MBL Lectures in Biology

Volume 11

PERSPECTIVES ON CELLULAR REGULATION: FROM BACTERIA TO CANCER

MBL LECTURES IN BIOLOGY

Volume 1
The Origins of Life and Evolution
Harlyn O. Halvorson and K.E. Van Holde, *Editors*

Volume 2
Time, Space, and Pattern in Embryonic Development
William R. Jeffery and Rudolf A. Raff, *Editors*

Volume 3
Microbial Mats: Stromatolites
Yehuda Cohen, Richard W. Castenholz, and Harlyn O. Halvorson, *Editors*

Volume 4
Energetics and Transport in Aquatic Plants
John A. Raven

Volume 5
The Visual System
Alan Fein and Joseph S. Levine, *Editors*

Volume 6
Blood Cells of Marine Invertebrates: Experimental Systems in Cell Biology and Comparative Physiology
William D. Cohen, *Editor*

Volume 7
The Origin and Evolution of Sex
Harlyn O. Halvorson and Alberto Monroy, *Editors*

Volume 8
Development as an Evolutionary Process
Rudolf A. Raff and Elizabeth C. Raff, *Editors*

Volume 9
The Biology of Parasitism: A Molecular and Immunological Approach
Paul T. Englund and Alan Sher, *Editors*

Volume 10
Perspectives in Neural Systems and Behavior
Thomas J. Carew and Darcy B. Kelley, *Editors*

Volume 11
Perspectives on Cellular Regulation: From Bacteria to Cancer
Judith Campisi, Dennis D. Cunningham, Masayori Inouye, and Monica Riley, *Editors*

PERSPECTIVES ON CELLULAR REGULATION: FROM BACTERIA TO CANCER

Essays in Honor of Arthur B. Pardee

Editors

Judith Campisi
Division of Cell and Molecular Biology
Lawrence Berkeley Laboratory
University of California
Berkeley, California

Dennis D. Cunningham
Department of Microbiology and Molecular Genetics
University of California College of Medicine
Irvine, California

Masayori Inouye
Department of Biochemistry
Robert Wood Johnson Medical School
University of Medicine and Dentistry of New Jersey
Piscataway, New Jersey

Monica Riley
Marine Biological Laboratory
Woods Hole, Massachusetts

WILEY-LISS

A JOHN WILEY & SONS, INC., PUBLICATION
New York • Chichester • Brisbane • Toronto • Singapore

Address all Inquiries to the Publisher
Wiley-Liss, Inc., 605 Third Avenue, New York, NY 10158-0012

Printed in United States of America

Library of Congress Cataloging-in-Publication Data

Perspectives on cellular regulation : from bacteria to cancer : essays in honor of Arthur B. Pardee / editors, Judith Campisi . . . [et al.].
p. cm. — (MBL lectures in biology ; v. 11)
Includes bibliographical references and index.
ISBN 0-471-56090-1
1. Cells—Growth—Regulation. 2. Cancer cells—Growth—Regulation. 3. Cell cycle. 4. Pardee, Arthur Beck, 1921– . I. Pardee, Arthur Beck, 1921– . II. Campisi, Judith III. Series.
QH604.P44 1991
574.87′61—dc20 90-26568
CIP

Cover:

Upper left: Segment of videomicroscopic sequence of caffeine-induced mitotic events in BHK cells arrested in early S phase. See Schlegel and Craig, page 240.

Upper right: Primer extension analysis of cytoplasmic (C) and nuclear (N) RNAs transcribed from the murine thymidine kinase gene. Arrows indicate the major transcriptional start sites for these RNAs. Figure courtesy of Arthur B. Pardee.

Lower left: Nucleotide sequences of the wild-type and Sp1 single-base mutated MT1 regions of the murine thymidine kinase promoter. Figure courtesy of Arthur B. Pardee and the National Academy of Sciences U.S.A.

Center: A G1/S phase-specific protein complex, Yi, that binds to the murine thymidine kinase promoter, is detected by band-shift analysis using a synthetic oligonucleotide containing the Yi binding site. Figure courtesy of Arthur B. Pardee and the National Academy of Sciences U.S.A.

Lower right: Differentiation of 3T3-F442A mouse preadipocytes in serum-supplemented medium. See Cherington, pages 170–171.

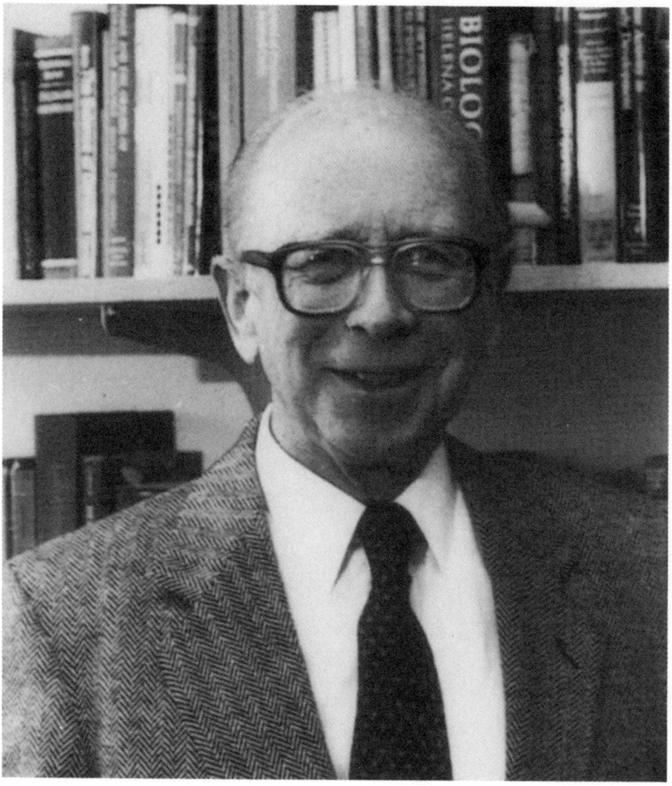

DEDICATION

This book is dedicated with admiration and affection to Arthur B. Pardee on the occasion of his 70th birthday. It was conceived as an adjunct to a symposium organized by his students and postdoctoral fellows, some of whom have contributed chapters summarizing their current research activities.

One part of a scientist's life work is embodied in contributions of original research. Many authors in this book trace the origins of their present research to seminal experiments and ideas that were generated in Arthur Pardee's laboratory. The younger authors describe exciting projects that are currently under way in his laboratory.

Another facet of a scientist's life is reflected in the influence exerted on other scientists, both budding and mature. The entire book is a tribute to Art's constructive and lasting influence on his students, in their enthusiastic and expert pursuit of biological principles, especially molecular mechanisms of cellular regulation.

Contents

Contributors

Mitchell S. Abrahamsen, Department of Biochemistry, University of Washington, Seattle, WA 98195 **[107]**

Jon Beckwith, Department of Microbiology and Molecular Genetics, Harvard Medical School, Boston, MA 02115 **[85]**

Lucas D. Bowler, Microbial Genetics Group, School of Biological Sciences, University of Sussex, Falmer, Brighton BN1 9QG, United Kingdom **[73]**

James A. Brannigan, Microbial Genetics Group, School of Biological Sciences, University of Sussex, Falmer, Brighton BN1 9QG, United Kingdom **[73]**

Judith Campisi, Division of Cell and Molecular Biology, Lawrence Berkeley Laboratory, University of California, Berkeley, CA 94720 **[153]**

John J. Castellot, Jr., Department of Anatomy and Cellular Biology, Tufts University Health Science Schools, Boston, MA 02111 **[201]**

Rinku Chatterjee, Department of Physiology, Tufts University School of Medicine, New England Medical Center, Boston, MA 02111 **[165]**

Van Cherington, Departments of Pathology, Anatomy and Cellular Biology, and Physiology, Tufts University School of Medicine, New England Medical Center, Boston, MA 02111 **[165]**

Ruth W. Craig, Department of Physiology, Johns Hopkins University School of Medicine, Baltimore, MD 21218 **[235]**

Dennis D. Cunningham, Department of Microbiology and Molecular Genetics, University of California College of Medicine, Irvine, CA 92717 **[175]**

Bruce J. Dezube, Division of Cell Growth and Regulation, Department of Biological Chemistry and Molecular Pharmacology and Department of Medicine, Dana-Farber Cancer Institute, and Division of Medical Oncology, Department of Medicine, Beth Israel Hospital, Boston, MA 02115 **[251]**

William D. Donachie, Institute of Cell and Molecular Biology, University of Edinburgh, Edinburgh EH9 3JR, Scotland **[51]**

Qing-Ping Dou, Department of Biological Chemistry and Molecular Pharmacology, Harvard Medical School, and Division of Cell Growth and Regulation, Dana-Farber Cancer Institute, Boston, MA 02115 **[265]**

Christopher G. Dowson, Microbial Genetics Group, School of Biological Sciences, University of Sussex, Falmer, Brighton BN1 9QG, United Kingdom **[73]**

Howard J. Fingert, Division of Hematology and Oncology, St. Elizabeth's Hospital, Boston, MA 02135 **[251]**

Judith L. Fridovich-Keil, Department of Biological Chemistry and Molecular Pharmacology, Harvard Medical School, and Division of Cell Growth and Regulation, Dana-Farber Cancer Institute, Boston, MA 02115 **[265]**

The numbers in brackets are the opening page numbers of the contributors' articles.

John Gerhart, Department of Molecular and Cell Biology, University of California, Berkeley, CA 94720 **[185]**

Jean M. Gudas, Department of Biological Chemistry and Molecular Pharmacology, Harvard Medical School, and Division of Cell Growth and Regulation, Dana-Farber Cancer Institute, Boston, MA 02115 **[265]**

Lorraine J. Gudas, Department of Biological Chemistry and Molecular Pharmacology, Harvard Medical School, and Division of Cellular and Molecular Biology, Dana-Farber Cancer Institute, Boston, MA 02115 **[121]**

Joyce L. Hamlin, University of Virginia School of Medicine, Charlottesville, VA 22908 **[297]**

Agnes Hutchison, Microbial Genetics Group, School of Biological Sciences, University of Sussex, Falmer, Brighton BN1 9QG, United Kingdom **[73]**

Masayori Inouye, Department of Biochemistry, Robert Wood Johnson Medical School, University of Medicine and Dentistry of New Jersey, Piscataway, NJ 08854 **[95]**

Ron Krieser, Department of Pathology, Tufts University School of Medicine, New England Medical Center, Boston, MA 02111 **[165]**

Peter Kuempel, Department of Molecular, Cellular and Developmental Biology, University of Colorado, Boulder, CO 80309 **[39]**

Ching C. Lau, Laboratory of Gynecologic Oncology, Brigham and Women's Hospital, Boston, MA 02115 **[251]**

Millicent Masters, Institute of Cell and Molecular Biology, University of Edinburgh, Edinburgh EH9 3JR, Scotland **[17]**

Estela E. Medrano, Department of Dermatology, University of Cincinnati College of Medicine, Cincinnati, OH 45267 **[213]**

David R. Morris, Department of Biochemistry, University of Washington, Seattle, WA 98195 **[107]**

G. Prem Veer Reddy, Department of Obstetrics and Gynecology and Department of Biochemistry, Health Sciences Center, University of Virginia, Charlottesville, VA 22908 **[315]**

Monica Riley, Marine Biological Laboratory, Woods Hole, MA 02543 **[3]**

Enrique Rozengurt, Imperial Cancer Research Fund, Lincoln's Inn Fields, London WC2A 3PX, United Kingdom **[129]**

Robert Schlegel, Laboratory of Toxicology, Harvard School of Public Health, Boston, MA 02115 **[235]**

Helene S. Smith, Geraldine Brush Cancer Research Institute, Pacific Presbyterian Medical Center, San Francisco, CA 94115 **[225]**

Brian G. Spratt, Microbial Genetics Group, School of Biological Sciences, University of Sussex, Falmer, Brighton BN1 9QG, United Kingdom **[73]**

Robert W. Tucker, The Johns Hopkins Oncology Center, Baltimore, MD 21205 **[143]**

James R. Walker, Department of Microbiology, The University of Texas, Austin, TX 78712 **[31]**

Allan C. Wilson, Division of Biochemistry and Molecular Biology, University of California, Berkeley, CA 94720 **[331]**

Andrew Yen, Department of Pathology, Cornell University Veterinary College, Ithaca, NY 14853 **[279]**

Qian-yun Zhang, Microbial Genetics Group, School of Biological Sciences, University of Sussex, Falmer, Brighton BN1 9QG, United Kingdom **[73]**

David R. Zusman, Department of Molecular and Cell Biology, University of California, Berkeley, CA 94720 **[65]**

Foreword

The Marine Biological Laboratory is honored to participate in this timely tribute to Arthur Pardee. Early workers at the MBL, such as C.O. Whitman, E.B. Wilson, T.H. Morgan, and Ross Harrison, pioneered in solving fundamental problems of heredity, genetics, and embryology and in exploring the advantages of utilizing diverse biological systems. Arthur Pardee was one of those who recognized that common fundamental mechanisms regulate cellular processes from bacteria to human cells. With the availability of present molecular techniques, new frontiers are now available, such as a study of the regulation of the cell cycle.

Major advances in biological science arise from a few key leaders. Isaac Newton has been credited with the aphorism, "If I have seen farther, it is by standing on the shoulders of giants." In a literary detective story, "On the Shoulders of Giants," Robert K. Merton traces this saying to Bernard of Chartres (1126). If you were to visit Chartres, you could observe this phenomenon in the stained glass windows or in the carved figures outside the cathedral. Arthur is one of those giants upon whose shoulders future scientists will stand.

Having shared with Arthur a common interest in the regulation of the cell cycle, as well as parallel experiences in Monod's laboratory at the Pasteur Institute, I have followed Arthur's career with great interest and admiration. He has repeatedly contributed landmark papers that have generated more than one area of study. In recognition of these achievements, Arthur was one of the early recipients of the Rosenstiel Award in Basic Biomedical Research in the early 1970s and has gone on to many other significant awards. We join his many students and postdoctoral fellows in a fitting tribute.

Harlyn O. Halvorson
President/Director
Marine Biological Laboratory

Preface

Biologists have been fascinated with the processes that control cell division, cell function, and the ability to respond to environmental cues, starting as long ago as 300 B.C. when Aristotle studied the development of a fertilized chick egg. However, only recently has there been a virtual explosion of knowledge about how basic cellular processes are controlled. Although formidable challenges remain, we now have powerful new concepts and tools for understanding cell function—in both normal and diseased states—at a molecular level. This book presents a sampling of these new developments.

Modern biochemistry—or what is now known as molecular and cellular biology—owes much of its impressive progress to the key insights and unfailing dedication of a relatively small group of intellectual leaders in the biological sciences. One such scientist is Arthur Pardee. Early in 1988, several of Art's former students and postdoctoral fellows suggested the possibility of marking his seventieth birthday in a way that would acknowledge the importance of his scientific contributions and the influence that he has had on the many young scientists who have worked in his laboratory. The enthusiasm for doing this gained momentum as more of his former students and postdoctoral fellows were contacted. What became clear as these contacts were made was that—in addition to making seminal and important discoveries in his own lab—Art had trained a cohort of first-class scientists whose current interests and contributions spanned many of the important areas in modern molecular and cellular biology. To enable Art's colleagues over the years to celebrate his seventieth birthday with him, we decided to convene a symposium, "Perspectives on Cellular Regulation: From Bacteria to Cancer." To provide a long-lasting memento of this occasion, we also decided to dedicate to him this book containing summaries of the research accomplishments and interests of some of his former students and postdoctoral fellows. It is therefore a sampling of leading research in a number of diverse areas of cell and molecular biology. The book is a striking testimony to the strong and lasting influence that Art Pardee has had on so many scientific careers. To put this influence into perspective, it is important to summarize some of Art's own scientific interests and accomplishments, which themselves span more than one area of biological regulation.

Arthur Pardee's scientific career can be divided into three periods: the Berkeley period (1949–1961), the Princeton period (1961–1975), and the Harvard period (1975–present). Although this progression reflects his professional advancement, Art developed distinct research interests that characterize each period. Notably, during the Berkeley period he spent a sabbatical (1957–1958) at the Pasteur Institute where he carried out the famous PaJaMo experiment, which defined the repressor concept. Remarkably, as illustrated in this symposium, the majority of his students continued their careers with research that grew out of their years with Art, while encompassing forefront achievements in their varied approaches to the study of regulatory mechanisms.

Art Pardee has had a long-standing interest in the biochemistry and molecular genetics of cellular regulatory processes. Trained as a chemist, he first studied antibody–antigen interactions in vitro with Linus Pauling, and then studied respiratory reactions in animal tissues with Van Potter. In his own laboratory, in Berkeley, Art blended his knowledge of biochemistry and physiology with the then-emerging field of molecular genetics. His published a series of landmark papers on the control of DNA synthesis in bacteria and the enzymes needed for its metabolism. Collectively, these studies crystallized some of the fundamental concepts of how genetic and environmental information is processed by living organisms. His work in these areas led to some now-classic paradigms of gene regulation: the control of protein function by allosteric regulators and the control of gene expression by cis-acting sequences and trans-acting proteins.

Pardee's laboratory also pioneered studies on the induction and repair of gene mutations and their consequences for cell function and survival. Art understood, long before it was obvious or fashionable, that a molecular description of how DNA synthesis and cell growth was controlled would provide us with key insights into the causes and treatment of cancer—and a number of other clinically important diseases. That understanding resulted in another series of landmark papers on the control of animal cell proliferation, which began during his Princeton period. Over the past two decades, Art's laboratory has contributed many of the critical studies on the molecular basis for control points in the animal cell cycle; on the concepts of metabolic channeling and a multienzyme complex for DNA replication; on the repair and cellular consequences of DNA damage; and on the connections between growth factors, intracellular signal transduction mechanisms, and the expression of protooncogenes. Always, Art nurtured an awareness that cancer cells are, in a sense, growth control mutants and that their behavior can provide important information on the regulation of normal processes, as well as strategies for clinical intervention.

Pardee's laboratory at Harvard continues to work at the frontiers of cell and molecular biology, particularly in dissecting the molecular controls that govern progression through the animal cell cycle . We anticipate another series of landmark papers on how genes governing the cell cycle are switched on and off, and how a select group of cellular genes and protooncogenes influence cell cycle progression. Art has begun to realize one of his long-standing goals—to make our knowledge about cell cycle control and DNA repair mechanisms clinically relevant.

In reviewing the hundreds of papers originating from Art Pardee's lab over the past 40 years, it is clear that his research interests have spanned a great many aspects of cellular regulatory mechanisms. It is evident from the studies described in this book that Art created a fertile intellectual climate in which many minds and many areas of cell and molecular biology took root and began to grow.

The editors extend a warm acknowledgment to Ruth Sager for many insights during the preparation of this tribute to Arthur Pardee. She provided personal and scientific perspectives that greatly aided the preparation of this volume. We also thank her for helpful suggestions regarding the symposium that was organized to honor Art and that provided the occasion to present this volume to him.

Judith Campisi
Dennis D. Cunningham
Masayori Inouye
Monica Riley
January 1991

I. FROM CHROMOSOME STRUCTURE TO MORPHOGENESIS IN PROKARYOTES

Arthur Pardee began his independent scientific career at Berkeley, with an insightful, creative, and multifaceted inquiry into the physiology, biochemistry, and molecular genetics of bacteria. He worked on a surprisingly broad range of problems, including the metabolism of nucleic acids and amino acids, mutations induced by bromouracil and ultraviolet light, mechanisms of regulation of DNA synthesis, and most famously, feedback inhibition of protein synthesis and allosteric regulation of protein activity.

His studies of prokaryotes continued when he moved to Princeton, where his research became progressively more versatile and creative. Art is a born scientist—imaginative and able to open new fields effortlessly, simply guided by a deep sense of curiosity and keen insight. During this period, Art continued to work on allosteric proteins and began studies of suppressor regulation, sulfate binding protein, and RecA protein. Concomitantly, while working on the prokaryotic systems, he initiated work on animals cells. His interest gradually shifted toward more complex systems concerning cell cycle, cell division, and cancer. Subsequently, after his move to Harvard, he devoted his time entirely to the animal cell system.

In this prokaryotic section, nine authors contribute chapters, two of whom were graduate students (M. Riley and M. Masters) in the Berkeley period, and the rest of whom were postdoctoral fellows in the Princeton period. Remarkably, these authors have maintained the research interests developed during their training in Art's laboratory. Regrettably, because of space limitation, it is not possible to have every scientist who worked in Art's laboratory contribute a chapter. Nevertheless, the contributions of this section represent the breadth of their interests and the versatility of their mentor.

Perspectives on Cellular Regulation:
From Bacteria to Cancer, pages 3–15

The *Escherichia coli* Chromosome: A Mosaic

Monica Riley

Marine Biological Laboratory, Woods Hole, Massachusetts 02543

Introduction

In 1957 I was a graduate student in the "Comparative Biochemistry" program at Berkeley, taking courses and looking into possibilities for a thesis project and mentor. I knew that I wanted to work on some aspect of bacterial genetics because I loved the course in bacterial and phage genetics given by Gunther Stent and Ed Adelberg (and also the course in bacterial metabolism given by Roger Stanier, Michael Douderoff, and H. A. Barker.) When I began to talk with people about possible thesis projects and mentors, Gunther Stent was very helpful and told me about an absent faculty member, Arthur Pardee, then on sabbatical leave at the Pasteur Institute in Paris. Gunther described the PaJaMo experiment to me, then in progress. It sounded like the work was in the right ballpark for me, using bacterial genetics to answer questions about how the gene works, so I wrote Art (of course more formally "Dr. Pardee" at the time and throughout graduate school) to inquire about a place. We exchanged a few letters on the strength of which I applied for an NIH predoctoral fellowship (necessary to support myself and my three small children while going to school).

While waiting to hear about the fellowship and continuing with courses, I remember that Art suggested I read as background material the article in the Cold Spring Harbor Symposium 1956 about conjugation in *Escherichia coli* (Wollman et al., 1956). That was the paper that sorted out Hfrs from F^+s, defined a merozygote as having only a part of the Hfr DNA, all of the F^- DNA, and analyzed kinetics of marker transfer in both interrupted and uninterrupted matings. It was a landmark paper for me.

The ^{32}P Suicide Experiment

By the time Art returned to Berkeley, I had good news about the fellowship so that I was able to quit my job as a lab tech at the Naval Biological Lab in Oakland to become a full-time student, and I was raring to go on starting a research project. An experiment had been outlined in discussion among Pardee, Jacob, and Monod that used the inducible *lac* genes and the conjugation system in *E. coli* to determine whether a gene transferred its encoded information to a stable intermediate or whether the presence of an intact gene was necessary for continued production of a gene product. We planned to transfer *lac* genes from lac^+ Hfrs to lac^- F^-s. After mating, Hfrs could be eliminated and synthesis of β-galactosidase could be induced in the merozygotes. The Hfr DNA had been heavily labeled with ^{32}P so that with time the lac^+ Hfr DNA in the merozygotes was destroyed by ^{32}P decay. By following the rate of β-galactosidase production as a function of ^{32}P suicide of the lac^+ DNA, we could determine whether the gene passed on its encoded information to a stable intermediate such as stable RNA, or whether the gene was required to continue to play an active role to maintain the capacity for enzyme synthesis. The latter proved to be the case. There was no stable intermediate. The integrity of the lac^+ gene was required to continue to produced β-galactosidase, most likely, we concluded because the gene needed to renew continually an unstable intermediate. What was this intermediate? Not ribosomal RNA, since it is stable; not transfer RNA (then called soluble RNA), since it is too short to carry the requisite information. At that point we left an open question (Riley et al., 1960). In retrospect, I see the ^{32}P mating experiment as characteristic of a Pardee experiment in its essential simplicity of design and in the biological importance of the question it asked. This was at the beginning of formulating the idea of an unstable, fast turning-over type of RNA, now known as messenger RNA, and it was an exciting time to enter into the world of bacterial genetics and molecular biology.

The ^{32}P mating experiment is relatively quick to outline, but it took a long time to get the first result. It was a fussy procedure, especially the liquid nitrogen storage and thawing conditions and the elimination of all Hfrs. I remember countless Saturdays stopping by the Virus Lab with the children as preschoolers, driving a 1939 Plymouth (bought for $50 from someone at the Naval Biological Lab), traipsing upstairs to the lab to refill with liquid nitrogen dewars containing ^{32}P-labeled zygote samples. I had an association at the time: after the merry-go-round in Tilden Park, we'd stop by the lab to top up the liquid nitrogen. The children loved to have me spill a little on the floor so they could see the ''smoke'' and watch the little puddle of liquid dance around until it disappeared.

I remember well Art's kindness and sensitivity. He arranged to bridge an awkward interval between the end of my predoctoral fellowship and the beginning of my first academic job at UC Davis. To mark the completion of my time in his lab, he didn't have a conventional party for adults, but more appropriately treated me and my children to an afternoon at the circus.

Well, it has been a long time since those days. I finished up at Berkeley in

1960, so I'm thinking back to over 30 years ago. Our picture of the *E. coli* chromosome has changed since that day. I remember having a mental image at the time of a merozygote as a genetically static creature. In my mind it had a very round circle of chromosomal DNA in it plus a short, straight piece of Hfr DNA including the lac^+ genes, and I saw in my mind all members of the population of bacteria as being genetically identical. I understood the reasons for the spread in time of induction of enzyme synthesis in the bacterial population, but I did not think of the *E. coli* chromosome as being genetically dynamic, nor *E. coli* populations as continually giving rise to genetic variants. A great deal of work by many labs in the meantime has clarified and enriched our understanding of *E. coli* genetics. A recent book summarizes our current understanding structure and function of the bacterial chromosome (Drlica and Riley, 1990). I am taking this opportunity, the occasion of honoring Arthur Pardee, to recall some of the past work from my lab concerning *E. coli* conjugation and genetic variation in enterobacteria, and to summarize the current view of the *E. coli* chromosome as a dynamic and genetically active entity.

Recombination in *E. coli*

While I was in the Bacteriology Department at UC Davis, Amos Oppenheim and I did some work on the composition of the recombinant DNA formed by recombination in the zygote between Hfr and F^- parental DNAs (Oppenheim and Riley, 1966). We used density label and tritium to differentially label Hfr and F^- parental DNAs, mated them, then isolated a recombinant fraction from zygotes. We found no double-stranded unreplicated Hfr DNA in the joint molecules: all Hfr DNA had been replicated either before or in the course of incorporation into a recombinant molecule. Later Vapnek and Rupp (1970) showed that only one strand of preexisting Hfr DNA enters the zygote, suggesting that transferring Hfr DNA replicates by the rolling circle mode. We found that in the joint molecules there were places where both preexisting strands of F^- DNA were replaced by Hfr and newly synthesized DNA. In a day when breakage and reunion mechanisms were still rivalled by "copy choice" mechanisms, this was one vote for breakage and reunion. We also found that at an early stage the joint molecule was not covalently joined, but as the process of recombination continued, covalent union was established (Oppenheim and Riley, 1967).

Later when I was newly at Stony Brook (State University of New York), Aniko Paul and I examined the patchwork character of the joint molecules. Segments of parental DNAs and newly synthesized DNA on the order of 4×10^3 to 1.7×10^4 bp in size made up the joint molecules, and segments were separated by gaps. These gaps could be filled in in vitro by T4 DNA polymerase, and were no greater than 450 nucleotides long (Paul and Riley, 1973). This picture of the joint molecules, intermediates in recombination, is consistent with our present understanding of recombination and repair processes, and the patchwork additions of segments of Hfr DNA to the recipient DNA is entirely consistent with today's picture developed by Roger Milkman and colleagues of *E. coli* DNA as a patchwork of segments of DNA acquired over time by recombination and transmitted clonally to progeny (Stoltzfus et al., 1988; Milkman and Stoltzfus, 1988; Milkman and Bridges, 1990) (see further discussion below).

Comparisons of Bacterial DNA Sequences Give Information on Mosaic Composition

Some of our understanding of the dynamics of bacterial chromosomes has come from comparisons of *E. coli* K-12 genes and gene arrangements with other *E. coli* strains and with other enteric bacteria. Examples of such studies will be briefly summarized here.

The genetic makeup of a contemporary *E. coli* strain can be viewed as a mosaic, a patchwork consequence of a genetic history that includes mutational change and genetic recombination. The kinds of changes that occur over time in bacterial DNAs have been sorted out by comparing related strains with respect to nucleotide sequences of genes, restriction fragment length polymorphisms, physical maps, and genetic linkage maps. The kinds of changes undergone by *E. coli* DNA are changes in one or a few nucleotides (base substitutions or frameshifts), duplications and divergence, rearrangements (inversions, transpositions), recombination among *E. coli* strains, and acquisition of genetic material from other species (lateral transfer). Each type of change is superimposed on the others, each contributes to the mosaic character of the evolving genome. Specific examples of each of these types of change follow.

Variation Among *E. coli* strains

A collection of "wild" *E. coli* strains put together in the early 1970s by Roger Milkman, derived from humans, geographically disperse domestic animals, and zoo animals (Milkman, 1973). A small subset of 24 of these strains from 8 sources as well as a group of laboratory strains of *E. coli* was surveyed in my lab for variation in sizes of restriction fragments in and around 4 gene loci. The variation was on the order of 3% nucleotide substitution between strains, no more among the wild strains than among laboratory strains nor between the two groups (Anilionis and Riley, 1980; Harshman and Riley, 1980). Much more precise and detailed information has been obtained since by determining full nucleotide sequences of chosen genes in a group of *E. coli* strains. This information was gathered for *trp* (Milkman and Crawford, 1983), *gnd* (Sawyer et al., 1987), and *phoA* (DuBose et al., 1988) genes, and painted a picture of sporadic genetic change interspersed with substantial regions of genetic stability. Typically sequences were identical for much of a gene, but interspersed with clusters of differences. For instance, in the *trpC* gene, one strain, 45E, has a cluster of 4 single nucleotide differences from strain K-12 in 5 consecutive codons. Strain 202I has the same set and another cluster of 3 base substitutions further along in the *trpB* gene, as well as other substitutions not shared with the strains analyzed (Milkman and Crawford, 1983). The sequences of the *phoA* genes from 8 *E. coli* strains showed that nucleotide differences clustered and the clusters were shared between strains in patterns that are most readily understood as reflecting recombination events within the gene (DuBose et al., 1988). The mosaic character of *E. coli* DNA is seen within as well as between genes.

When comparative sequence analysis was extended to a series of open reading frames lying between the *trp* operon and the *tonB* locus (about 4 kb in all), the differences found were not only nucleotide substitutions, but small additions or deletions of one or a few bases (indels), and also large rearrangements of hundreds of nucleotides (Stoltzfus et al., 1988; Milkman and Stoltzfus, 1988). Recombination clearly played a role in generating some of the genetic variety since some strains had sequences that were combinations of the sequences found in other strains.

Recently, Milkman's lab has extended the comparative analysis to a stretch of 40 kb around the *trp* operon, using a combination of nucleotide sequence and RFLP analysis (Milkman and Bridges, 1990). The relationships between the *E. coli* strains examined leads to a model of the chromosome as a mosaic of clonally inherited segments of DNA. The model proposes that a clonal segment is acquired by virtue of containing a favorable mutation that confers increased fitness, and it spreads through the *E. coli* population on a worldwide scale, carrying the rest of the chromosome with it, resulting in the observed small number of sequence types in wild strains of *E. coli* of diverse origins.

The mosaic character pictured in this model of *E. coli* population biology has as its unit of change the clonal segments of the chromosome acquired by recombination between *E. coli* strains. The range of plausible sizes of segments proposed by Milkman and Bridges is 10^3–10^5 bp. This range includes the sizes of segments observed by Aniko Paul and myself during recombination in *E. coli* by conjugation: 4×10^3 to 1.7×10^4 bp (Paul and Riley, 1973). Thus today's estimates of the unit of recombinational transmission based on a population genetics approach are consistent with earlier molecular findings of the unit of recombination in bacterial matings.

Comparisons of similar genes within the *E. coli* genome gives further information on the nature of changes that accumulate in the DNA of the bacterial chromosome over time. For instance, the *E. coli* genes for transcarbamylases seem to be good examples of a history of duplication and divergence. The *argI* gene encodes ornithine transcarbamylase; the adjacent *pyrB* gene encodes aspartate transcarbamylase. The sequences of the two genes bear a similarity that suggests common ancestry, possibly by tandem duplication and divergence (Van Vliet et al., 1984; Riley, 1985). The changes that have occurred in the sequences go beyond single nucleotide substitutions and include small additions or deletions (indels), often resulting in reading frame shifts for the same gene products.

Variation Among Enteric Bacteria

Comparisons of similar genes among evolutionarily related bacterial relatives also gives information on the nature of changes that accumulate in the bacterial chromosome over time, and on the reasons for lack of uniformity in extent of change throughout the genome. The *lac* genes provides good examples of other kinds of genetic variation. In my lab *lac* genes from the enterobacterial relative, *Klebsiella pneumoniae*, were cloned and sequenced (MacDonald and Riley, 1983; Buvinger and Riley, 1985a,b). As compared with the sequences of the corresponding *lac* genes of *E. coli*, one sees that in the course of divergence from presumed common ancestral genes there has been an accumulation of nucleotide substitutions, small indels, and a gross rearrangement.

Aligning the 3152 bases of the *K. pneumoniae lacZ* gene with the homologous *E. coli lacZ* gene sequence, one sees that there are altogether 21 places where there seems to be a gap or indel in one sequence relative to the other: either a deletion in one sequence or an addition to the other (Fig. 1). The numbers of nucleotides at each gap ranges from 1 to 23 and is either a multiple of 3 so that no shift in codon reading frame results, or if it is not, the algebraic sum of the nucleotides of clustered gaps becomes a multiple of 3 over a short distance (18, 37, and 44 nucleotides in three instances), and thus only a short span of unrelated amino acids in the two enzymes. Presumably the frame shifts occur in parts of the protein that can tolerate change in the amino acid sequence. Leaving aside the indels, the paired nucleotides of the two *lacZ* genes were 65% identical overall. Local heterogeneity in degree of match may reflect differences in functional constraints in different parts of the proteins.

A set of cryptic genes exists in the *E. coli*, related to the *lac* genes and

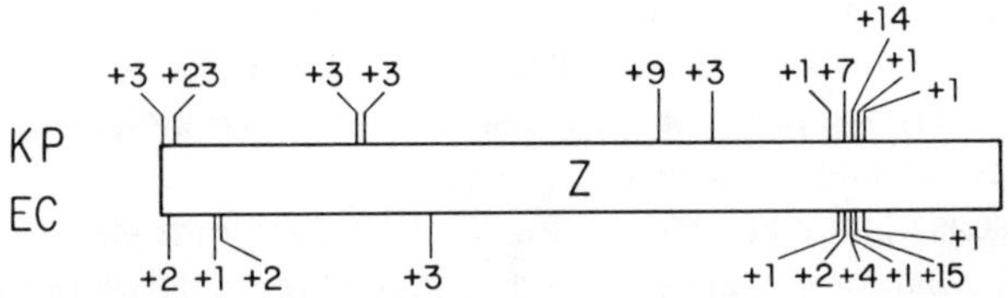

Fig. 1. Schematic representation of locations of additions and/or deletions (gaps) in the *K. pneumoniae lacZ* gene relative to the *E. coli lacZ* gene. At each position indicated, the numbers indicate the number of nucleotides for which there appears to be no counterpart in the other sequence. KP, *K. pneumoniae*; EC, *E. coli*. Reprinted from Buvinger and Riley (1985a).

TABLE 1. Similarities of Gene Products: Percent Identity of Inferred Amino Acid Sequences

	β-Galactosidases		
	E.c.[a]*lac*	E.c.*ebg*	
K.p. *lac*	61[b]	31[c]	
E.c. *lac*		34[c]	
	Repressor proteins		
	E.c.*lac*	E.c.*ebg*	E.c.*gal*
K.p. *lac*	40[b]	20[c]	23[b]
E.c. *lac*		25[c]	23[b]
E.g. *ebg*			23[c]

[a]K.p., *K. pneumoniae*; E.c., *E. coli*.
[b]From Buvinger and Riley (1985a).
[c]From Hall et al., (1989).

capable of mutation to active status and production of another β-galactosidase (Hall, 1982). The *ebg* genes have been sequenced and compared with the *E. coli* and *K. pneumoniae lacZ* sequences (Hall et al., 1989). Table 1 shows the percent similarity of amino acid sequences. The two enteric *lacZ* β-galactosidases are more closely related than either is to the *ebgA* β-galactosidase. One implication of the result is that the *ebg* gene split off from an ancestral *lac* gene before the organisms *E. coli* and *K. pneumoniae* and their *lacZ* genes diverged. As will be seen below, this is true of the phylogeny of repressor genes as well. The phylogeny of genes is not the same as the phylogeny of organisms.

Comparison of the nucleotide sequences of the *K. pneumoniae lacI* gene for the *lac* repressor (1161 bp) with the homologous *E. coli lacI* gene showed no gaps and less similarity in sequence (49%) than was found for the *lacZ* gene pair (65%). The similarities of deduced amino acid sequences of the *K. pneumoniae lac* repressor (Buvinger and Riley, 1985a), the *E. coli ebgR* repressor (Hall et al., 1989), and the *E. coli lac* repressor are shown in Table 1. The *K. pneumoniae* and *E. coli lac* repressor sequences are not as closely related as the β-galactosidase sequences.

Looking at the more distantly related *ebg* repressor sequence and the repressor of the *E. coli* galactose operon, amino acid sequences are all related in the range 20–25% (Table 1). Again, the separation of descent of the *lac*, *gal*, and *ebg* repressor genes from a common ancestral repressor gene predated divergence of the *E. coli* and *K. pneumoniae* organisms (Buvinger and Riley, 1985a; Hall et al., 1989). Genes have their own phylogenetic histories independent of organismal chromosomes (Fig. 2).

Another kind of change that takes place in bacterial chromosomes is illustrated in the comparison of the *K. pneumoniae* and *E. coli lac* operons: rearrangement of genes. The *lacI* gene is inverted in *K. pneumoniae* relative

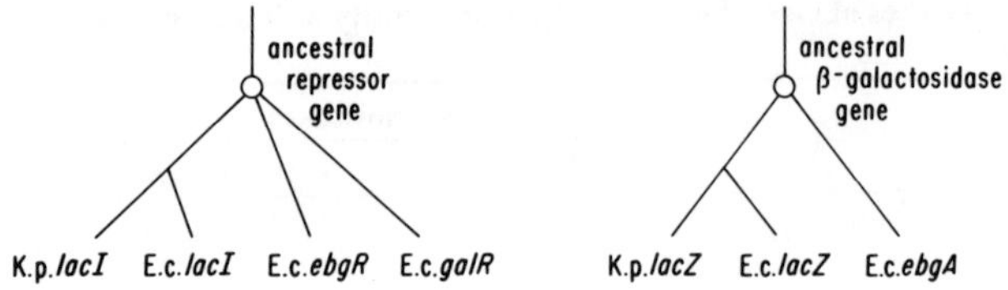

Fig. 2. Evolutionary branching order of *K. pneumoniae* and *E. coli lac* and *ebg* genes.

to the orientation in *E. coli*. The *lac* operon in *K. pneumoniae* is divergent, the *lacI* and *lacZ* genes being transcribed in opposite directions from a double-headed back-to-back promoter region between the coding segments (Buvinger and Riley, 1985b), whereas the *lac* operon in *E. coli* is not divergent, the *lacI* and *lacZ* genes arranged in tandem and transcribed in the same direction.

As genetic and physical maps of different strains of *E. coli* accumulate, including different substrains of *E. coli* K-12, differences in arrangements are discovered. The best known is probably the inversion of some 18% of the chromosome in a progenitor of *E. coli* K-12 W3110 relative to other K-12 strains. Many rearranged variants of *E. coli* have been isolated and mapped (for a summary see Riley, 1984). A question that has been before us for a long time is how the overall gene order of *E. coli* is preserved when rearrangements occur with apparent ease. How can a mosaic keep from becoming a perpetually turning kaleidoscope where every piece is free to move independently of every other piece in continuous free play? Comparison of *E. coli* and *Salmonella typhimurium* genetic maps shows how long the general outline of gene order has been maintained. The gene orders are nearly identical although the bacteria are estimated to have diverged some 120–160 million years ago (Ochman and Wilson, 1987). Some answers to this long-posed question have been proposed. It appears that more than one factor can be working to place limits on completely free internal rearrangement of the genes of the *E. coli* chromosome. Some of these factors are:

1. Maintenance of gene dosage ratios of genes close to origin relative to those farther away.
2. Maintenance of approximately equal lengths of replication arms.
3. Keep direction of transcription of most highly expressed genes oriented in the same direction as replication.
4. Mechanical barriers may prevent intrachromosomal recombination around the terminus, preventing formation of some types of rearrangements.
5. Interruption of two nondivisible zones (NDZs) on either side of the terminus loci seems to be prohibited.

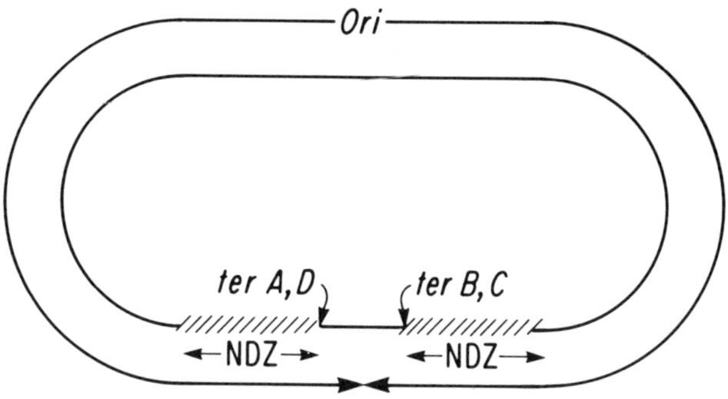

Fig. 3. Schematic representation of *E. coli* chromosome. Replication arms, terminating in arrows, are shown of equal length. Nondivisible zones (NDZs) defined by Rebollo et al. (1988) seem not to tolerate ends of inversions. Intrachromosomal recombination seems not to occur around the terminus (Segall and Roth, 1989). ori, origin of replication; ter, terminus loci.

The combination of these and possibly other limiting factors provides a rationale for believing that pieces of the mosaic can be shuffled around, but there are limitations to the choices, rules that govern what alternate arrangements can occur (see Fig. 3; Rebollo et al., 1988; Segall and Roth, 1989; discussion in Riley and Krawiec, 1990).

Lateral Transfer

The variations introduced in the bacterial chromosome by mutation, rearrangement, and genetic recombination within the species can be further enriched by an overlay of acquisition of genes from extrageneric sources. Lateral transfer of genes and incorporation of new genes into the recipient chromosome can occur over large taxonomic gulfs as is the case for bacterial genes transported into plants via the Ti plasmid of *Agrobacterium tumefaciens*. Promiscuous plasmids that carry genes in transposons and have broad host ranges seem likely vehicles for extrageneric transmission of genes. Composite plasmids are capable of transkingdom conjugation, *E. coli* to *Saccharomyces cerevisiae* (Heineman and Sprague, 1989). Some enterobacterial genes are often found on plasmids; among these are the *lac* genes (Guiso and Ullman, 1976; Michiels and Cornelis, 1984). It is conceivable that a part of the genetic material in bacteria has been derived by lateral transfer, enriching the mosaic further. It is not easy to make the distinction between genes that have evolved to large differences at a fast rate versus genes of foreign origin that were acquired by a horizontal transfer.

A detailed comparison of the genetic maps of *E. coli* and *S. typhimurium* showed that there were 15 places where there was appreciable excess map distance in one chromosome relative to the other (Fig. 4) (Riley and Krawiec, 1987). If these distance differences do not represent distortions in recombina-

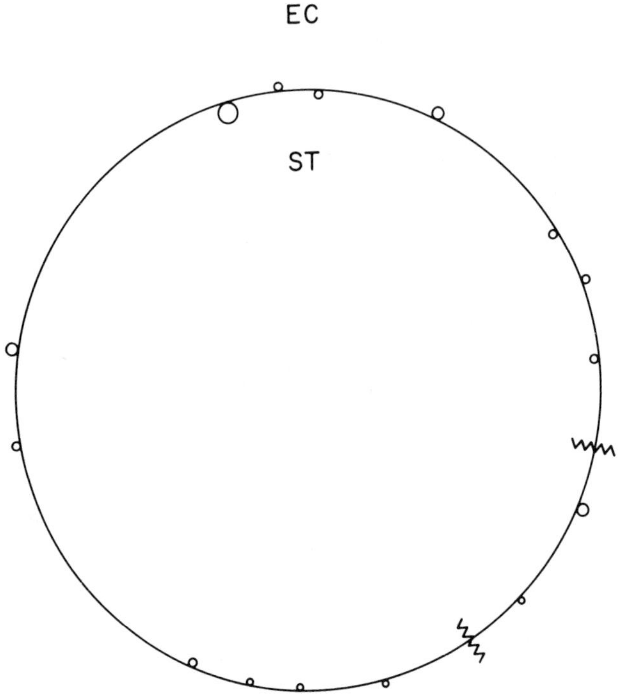

Fig. 4. Comparison of the genetic maps of *E. coli* and *S. typhimurium*. The large circle represents linkage data that are essentially congruent for the two maps at a resolution of 0.6 map units. Loops represent locations of excess genetic distance in one genome with respect to the other, based on the genetic maps in the 1987 volume on *E. coli* and *S. typhimurium* (Neidhardt et al., 1987). Sawtooth lines designate ends of an inversion in one map relative to the other. EC, *E. coli*; ST, *S. typhimurium*. Reprinted from Krawiec and Riley (1987).

tion frequencies but in fact represent differences in amounts of DNA at discrete loci on the two chromosomes, then these ''loops'' or ''bulges'' of extra DNA could represent independent uptake of DNA in the two genera, DNA acquired by lateral transmission.

One of the loops in *E. coli* seems to be quite complex and bears looking at in some detail, and this is the loop that contains the *lac* operon (Lampel and Riley, 1982). A compilation of information on the locations of IS elements in *E. coli* K-12 strains shows a disproportionately high number located in this region, 10 IS elements in a span of about 3 min of map (Birkenbihl and Vielmetter, 1989). Genetically, *E. coli* and *S. typhimurium* differ in a number of places in this area: *newD* and *supQ* are linked genes present in *S. typhimurium*, not in *E. coli; argF* is a second gene for an ornithine transcarbamylase, related to the *argI* gene, and is present in *E. coli* K-12, but not in *S. typhimurium* or other *E. coli* strains; the *lac* operon is present in *E. coli,* not in *S. typhimurium*; the *phoA* gene for alkaline phosphatase is present in *E. coli*, not in *S. typhimurium*. Have these genes been acquired by lateral transmission, perhaps via transposons brought into the cell on plasmids? A discontinuity in homology of DNA between *E. coli* and *S. typhimurium* was found on one side of the *lac* operon. Did that locate one end of remnants of a *lac* transposon (Lampel and Riley, 1982; Buvinger et al., 1984)?

The nucleotide sequence of the *argF* gene is 78% identical to the sequence of the *argI* gene, and the amino acid sequences are 86% identical (Van Vliet et al., 1984). Did the *argF* gene arise in *E. coli* K-12 by duplication of *argI*, transposition, and divergence? Or was *argF* transported as an ISl-flanked transposon? (The *argF* gene resides in the chromosome within 10 kb of DNA flanked by two ISl elements.) There are reasons to believe *argF* is of extrageneric origin. Close scrutiny of the nucleotide sequence of the *argF* gene showed that it is not typical of *E. coli* genes in base composition or codon usage (Van Vliet et al., 1988). With a G + C content of 59%, it is higher than other sequenced *E. coli* genes. Whereas the proportion of G + C in the third positions of codons in the *E. coli* genes surveyed (up to 1986) averages 56%, the value for *argF* is anomalously high, 76%. Van Vliet et al. (1988) conclude that in respect to total G + C content and third position codon G + C, *argF* is more like the DNA of *K. aerogenes* or *Serratia marcescens* than like *E. coli*, and may well have been acquired by lateral transfer. Additional sequence data for this genetic region in *E. coli* strains and other enterobacteria will no doubt provide a good picture of the complexity of past genetic events and will clarify the role that lateral transmission has played in introducing change.

Summary

One view of the *E. coli* chromosome is as a mosaic of genetic information derived from many sources. The *E. coli* DNA sequence of 4700 kb is subject to many avenues of genetic change: through mutations by base substitutions, small indels, and major rearrangements; through recombination and genetic exchange within the species with clonal inheritance of the acquired segments of DNA; through lateral transmission of segments from extrageneric sources. These several processes are superimposed on one another, providing rich possibilities for genetic change. A balance must be struck between these opportunities for change and the need for an essential genetic stability. Some of the genetically conservative factors that limit genetic change are now being identified. Models describing the genetic dynamics of populations of *E. coli* are emerging from studies in the area of population genetics: it is becoming possible to make reasonable estimates of rates of mutation, recombination, and fixation in natural populations of *E. coli*. Through the work of both molecular biologists and population geneticists, a picture of the dynamics of formation, maintenance and continuing change of the mosaic of the *E. coli* chromosome is emerging.

This article is too short to pass on all of the words of wisdom and helpful hints that I learned from Art Pardee. It is sometimes hard for me to wrap up a seminar or an article, particularly when the topic is still unfinished, with much more to be learned. After I had presented a seminar as a student with a closing that just feebly trailed off, Art told me that one should have in mind before giving a talk what the ending is going to be, and that if nothing else, at least one can fall back on an upbeat closing remark, his parody being "the future lies ahead." So, in honor of this memorable occasion, let me close with the resounding statement that The Future Lies Ahead!

References

Anilionis A, Riley M (1980): Conservation and variation of nucleotide sequences. I. *E. coli* strains. J Bacteriol 143:355–376.

Birkenbihl RP, Vielmetter W (1989): Complete maps of IS1, IS2, IS3, IS4, IS5, IS30 and IS150 locations in *Escherichia coli* K-12. Mol Gen Genet 220:147–153.

Buvinger WE, Lampel KA, Bojanowski RJ, Riley M (1984): Location and analysis of nucleotide sequences at one end of a putative *lac* transposon in the *Escherichia coli* chromosome. J Bacteriol 159:618–623.

Buvinger WE, Riley M (1985a): Nucleotide sequence of *Klebsiella pneumoniae lac* genes. J Bacteriol 163:850–857.

Buvinger WE, Riley M (1985b): Regulatory region of the divergent *Klebsiella pneumoniae lac* operon. J Bacteriol 163:858–862.

DuBose RF, Dykhuizen DE, Hartl DL (1988): Genetic exchange among natural isolates of bacteria: Recombination within the *phoA* gene of *Escherichia coli*. Proc Natl Acad Sci USA 85:7036–7040.

Drlica K, Riley M (eds.) (1990): The Bacterial Chromosome. Washington, DC: American Society for Microbiology.

Guiso N, Ullman A (1976): Expression and regulation of lactose genes carried by plasmids. J Bacteriol 127:691–697.

Hall B (1982): Evolution of a regulated operon in the laboratory. Genetics 101:335–344.

Hall BG, Betts PW, Wootton JC (1989): DNA sequence analysis of artificially evolved ebg enzyme and ebg repressor genes. Genetics 123:635–648.

Harshman L, Riley M (1980): Conservation and variation of nucleotide sequences in *Escherichia coli* strains isolated from nature. J Bacteriol 144:560–568.

Heineman JA, Sprague GV Jr (1989): Bacterial conjugative plasmids mobilize DNA transfer between bacteria and yeast. Nature (London) 340:205–209.

Krawiec S, Riley M (1990): Organization of the bacterial chromosome. Microbiol Rev 54:502–539.

Lampel KA, Riley M (1982): Discontinuity of homology of *Escherichia coli* and *Salmonella typhimurium* DNA in the *lac* region. Mol Gen Genet 186:82–86.

MacDonald C, Riley M (1983): Cloning chromosomal *lac* genes of *Klebsiella pneumoniae*. Gene 24:347–351.

Michiels T, Cornelis G (1984): Detection and characterization of Tn2501, a transposon included within the lactose transposon Tn951. J Bacteriol 158:866–871.

Milkman R (1973): Electrophoretic variation in *Escherichia coli* from natural sources. Science 182:1024–1026.

Milkman R, Stoltzfus A (1988): Molecular evolution of the *Escherichia coli* chromosome. II. Clonal segments. Genetics 120:378–380.

Milkman R, Crawford IP (1983): Clustered third-base substitutions among wild strains of *Escherichia coli*. Science 221:378–380.

Milkman R, Bridges MM (1990): Molecular evolution of the *Escherichia coli* chromosome. III. Clonal frames. Genetics 126:505–517.

Neidhardt FC, Ingraham JL, Low KB, Magasanik B, Schaechter M, Umbarger HE (eds.) (1987): *Escherichia coli* and *Salmonella typhimurium*: Cellular and Molecular Biology. Washington, DC: American Society for Microbiology, pp. 808–809, 892–893.

Ochman H, Wilson AC (1987): Evolutionary history of enteric bacteria. In F.C. Neidhardt, J.L. Ingraham, K.B. Low, B. Magasanik, M. Schaechter, and H.E. Umbarger (eds.): *Escherichia coli* and *Salmonella typhimurium*: Cellular and Molecular Biology. Washington, DC: American Society for Microbiology, pp. 1649–1654.

Oppenheim AB, Riley M (1966): Molecular recombination following conjugation in *Escherichia coli*. J Mol Biol 20:331–357.

Oppenheim AB, Riley M (1967): Covalent union of parental DNA's following conjugation in *Escherichia coli*. J Mol Biol 28:503–511.

Paul AV, Riley M (1973): Joint molecule formation following conjugation in wild type and mutant *Escherichia coli* recipients. J Mol Biol 81:1–22.

Rebollo JE, Francois V, Louarn JM (1988): Detection and possible role of two large nondivisible zones on the *E. coli* chromosome. Proc Natl Acad Sci USA 85:9391–9395.

Riley M (1984): Arrangement and rearrangement of bacterial genomes. In Mortlock (ed.): Microorganisms as Model Systems for Studying Evolution. New York: Plenum, pp. 285–315.

Riley M (1985): Discontinuous processes in the evolution of the bacterial genome. In M. Hecht, G. Prance, and B. Wallace (eds.): Evolutionary Biology, Vol. 19. New York: Plenum, pp. 1–36.

Riley M, Krawiec S (1987): Genome organization. In F.C. Neidhardt, J.L. Ingraham, K.B. Low, B. Magasanik, M. Schaechter, and H.E. Umbarger (eds.): *Escherichia coli* and *Salmonella typhimurium*: Cellular and Molecular Biology. Washington, DC: American Society for Microbiology, pp. 967–981.

Riley M, Pardee AB, Jacob F, Monod J (1960): On the expression of a structural gene. J Mol Biol 2:216–255.

Sawyer SA, Dykhuizen DE, Hartl DL (1987): Confidence interval for the number of selectively neutral amino acid polymorphisms. Proc Natl Acad Sci USA 84:6225–6228.

Segall AM, Roth JR (1989): Recombination between homologies in direct and inverse orientation in the chromosome of Salmonella: intervals which are nonpermissive for inversion formation. Genetics 122:737–747.

Stoltzfus A, Leslie JF, Milkman R (1988): Molecular evolution of the *Escherichia coli* chromosome. I. Analysis of structure and natural variation in a previously uncharacterized region between *trp* and *tonB*. Genetics 120:345–358.

Van Vliet F, Cunin R, Jacobs A, Piette J, Gigot D, Pierard A. Glansdorff N (1984): Evolutionary divergence of genes for ornithine and aspartate carbamoyl-transferases—complete sequence and mode of regulation of the *Escherichia coli argF* gene; comparison of *argF* with *argI* and *pyrB*. Nucleic Acids Res 12:6277–6289.

Van Vliet F, Boyen A, Glansdorff N (1988): On interspecies gene transfer: The case of the *argF* gene of *Escherichia coli*. Ann Inst Pasteur/Microbiol 139:493–496.

Vapnek D, Rupp D (1970): Asymmetric segretation of the complementary sex- factor DNA strands during conjugation in *E. coli*. J Mol Biol 53:287–303.

Wollman EL, Jacob F, Hayes W (1956): Conjugation and genetic recombination in *Escherichia coli* K-12. Cold Spring Harbor Symp Quant Biol 21:142–162.

**Perspectives on Cellular Regulation:
From Bacteria to Cancer, pages 17–29**

The *Escherichia coli* Chromosome: Organization, Replication, and Reminiscence

Millicent Masters

Institute of Cell and Molecular Biology, University of Edinburgh, Edinburgh EH9 3JR, Scotland

My contribution to this volume will be in two parts. Reminiscence and reflection seem appropriate in a Festschrift volume and the first part of my contribution will contain both, a reminiscence on my time in the Pardee group and a reflection on the work done in the late 1960s and early 1970s in many labs—my involvement in which grew directly out of my earlier work in the Pardee lab—to determine the sequence in which the genes on the *coli* chromosome are replicated. The straightforward interpretation of much of this latter work was made difficult by a generally accepted false premise (see below), and a shared desire to reach a consensus, consistent with this premise, which would also accommodate the results from a number of laboratories. This particular story provides an object lesson in the potential dangers of reaching truth through consensus, and is worth recalling for that reason.

The second part of my contribution will be a brief review of current thinking on the *Escherichia coli* chromosome and its replication.

From Berkeley to Princeton: 1959–1964

I entered the Biochemistry Department at the University of California, Berkeley in 1959 as a graduate student, a naive ex-chemistry major from a New York City college, unfamiliar with even the rudiments of DNA. This ignorance was soon remedied. Excited by G. Stent's lectures on bacteriophage and the origins of molecular biology, I sought a research supervisor in this then new area and was directed towards Art's group. Fortunately (as Monica Riley was just about to depart for a position at Davis) there was space available in the crowded lab on the top floor of the Virus Lab. After an apprenticeship project, in which I showed that UVed cells that produce no active β-galactosidase on induction, produce no inactive enzyme either (Masters and Pardee,

1962), a longer term Ph.D. project was required. Steve Zamenhof was a visitor in the lab in the summer of 1961 and he was able to initiate me into the mysteries of the recently developed *Bacillus subtilis* transformation system: DNA at last! My project was to look at gene expression after transformation (although this eventually proved unfeasible as competent cells were not metabolically active). By this time the lab had moved to Princeton and there were frequent contacts between our group and that of Sueoka. Sueoka and Yoshikawa were in the process of doing their seminal work, which showed that the *B. subtilis* chromosome replicates sequentially from a fixed origin (Yoshikawa and Sueoka, 1963) and, later, that in rapidly growing cells, there are overlapping rounds of replication (Yoshikawa et al., 1964). Peter Kuempel, in Pardee's lab, was engaged in trying to synchronize *E. coli* cells to reexamine the surprising earlier finding of Abbo and Pardee (1960) that, in synchronously dividing cells of *E. coli*, nothing else appeared to be phased. With synchrony in the air, I soon noticed that my cultures of *B. subtilis*, without much in the way of special treatment, appeared to be dividing synchronously. Having by this time adapted some enzyme assays for use in *B. subtilis* extracts, and taking advantage of *subtilis*'s propensity for synchronous growth, I showed that not only did enzyme inducibility double at characteristic times in the cell cycle (confirming that DNA replication was sequential, most likely from a fixed origin) but that a range of biosynthetic and degradative enzymes were synthesized discontinuously, each at a characteristic time in the cell cycle. Peter had obtained similar results with *E. coli* and we published two papers in 1964 and 1965 (Masters et al., 1964; Kuempel et al., 1965). Art was excited by these results and arranged for the latter paper to include some computer modeling. Starting with the premise that DNA replication was sequential and phased in our synchronous cultures, steps in enzyme synthesis could be generated, with periodic gene doublings as the driver and overshoots and repression acting in opposition to stabilize stepwise syntheses.

Meanwhile Sueoka and his colleagues had developed a method of genetic mapping in *B. subtilis* based on their findings that gene frequency distributions in growing populations were skewed, with genes replicated early in the replication cycle more abundant than genes replicated late. I was able to apply their methods to show that the order of discontinuous enzyme synthesis corresponded to the replication order of the relevant genes (Masters and Pardee, 1965).

Having by then completed my Ph.D. work, I departed first to Edinburgh and then to William Hayes' Microbial Genetics Research Unit in London to work in collaboration with Willie Donachie. Between us we showed that although increases in rates of enzyme induction required continuing DNA replication, cyclic enzyme synthesis steps did not, and that, moreover, timing of enzyme synthesis could be shifted relative to division. Also, the observed steps appeared, at least in part, to be caused by enzyme instability; it thus seemed that their origin lay, most probably, in a complex entrainment process in which gene dosage was an important factor, but in which other factors were obviously also involved (Donachie and Masters, 1969). The potential complexities were daunting and, at this point, Willie transferred his interests to cell division and I turned to the study of the *E. coli* chromosome and its replication: DNA again.

E. coli Chromosome Replication: The 1960s and 1970s

There was much excitement regarding *coli* chromosome replication in the mid-1960s, sparked by Cairns' beautiful autoradiogram, which provided immediate visual evidence that the replicating *E. coli* chromosome is a circular structure with two forks (Cairns,1963). Cairns interpreted the labeling pattern of this autoradiogram as showing that one of the forks was the site and the other the origin of replication, and many were stimulated to try and demonstrate this to be so. Equally provocative were the experiments of Nagata (1963) from which he concluded that Hfr strains of *E. coli* replicated sequentially from a fixed origin (the insertion site of F) but that F^- strains did not. The false hares set off by these undoubtedly mistaken interpretations led to many papers being published between 1965 and 1970 (I will cite only some of these here; for a review see Kuempel, 1970) in which a wide variety of techniques were used to determine whether replication was from a fixed chromosomal origin in male and female cells and, if so, at what point on the genetic map it was located. Synchronously dividing cultures were used to measure the rate at which several enzymes (whose genes were widely spaced on the chromosome) could be induced (it was anticipated that these rates would double soon after gene duplication) (Donachie and Masters, 1966; Pato and Glaser, 1968); similarly, rates of nitrosoguanidine (NTG) induced mutagenesis for a selection of genes (NTG acts at the replicating fork) were determined (Cerda-Olmeda et al., 1968). Cultures aligned to initiate DNA replication synchronously were used in order to label [with the heavy thymine analog, bromouracil (BU)], either the origins or termini of replication; the labeled DNA was encapsidated in P1 transducing phages, which were separated by density and used in transduction assays to determine how the genetic markers were distributed among phages of different densities (Wolf et al., 1968; Caro and Berg, 1969).

It was quickly agreed that the majority of strains, both male and female, had the same fixed origin of replication. But although all the results obtained were either consistent with or only consistent with the interpretation that replication starts in the upper left-hand quadrant of the chromosome and terminates opposite, we were all so under the spell of mistaken preconception that we failed to see this. Instead we mostly came to the conclusion that replication was unidirectional from an origin in the lower left quadrant of the map. I well remember a Cell Cycle meeting in Oak Ridge in 1967 when a group of us convened to try and correlate our results and agreed on that conclusion, even though convoluted special pleading was needed in many cases to fit the data into this preconception (see Masters, 1970 for an example of this).

Not until 1970 did we find ourselves able to discard the unidirectional paradigm and conclude that replication was bidirectional from a fixed origin. First, Nishioka and Eisenstark (1970) (using synchronously growing *Salmonella*, BU labeling, and transduction to measure marker frequency) introduced a period of BU labeling into the middle of the replication cycle and found that markers on opposite sides of the chromosome were labeled. Then I, using transduction alone and capitalizing on the gene frequency differences expected in cultures growing at different rates concluded that replication in *coli* is bidirectional (Masters and Broda, 1971). This was quickly followed by molecular confirmation—Bird et al. (1972), with their classical analysis of gene frequency using hybridization, the autoradiographs of Prescott and Kuempel (1972), the demonstration from my lab that BU label incorporated at the time of initiation was in the middle of the earliest synthesized fragment (McKenna and Masters, 1972)—and then finally, years later, and using new technology, the demonstration that label present during initiation appears sequentially in the restriction fragments located to the left and right of *oriC* (Marsh and Worcel, 1977). Although there are undoubtedly still some unexplained anomalies in the late 1960s data, as a whole they so clearly demonstrate bidirectional replication that it is difficult to imagine, at this time, why it took some fine minds so long to arrive at that conclusion.

The *E. coli* Chromosome: 1990

The *E. coli* chromosome is a single circular molecule 4700 kb in length. Although its approximate size has been known for many years (viscosity, autoradiography measurements, etc.), the relatively accurate figure quoted above is the result of two fairly recent determinations. The first used restriction by enzymes with 8 bp recognition sequences followed by pulse-gel electrophoresis to cut the chromosome into a manageable number of large restriction fragments that could be separated and have their lengths summed (Smith et al., 1987). (This technique has now been applied to measure the length of several other bacterial chromosomes.) The second was a tour-de-force of concentrated effort (Kohara et al., 1987), which resulted in the creation of an ordered overlapping clone bank in λ and a concomitantly determined restriction map (using eight enzymes) that covers almost the entire length of the chromosome. The existence of these clones is already simplifying the study of the *E. coli* genetic complement immensely: they are being used for sequencing, as the raw material for subcloning, and as a molecular substitute for conventional genetic mapping—the chromosomal position of a cloned gene can quickly be determined by hybridization to gridded filters containing DNA of an overlapping subset of the phages (these filters are likely to be commercially available soon) or by comparison of restriction patterns.

The average *E. coli* protein has a molecular size of 45 kDa and thus requires 1.2 kb of DNA to encode it. If reading frames were packed end to end the chromosome could encode 3800 proteins of this size. Since there are some intragenic regions, perhaps 3500 seems a reasonable upper limit to the number of genes we can expect eventually to be able to identify. Using 2-D polyacrylamide gels, Neidhardt's group have already catalogued over 2000 different polypeptides produced by *E. coli* growing under a variety of conditions (Phillips et al., 1987). The last edition of the genetic map (Bachmann, 1990) lists over 1400 mapped genes. Although only about 200 of these have so far been identified with particular proteins on the 2-D gels, further correlations can certainly be expected soon. This information, combined with the expected results of macrosequencing projects (i.e., the whole chromosome) reputed to be underway, should soon define the chromosomal arrangement of all the genes in the *E. coli* genetic complement. Uncovering the functions of all these newly identified coding units should, however, keep many scientists busy for some time yet!

How are all these genes arranged on the chromosome? The chromosome replicates bidirectionally from an origin (at 84 min on the genetic map) to a terminus opposite and requires 40 min (at 37°C) to do so. As a consequence, in rapidly growing cultures, origin proximal genes can be up to four times as numerous as those near the terminus. Thus, it would be advantageous for genes encoding products required in high concentrations to be located in the origin region. Such an arrangement, broadly speaking, is apparent; genes whose products are required for the transcription and translation apparatus of the cell, which might be expected to be operating flat out at high growth rates, are all located within the chromosomal quadrant surrounding the origin. The genes encoding ribosomal RNA (rrn) are not only present in seven copies (the only *coli* genes repeated in this way) but five of the copies are close to the origin. A further adaptation for efficient translation also seems apparent (Brewer, 1988). All strongly transcribed genes are arranged on the chromosome such that their directions of transcription are away from the origin (i.e., in the same direction as they are replicated). This may be to minimize collisions between RNA and DNA polymerases, which could interfere with replication. This appears plausible as weakly or infrequently transcribed genes are not arranged directionally. Further support is provided by the observation that translocations, duplications, or inversions across the origin or terminus that would not alter direction of transcription are more easily isolated than are inversions within a single arm.

If the chromosomal positions of mapped genes are analyzed, it can be seen that although genes are not distributed equally along the length of the chromosome there are few gene sparse gaps. Two regions that do appear relatively gene sparse are at 76–80 min and, more strikingly, the region in which replication terminates at 29–35 min. This latter region, which contains directional stops for replication forks (see P. Kuempel's article and Masters, 1989 for references), has been much studied. It can be deleted from the chromosome without ill effect (as might be expected if the absence of mapped genes denotes an absence of essential functions) and is rich in DNA sequences of possible phage or plasmid origin. Even a portion of the region not previously identified as containing DNA of foreign origin is less conserved amongst related enterobacteria than are sequences from other chromosomal regions. Indeed some portions of the studied region are found only in *E. coli* (Masters and Oliver, unpublished). Despite this, cloned DNA from the region is fully capable of promoting protein synthesis, proving intact coding units to be present. It has been variously suggested that the terminus region contains DNA either on the way in or on the way out of the chromosome. It would be interesting to see if other species of bacteria possess similar inessential regions in the vicinity of their replication termini.

Physical State of Chromosomal DNA

The extended length of the unreplicated chromosome, at 1.4 mm, is about 1000 times the length of the cell containing it. The chromosome (Schmid, 1988), even when replicating, is visible as a well-defined mass in the center of the cell, and can be extracted intact using gentle lysis conditions (Drlica, 1987). Within the cell completed daughter chromosomes do not separate into two discrete bodies until decatenated by gyrase and until some post-termination protein synthesis occurs. They then rapidly move to positions such that when the cell later divides in the middle they will be located in the centres of the daughter cells (this process will be discussed further by W. Donachie, this volume).

What maintains the chromosome in a compact and folded state within the cell? The short answer to this question is that we do not know. The chromosome appears to be organized into perhaps 50 domains of supercoiling, but supercoiling alone is insufficient to account for the compact size of the nuclear body. A number of small basic proteins, akin to histones, have been identified in *E. coli* (Drlica and Rouviere-Yaniv, 1987) but no compelling evidence indicates that they are responsible for maintaining *coli* DNA in a nucleosome-like structure or, indeed, that they are even associated with the nuclear body (Dürrenberger et al., 1988). It is furthermore not clear what determines the position of the nuclear body in the cell. There is evidence that indicates that the replication origin is attached to the cell membrane (Firshein, 1989), but whether this attachment is instrumental in determining the location of the nucleus is not known.

How can the DNA in a tightly packaged nucleus serve as a template for transcription and, since transcription and translation are coupled in *E. coli*, how can extranuclear ribosomes translate an intranuclear message? The answer of course is that it cannot and that genes in the process of transcription must therefore be on the outside of the nucleus. The nucleus is indeed more

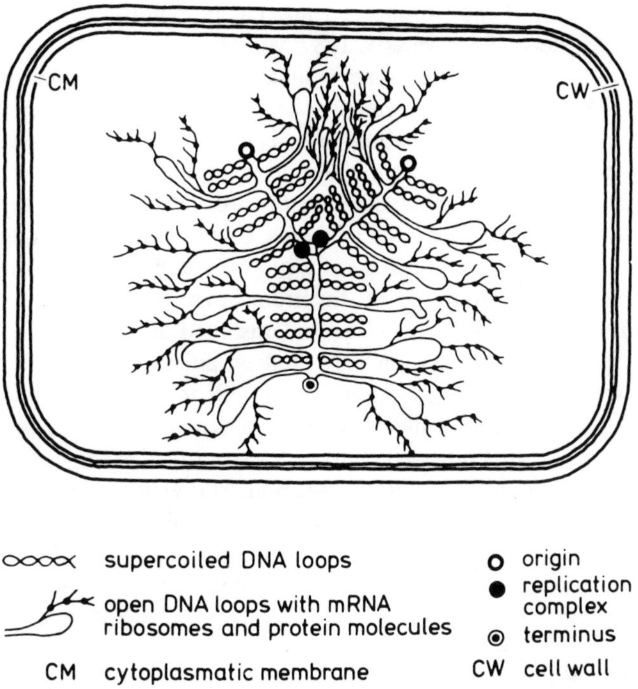

Fig. 1. Schematic drawing showing possible nuclear structure and flow of information in an *E. coli* cell. Note that compacted, inactive DNA is on the inner part of the structure while active DNA is at the edges. The two oppositely moving replication forks are shown as close together; there is no evidence for or against this. The symmetrical arrangement of loops in the unreplicated portion of the chromosome depicted is not meant to reflect the in vivo situation. The number of loops shown is about one-third to one-half the number believed to exist in vivo. Reproduced, with permission, from Hobot et al. (1985), from a drawing by J. Gumpert.

diffuse in actively growing cells than in those inhibited from making protein and there is also evidence that loops of DNA extend into the ribosome-filled cytoplasm (Hobot et al., 1987); it is these that may be in the process of active transcription. The chromosome in the cell may well adapt something like the form represented in the schematic diagram (Fig. 1) with a central portion consisting of restrained inactive DNA while DNA in the process of transcription and translation is located at the periphery.

Chromosome Replication

This subject has been reviewed by several different sets of authors recently and therefore I will try here to give only a very general overview of the current state of knowledge and outstanding problems; further details can be found in the reviews (Bramhill and Kornberg, 1988; Georgopolous, 1989; McMacken et al., 1987; Masters, 1989; Messer, 1987; von Meyenburg and Hansen, 1987).

Initiation of *E. coli* chromosome replication occurs at a single intergenic sequence, 245 bp in length, called *oriC*. This sequence has a number of sites, conserved in related bacteria, which are essential for initiation activity. They include binding sites for the essential initiation protein DnaA (4–5 in *E. coli*), GATC sites at which deoxyadenosine methylase (DAM) adds methyl groups (11 in *E. coli*), and 13-mer sites at which DnaB helicase is thought to be loaded onto the DNA. The spacing, as well as the sequence, of most of these sites appears to be essential for *oriC* function. In vivo preconditions for initiation include transcription by RNA polymerase and attainment of a specific cell mass, the initiation mass. In vitro studies, although they have not been able to illuminate these two requirements, have led to the development of the model shown in Figure 2. A supercoiled template in the presence of the DNA binding protein HU (not required in vivo; deletion mutants are viable) (Ogawa et al., 1989) the initiation protein DnaA, and ATP yields a structure, visible by electron microscopy, in which 20–40 monomers of DnaA are bound to 225 bp of DNA introducing a bend into the DNA. DnaC can then deliver DnaB to this complex, forming a larger structure that incorporates a further 55 bp (containing 3 tandem 13-mer repeats) from the left hand side of *oriC*. DnaB but not DnaC is retained in the complex. DnaB is a helicase that promotes unwinding in both directions from the site of its loading at *oriC*. It has been suggested that this bidirectionality may result from the fact that the two pairs of DnaA binding sites in the origin are oppositely oriented and that this serves to direct DnaB to one strand or the other (Georgopoulos, 1989); this remains to be proven. Following strand separation and helicase unwinding at the origin, DnaG primase and the complex DNA polymerase holoenzyme may then enter the nascent fork and replication proceed. The recent purification of DNA polymerase III (which contains at least 10 different sorts of subunit), as a 900-kDa particle, which is judged to be an asymmetric dimer, lends support to the suggestion that a single replisome concurrently copies both leading and lagging strand templates (Maki et al., 1988). For this to occur the lagging strand template must be looped out so that synthesis can proceed from 3′ to 5′ in both directions (Fig. 3). It is not yet clear whether or not replication begins simultaneously at the two forks that proceed around the chromosome in opposite directions.

Initiation of Replication: Control

Initiation of replication occurs simultaneously at all copies of *oriC* present in the cell (Leonard and Helmstetter, 1986) once per cell cycle at the time when the fixed initiation mass is reached. The signals that trigger this initiation event, despite years of effort on the part of many, still remain elusive.

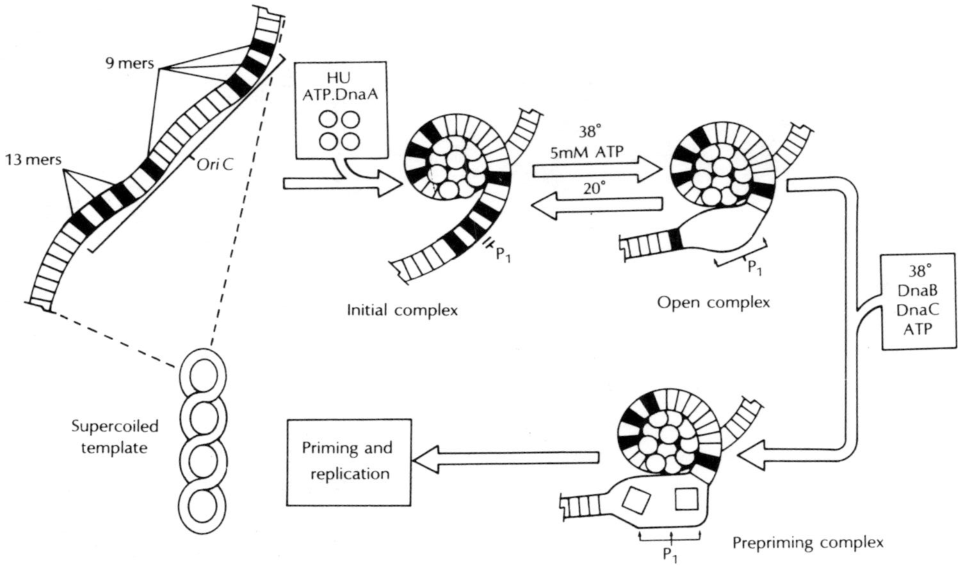

Fig. 2. A scheme for initiation at *oriC*. The DnaA protein binds the four 9-mers, organizing *oriC* around a protein core to form the initial complex. The three 13-mers are then melted serially by DnaA protein to create the open complex. The DnaB–DnaC complex can now be directed to the 13-mer region to extend the duplex opening and generate a prepriming complex, which unwinds the template for priming and replication. Adapted from Bramhill and Kornberg (1988) and reproduced from Masters (1989).

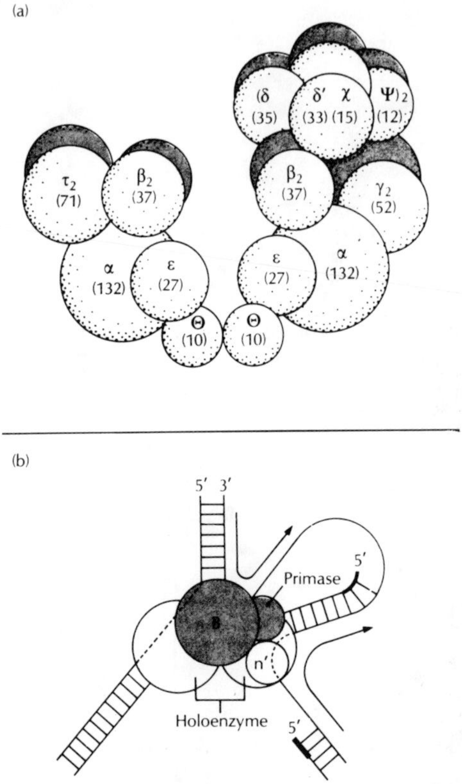

Fig. 3. **(a)** Provisional structure of DNA polymerase III holoenzyme, showing the asymmetry that may allow concomitant replication of both strands at a single fork. Figures in parentheses are the molecular weights of each subunit $\times\ 10^{-3}$. **(b)** Model for the primosome at a replication fork. The bars represent RNA primers, and arrows indicate the direction DNA moves through the protein complex. Note that both strands are being replicated at once. Adapted from Maki et al. (1988) and Lasken and Kornberg (1988) and reproduced from Masters (1989).

According to current thinking, the initiator protein DnaA, postreplicational DAM-methylation at GATC sites in *oriC*, and cell membrane attachment are all of central importance in the timing process. Mutations in either the *dnaA* (Skarstad et al., 1988) or *dam* (Bakker and Smith, 1989) genes disturb the synchrony of initiation and overproduction of either DnaA (Skarstad et al., 1989) or DAM-methylase (Messer et al., 1985) can reduce initiation mass, confirming the central role of these activities in timing. There is a long history of reports of *oriC*-membrane association (Firshein, 1989; Gudas et al., 1976), and recently it has been shown that hemimethylated DNA has a particular affinity for cell membrane (Ogden et al., 1988). DnaA protein is reasonably plentiful in the cell and thus its total amount it unlikely to be rate-limiting for initiation. It occurs however in both an ATP bound and an ADP bound form, only the former of which is active in initiation (Sekimizu et al., 1987). Thus conversion by whatever means to the active DnaA.ATP form could be a controlling event in initiation timing (Hwang and Kaguni, 1988). It has been suggested that membrane lipids may be active in the conversion process, (Sekimizu and Kornberg, 1988; Yung and Kornberg, 1988), consistent with the report that a considerable fraction of DnaA protein is membrane associated (Sekimizu et al., 1988).

A tentative model, based on these considerations, might be as follows. Initiation occurs at origins that are not membrane attached. The hemimethylated initial product of replication binds to the membrane and is sequestered from acting as a substrate for further initiations. It becomes fully methylated after about 10 min but does not detach until initiation mass is again reached. Detachment then occurs by a "black-box" mechanism which could involve *oriC* binding of newly activated DnaA, ATP binding of DnaA already attached at *oriC* (Samitt et al., 1989), or a change in state of the membrane attachment site. Convincing evidence to support this or any other model is eagerly awaited—the problem of how initiation of replication is timed, so central to our understanding of how the bacterial cell works as an organism, is a question that, although formulated over 20 years ago, has remained stubbornly refractory to solution.

References

Abbo FE, Pardee AB (1960): Synthesis of macromolecules in synchronously dividing bacteria. Biochim Biophys Acta 39:478–485.

Bachmann BJ (1990): Linkage map of *Escherichia coli* K-12, Edition 8. Microbiol Rev 54:130–197.

Bakker A, Smith DW (1989): Methylation of GATC sites is required for precise timing between rounds of DNA replication in *Escherichia coli*. J Bacteriol 171:5738–5742.

Bird R, Louarn J, Martuscelli J, Caro L (1972): Origin and sequence of chromosome replication in *Escherichia coli*. J Mol Biol 70:549–566.

Bramhill D, Kornberg A (1988): A model for initiation at origins of DNA replication. Cell 54:915–918.

Brewer BJ (1988): When polymerases collide: Replication and the transcriptional organization of the *E. coli* chromosome. Cell 53:679–686.

Cairns J (1963): The chromosome of *Escherichia coli*. Cold Spring Harbor Symp Quant Biol 28:43–46.

Caro L, Berg CM (1969): Chromosome replication in *Escherichia coli*. II. Origin of replication in F^- and F^+ strains. J Mol Biol 45:325–336.

Cerda-Olmeda E, Hanawalt PC, Guerola N (1968): Mutagenesis of the replication point by nitrosoguanidine: Map and pattern of replication of the *Escherichia coli* chromosome. J Mol Biol 33:705–719.

Donachie WD, Masters M (1966): Evidence for polarity of chromosome replication in F^- strains of *Escherichia coli*. Genet Res Camb 8:119–124.

Donachie WD, Masters M (1969): Temporal control of gene expression in bacteria. In G.M. Padilla, G.L. Whitson, and I.L. Cameron (eds): The Cell Cycle. Gene-Enzyme Interactions. New York: Academic Press, pp. 37–76.

Drlica K (1987): The nucleoid. In F.C. Neidhardt (ed): *Escherichia coli* and *Salmonella typhimurium*: Cellular and Molecular Biology. Washington: American Society of Microbiology, pp. 91–103.

Drlica K, Rouviere-Yaniv (1987): Histone-like proteins of bacteria. Microbiol Rev 51:301–319.

Dürrenberger M, Bjornsti MA, Uetz T, Hobot JA, Kellenberger E (1988): Intracellular location of the histone-like protein HU in *Escherichia coli*. J Bacteriol 170:4757–4768.

Firshein W (1989): Role of the DNA/membrane complex in procaryotic DNA replication. Annu Rev Microbiol 43:89–120.

Georgopoulos C (1989): The *E. coli* dnaA initiation protein: A protein for all seasons. Trends Genet 5:319–321.

Gudas LJ, James R, Pardee AB (1976): Evidence for the involvement of an outer membrane protein in DNA initiation. J Biol Chem 251:3470–3479.

Hobot JA, Bjornsti MA, Kellenberger E (1987): Use of on-section immunolabeling and cryosubstitution for studies of bacterial DNA distribution. J Bacteriol 169:2055–2062.

Hobot JA, Villiger W, Escaig J, Maeder M, Ryter A, Kellenberger E (1985): Shape and fine structure of nucleoids observed on sections of ultrarapidly frozen and cryosubstituted bacteria. J Bacteriol 162:960–971.

Hwang DS, Kaguni JM (1988): Interaction of *dna*A46 protein with a stimulatory protein in replication from the *Escherichia coli* chromosomal origin. J Biol Chem 263:10633–10640.

Kohara Y, Akiyama K, Isono K (1987): The physical map of the whole *E. coli* chromosome: Application of a new strategy for rapid analysis and sorting of a large genomic library. Cell 50:495–508.

Kuempel P (1970): Bacterial chromosome replication. In D.M. Prescott et al. (eds): Advances in Cell Biology, Vol 1. New York: Appleton-Century-Crofts, pp. 3–56.

Kuempel PL, Masters M, Pardee AB (1965): Bursts of enzyme synthesis in the bacterial duplication cycle. Biochem Biophys Res Commun 18:858–867.

Lasken RS, Kornberg A (1988): The primosomal protein n′ of *Escherichia coli* is a DNA helicase. J Biol Chem 263:5512–5518.

Leonard AC, Helmstetter CE (1986): Cell cycle-specific replication of *Escherichia coli* minichromosomes. Proc Natl Acad Sci USA 83:5101–5105.

Lobner-Olesen A, Skarstad K, Hansen FG, von Meyenburg K, Boye E (1989): The DnaA protein determined the initiation mass of *Escherichia coli* K-12. Cell 57:881–889.

Maki H, Maki S, Kornberg A (1988): DNA polymerase II holoenzyme of *Escherichia coli* IV. The holoenzyme as an asymmetric dimer with twin active sites. J Biol Chem 263:6570–6578.

Marsh RC, Worcel A (1977): A DNA fragment containing the origin of replication of the *Escherichia coli* chromosome. Proc Natl Acad Sci USA 74:2720–2724.

Masters M (1970): Origin and direction of replication of the chromosome of *E. coli* B/r. Proc Natl Acad Sci USA 65:601–608.

Masters M (1989): The *Escherichia coli* chromosome and its replication. Curr Opinion Cell Biol 1:241–249.

Masters M, Broda P (1971): Evidence for the bidirectional replication of the *Escherichia coli* chromosome. Nature New Biol 232:137–140.

Masters M, Kuempel PL, Pardee AB (1964): Enzyme synthesis in synchronous cultures of bacteria. Biochem Biophys Res Commun 15:38–42.

Masters M, Pardee AB (1962): Failure of ultraviolet-irradiated *Escherichia coli* to produce a cross-reacting protein. Biochim Biophys Acta 56:609–611.

Masters M, Pardee AB (1965): Sequence of enzyme synthesis and gene replication during the cell cycle of *Bacillus subtilis*. Proc Natl Acad Sci USA 54:64–70.

McKenna WG, Masters M (1972): Biochemical evidence for the bidirectional replication of DNA in *Escherichia coli*. Nature (London) 240:536–539.

McMacken R, Silver L, Georgopoulos C (1987): DNA replication. In F.C. Neidhardt (ed): *Escherichia coli* and *Salmonella typhimurium*: Cellular and Molecular Biology. Washington: American Society of Microbiology, pp. 564–610.

Messer W (1987): Initiation of DNA replication in *Escherichia coli*. J Bacteriol 169:3395–3399.
Messer W, Bellekes U, Lother H (1985): Effect of *dam* methylation on the activity of the *E. coli* replication origin. EMBO J 4:1327–1332.
Nagata T (1963): The molecular synchrony and sequential replication of DNA in *Escherichia coli*. Proc Natl Acad Sci USA 49:551–559.
Nishioka Y, Eisenstark A (1970): Sequence of genes replicated in *Salmonella typhimurium* as examined by transduction techniques. J Bacteriol 102:320–333.
Ogawa T, Wada M, Kano Y, Imamoto F, Okazaki T (1989): DNA replication in *Escherichia coli* mutants that lack protein HU. J Bacteriol 171:5672–5679.
Ogden GB, Pratt MJ, Schaechter M (1988): The replicative origin of the *E. coli* chromosome binds to cell membranes only when hemimethylated. Cell 54:127–135.
Pato ML, Glaser DA (1968): The origin and direction of replication of the chromosome of *Escherichia coli* B/r. Proc Natl Acad Sci USA 60:1268–1274.
Phillips TA, Vaughn V, Block PL, Neidhardt FC (1987): Gene-protein index of *Escherichia coli*. In F.C. Neidhardt (ed): *Escherichia coli* and *Salmonella typhimurium*: Cellular and Molecular Biology. Washington: American Society of Microbiology, pp. 919–966.
Prescott DM, Kuempel PL (1972): Bidirectional replication of the chromosome in *Escherichia coli*. Proc Natl Acad Sci USA 69:2842–2845.
Samitt CE, Hansen FG, Miller JF, Schaechter M (1989): *In vivo* studies of DnaA binding to the origin of replication of *Escherichia coli*. EMBO J 8:989–993.
Schmid M (1988): Structure and function of the bacterial chromosome. Trends Biol Sci 13:131–135.
Sekimizu K, Bramhill D, Kornberg A (1987): ATP activates dnaA protein in initiating replication of plasmids bearing the origin of the *E. coli* chromosome. Cell 50:259–266.
Sekimizu K, Kornberg A (1988): Cardiolipin activation of dnaA protein, the initiation protein of replication in *Escherichia coli*. J Biol Chem 263:7131–7135.
Sekimizu K, Yung BY, Kornberg A (1988): The dnaA protein of *Escherichia coli*. Abundance, improved purification, and membrane binding. J Biol Chem 263:7136–7140.
Skarstadt K, Lobner-Olesen A, Atlung T, von Myenburg K, Boye E (1989): Initiation of DNA replication in *Escherichia coli* after overproduction of the DnaA protein. Mol Gen Genet 57:881–889.
Skarstad K, von Meyenburg K, Hansen FG, Boye E (1988): Coordination of chromosome replication initiation in *Escherichia coli*: Effects of different *dnaA* alleles. J Bacteriol 170:852–858.
Smith CL, Econome J, Schutt A, Klco S, Cantor CR (1987): A physical map of the *E. coli* K12 genome. Science 236:4481–4490.
von Meyenburg K, Hansen FG (1987): Regulation of chromosome replication. In F.C. Neidhardt (ed): *Escherichia coli* and *Salmonella typhimurium*: Cellular and Molecular Biology. Washington: American Society of Microbiology, pp. 1555–1577.
Wolf B, Newman A, Glaser DA (1968): On the origin and direction of replication of the *Escherichia coli* K12 chromosome. J Mol Biol 32:611–629.
Yoshikawa H, Sueoka N (1963): Sequential replication of *Bacillus subtilis* chromosome I. Comparison of marker frequencies in exponential and stationary growth phases. Proc Natl Acad Sci USA 49:559–566.
Yoshikawa N, O'Sullivan A, Sueoka N (1964): Sequential replication of the *Bacillus subtilis* chromosome III. Regulation of initiation. Proc Natl Acad Sci USA 52:973–980.
Yung BY, Kornberg A (1988): Membrane attachment activates dnaA protein, the initiation protein of chromosome replication in *Escherichia coli*. Proc Natl Acad Sci USA 85:7202–7205.

Perspectives on Cellular Regulation:
From Bacteria to Cancer, pages 31–38

Organization of *Escherichia coli* DNA Replication Genes

James R. Walker

Department of Microbiology, The University of Texas, Austin, Texas 78712

Replication of the *Escherichia coli* chromosome, the single most important controlling step of the cell cycle, occurs in three complex stages. Polymerization initiates near the *oriC* region (Tabata et al., 1983) and proceeds bidirectionally around the 4.55 megabase pair (Kohara et al., 1987) chromosome. Initiation depends on ATP-activated DnaA protein binding to four 9-bp DnaA Boxes of *oriC* in the presence of HU protein and unwinding three adjacent AT-rich 13-mers (Fuller et al., 1984; Sekimizu et al., 1987; Bramhill and Kornberg, 1988; Seufert et al., 1988; Samitt et al., 1989; Kowalski and Eddy, 1989). More extensive unwinding by the helicase activity of DnaB protein (LeBowitz and McMacken, 1986) in the presence of single-strand binding protein and gyrase prepares the template for priming by primase and polymerization by DNA polymerase III (Baker et al., 1987; Sekimizu et al., 1988). The combined action of DnaB helicase and DnaG primase continues to unwind and also to prime lagging strand polymerization. To complete the model, the lagging strand from the fork moving counterclockwise becomes the leading strand of the fork moving clockwise (Seufert and Messer, 1986; Hirose et al., 1983).

Once initiated, replication forks are thought to be propagated by the 7- protein primosome, originally identified by its ability to prime ϕX174 complementary strand synthesis (Arai and Kornberg, 1981). Factor n′ (also called Factor Y) binds unique sites on single strands coated with single strand binding protein and serves to assemble the preprimosome-containing proteins n, n″, DnaT, DnaB, and DnaC. Primase, the product of *dnaG*, binds to complete the primosome (Shlomai and Kornberg, 1980; Zipursky and Marians, 1981; Lee and Marians, 1987; Lasken and Kornberg, 1988; Lee and Marians, 1989).

DNA polymerase III holoenzyme contains 10 subunits (Maki and Kornberg, 1988a) and is thought to coordinate leading and lagging strand synthesis as an asymmetric dimer (Sinha et al., 1980; Johanson and McHenry, 1984; H. Maki et al., 1988). The core subunits are α, ϵ, and θ. α is the $5'\rightarrow3'$ polymerase—the product of the *dnaE* gene (Welch and McHenry, 1982). *dnaQ* encodes ψ, the $3'\rightarrow5'$ proofreading exonuclease and processivity factor (Scheurmann et al., 1983; Studwell and O'Donnell, 1990). The function and origin of θ are unknown. Auxiliary factors include τ, γ, δ, δ′, β, χ, and Ψ. β is the *dnaN* product (Burgers et al., 1981). Both τ and γ are products of one gene, *dnaX* (Yin et al., 1986; Flower and McHenry, 1986; Lee and Walker, 1987; Hawker and McHenry, 1987; Maki and Kornberg, 1988b). The shorter γ is generated from within the τ reading frame by a programmed ribosomal frameshift followed by a stop codon in the new frame (Blinkowa and Walker, 1990).

The 5′→3′ exonuclease activity of DNA polymerase I and DNA ligase is necessary for joining the lagging strand Okazaki fragments and for viability (Konrad and Lehman, 1974; Olivera and Bonhoeffer, 1974; Gottesman et al., 1973; Nagata and Horiuchi, 1974).

Termination is an active process by which helicase movement of forks is arrested by the action of at least one protein, designated Tus, acting at specific sequences in the terminus region (Khatri et al., 1989; Lee et al., 1989; Hill et al., 1989).

The genes that encode these replication factors are located around the chromosome (Fig. 1). Some occur as single genes but others are organized in groups of related genes. The organization of four such related groups and possible advantages of their organization will be reviewed.

One group, the macromolecular synthesis operon, includes three cistrons

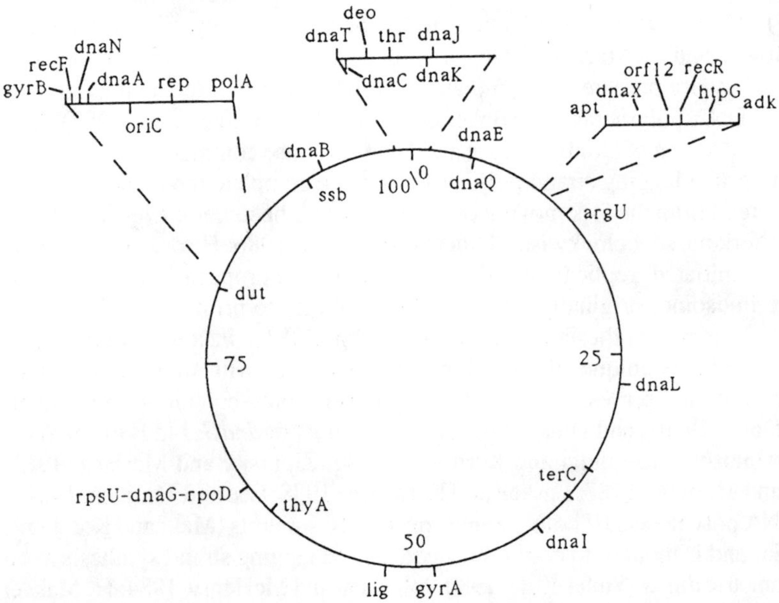

Fig. 1. The *E. coli* chromosome map showing the locations of some DNA replication (*dna*) genes (adapted from Bachmann, 1987).

encoding translation, replication, and transcription factors. This *rpsU–dnaG–rpoD* operon encodes 30 S ribosomal protein S21, primase, and the sigma-70 subunit of RNA polymerase (Lupski and Godson, 1984) and is controlled by complex regulatory mechanisms. Discoordinate cistron expression effected by an internal terminator upstream of *dnaG* and internal promoters upstream of *rpoD* establish the steady-state level of the three proteins of 50,000:50:5,000 molecules per cell (Lupski and Godson, 1984). These levels reflect the demand for the proteins in the three processes. The relative amount of sigma-70 is varied under stress conditions by a heat shock promoter (Lupski and Godson, 1984). This organization suggests that this transcription unit most likely plays a role in regulating cell growth (Nesin et al., 1988; Almond et al., 1989).

A second group is the *dnaA–dnaN–recF–gyrB* region. DnaA seems to be the critical controlling factor in initiation of replication (Georgopoulos, 1989). *dnaN* encodes the DNA polymerase III subunit β (Burgers et al., 1981), the *recF* product is required for UV repair, and the RecF pathway of recombination (Horii and Clark, 1973), and *gyrB* encodes a topoisomerase II subunit (Adachi et al., 1987). All are contiguous on a 7-kb region and transcribed in the same direction (Fig. 2A). *dnaA* is transcribed from two prmoters (F.G. Hansen, et al., 1982; E. B. Hansen, et al., 1982; Ohmori et al., 1984) and negatively regulated by DnaA protein binding to two DnaA-binding sites. Binding to the site between the promoters represses transcription (Stuitje et al., 1986; Atlung et al., 1985; Braun et al., 1985; Wang and Kaguni, 1987; Kucherer et al., 1986; Hansen et al., 1987); binding to the site within the *dnaA* coding sequence terminates transcription (Schaefer and Messer, 1989). The *dnaA* promoter region contains an unusually large number of Dam methylation sites and maximum expression from the more proximal promoter requires Dam methylation enzyme activity (Braun and Wright, 1986). *dnaA* is subject also to growth rate and stringent control (Chiaramello and Zyskind, 1989; Rokeach and Zyskind, 1986). In addition, DnaA represses other genes involved in nucleic acid metabolism, including *uvrB* (Van den Berg et al., 1985) and *rpoH* (Wang and Kaguni, 1989). Thus, the *dnaA* gene product seems to be a protein important in coordinating replication with growth.

The *dnaN* reading frame begins five nucleotides from *dnaA* and is thought to be part of the *dnaA* operon (E.B. Hansen, et al., 1982; Ohmori et al., 1984; Sako and Sakakibara, 1980; Sakakibara et al., 1981). Although *dnaA* transcripts extend into *dnaN*, there are at least three *dnaN* promoters located within the *dnaA* coding sequence (Quiñones and Messer, 1988; Armengod et al., 1988). The most distal promoter would produce messenger with a long untranslated region of about 400 nucleotides (Armengod et al., 1988). Although *dnaN* is part of the *dnaA* operon, the *dnaN* prmoters must be responsible for maintenance of *dnaN* expression when *dnaA* is decreased or for differential regulation (Quiñones and Messer, 1988; Armengod et al., 1988).

The *recF* cistron also has separate promoters located far upstream in the

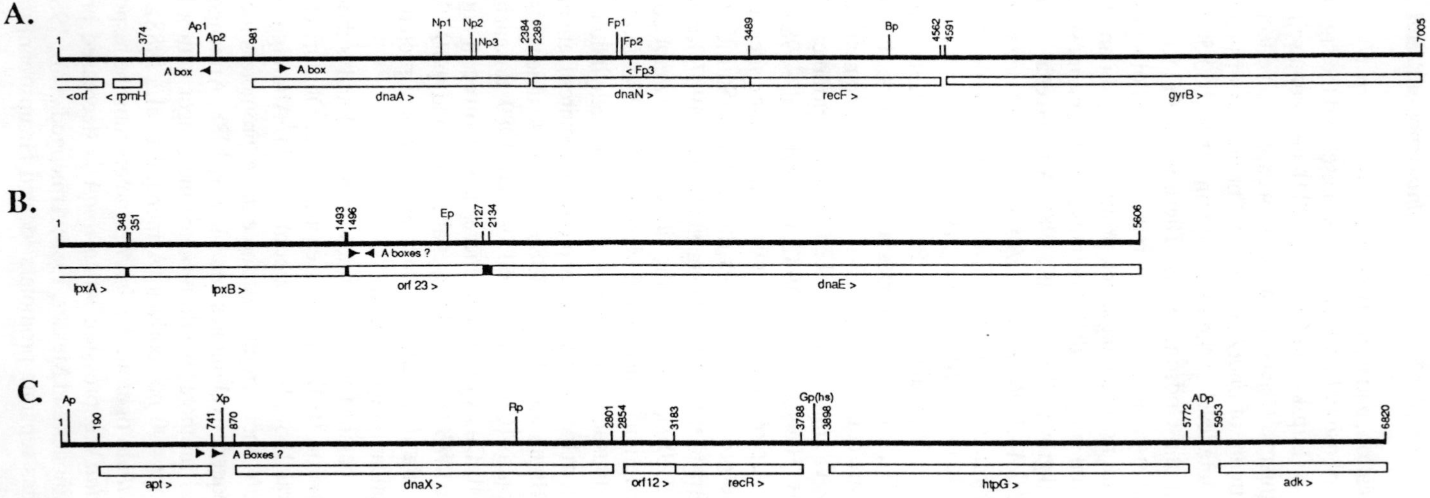

Fig. 2. Organization of *dna* replication genes. **(A)** The *dnaA* operon; **(B)** the *lpx–dnaE* operon; **(C)** the *dnaX* region. Heavy bars are chromosomal fragments with the nucleotides numbered. Open reading frame (open bars) beginning and ending nucleotides are numbered. Translation direction is indicated by the open arrow points. Promoters are indicated by a gene symbol plus p and a number where there are multiple promoters. Promoters above the DNA bar direct rightward transcription. The *recF* promoter (Fp3) directs leftward transcription. DnaA boxes and potential DnaA boxes (?) are represented by filled arrow points. *lpxA* is only a portion of the *orf*.

middle of the preceding cistron, *dnaN* (Armengod and Lambies, 1986). Maximum *recF* expression, however, requires *dnaN* expression and *recF* is, therefore, part of the *dnaA–dnaN* operon (Armengod and Lambies, 1986). In addition, *dnaN* and *recF* are probably translationally coupled (Das and Yanofsky, 1984). Their reading frames overlap by one nucleotide (Blanar et al., 1984) and interruption of *dnaN* translation reduced *recF* expression (Armengod et al., 1988). Transcription from *recF* promoters also is controlled negatively by downstream sequences in the *dnaN* reading frame by a mechanism not yet understood. Perhaps this is related to the finding of a promoter that directs an antisense transcript of *recF* (Armengod and Lambies, 1986) (Fig. 2A).

Finally, *gyrB* is located only 29 basepairs downstream of *recF* (Adachi et al., 1987; Blanar et al., 1984; Yamagishi et al., 1986), but there seems to be no evidence that *gyrB* is part of the *dnaA-dnaN-recF* operon.

A third operon contains *dnaE* (α subunit of DNA polymerase II (Welch and McHenry, 1982)), two genes involved in outer membrane lipid A biosynthesis *(lpxAB)*, and a fourth cistron (*orf*23) of unknown function (Fig. 2B) (Tomasiewicz and McHenry, 1987; Crowell et al., 1987). *lpxA* and *B* encode UDP-*N*-acetylglucosamine acyltransferase (Crowell et al., 1987) and lipid A disaccharide synthase (Tomasiewicz and McHenry, 1987; Crowell et al., 1987), The *lpxA* and *lpbB* termination codons overlap the *lpxB* and *orf*23 presumptive initiative codons (Tomasiewicz and McHenry, 1987; Crowell et al., 1987), respectively, and the *orf*23 and *dnaE* reading frames overlap by five base pairs. The *lpxA* and *lpxB* termination codons overlap the *lpxB* and *orf*23 presumptive initiative codons (Tomasiewicz and McHenry, 1987; Crowell et al., 1987), This suggests that the four are cotranscribed and translationally coupled (Tomasiewicz and McHenry, 1987; Crowell et al., 1987). [A fifth *orf*, also of unknown function, is located upstream of and overlaps *lpxA* (Crowell et al., 1987).] Inasmuch as lipid A biosynthesis is essential for growth (Nishijima et al., 1981), it seems possible that clustering of the replication and outer membrane component genes into an operon may serve to coordinate replication with growth. Two potential DnaA boxes (8 of 9 matches) are located upstream of the *dnaE* promoter within the *orf*23 coding region (Fig. 2B) (Tomasiewicz and McHenry, 1987) and one of them is in the orientation to terminate transcription (Schaefer and Messer, 1989) from an upstream promoter reading into *dnaE*. Thus, DnaA protein could contribute to control also of *dnaE*.

A fourth region of interest is a 7 kilobase pair region containing six contiguous genes most, if not all, of which are involved in nucleic acid metabolism (Fig. 2C). *apt* encodes adenine phosphoribosyltransferase (Hershey and Taylor, 1986); *dnaX* encodes τ and γ subunits of DNA polymerase III (Yin et al., 1986; Flower and McHenry, 1986; Lee and Walker, 1987; Hawker and McHenry, 1987; Maki and Kornberg, 1988b); *orf*12 function is unknown; *recR* encodes a recombinational UV repair protein (Yeung et al., 1990; Mahdi and Lloyd, 1989);*htpG* is a heat shock gene of unknown function (Bardwell and Craig, 1987); *adk* encodes the essential adenylate kinase (Cousin and

Belaich, 1966). Although there is no evidence that the *htpG* product is involved in nucleic acid metabolism, several other heat shock proteins are (Georgopoulos et al., 1982; Bardwell et al., 1986; Ang et al., 1986). Although most of these genes have separate promoters, all are transcribed in the same direction and are tightly packed with few nucleotides between reading frames. The presence of two potential DnaA boxes (7 of 9 matches) correctly oriented upstream of *dnaX* to terminate transcription proceeding toward *dnaX* suggests that DnaA protein could function also to keep *dnaX* expression limited. The *orf*12–*recR* region seems to be an operon transcribed from a promoter far upstream within *dnaX*. This messenger contains a long, untranslated leader of 199 nucleotides. The *orf*12 termination and *recR* initiation codons overlap by one nucleotide suggesting also translational coupling.

The clustering of replication genes suggests that replication "modules" were assembled and advantageously maintained during evolution. What are the advantages?

One potential advantage could be coordination of expression of related genes. For example, the *dnaA* operon seems to coordinate the synthesis of DNA polymerase III β subunit (*dnaN*) with that of the DnaA activator protein. However, there are in the macromolecular synthesis operon terminators and multiple promoters that permit discoordinate expression. In addition, the presence of the heat shock promoter in this operon provides flexibility for increasing sigma-70 expression under stress conditions.

Another potential advantage of coupling replication and other macromolecular synthesis genes is coordination of cell growth. The maximal expression of *dnaE* is likely to be dependent first on *lpx* expression, providing a mechanism to link replication to outer membrane synthesis. Once *lpx*–*dnaE* transcription is initiated upstream of *lpx*, that which extends into *dnaE* could be limited by DnaA protein. The macromolecular synthesis operon links translation, replication, and transcription.

Acknowledgments

I thank Dr. Arthur B. Pardee for the introduction to microbial genetics in his laboratory and for the opportunity to observe the benefits of combining genetic and biochemical approaches. The original work mentioned in this chapter was supported by American Cancer Society Grant MV429. The assistance of Todd C. Peters, Susan Crossland, and Judy Parker in the preparation of this manuscript is acknowledged.

References

Adachi T, Mizuuchi M, Robinson EA, Appella E, O'Dea MH, Gellert M, Mizuuchi K (1987): Nucleic Acids Res 15:771–784.

Almond N, Yajnik V, Svec P, Godson GN (1989): Mol Gen Genet 216:195–203.
Ang D, Chandrasekhar GN, Zylicz M, Georgopoulos C (1986): J Bacteriol 167:25–29.
Arai KI, Kornberg A (1981): Proc Natl Acad Sci USA 78:69–73.
Armengod M-E, Lambies E (1986): Gene 43:183–196.
Armengod M-E, García-Sogo M, Lambies E (1988): J Biol Chem 263:12109–12114.
Atlung T, Clausen ES, Hansen FG (1985): Mol Gen Genet 200:442–450.
Bachmann BJ (1987): In F.C. Neidhardt, J.L. Ingraham, K.B. Low, B. Magasanik, M. Schaechter, and H. E. Umbarger (eds): *Escherichia coli* and *Salmonella typhimurium* Cellular and Molecular Biology. Washington, DC: American Society of Microbiology, pp 807–876.
Baker TA, Funnell BE, Kornberg A (1987): J Biol Chem 262:6877–6885.
Bardwell JCA, Tilly K, Craig E, King J, Zylicz M, Georgopoulos C (1986): J Biol Chem 261:1782–1785.
Bardwell JCA, Craig EA (1987): Proc Natl Acad Sci USA 84:5177–5181.
Blanar MA, Sandler SJ, Armengod ME, Ream LW, Clark AJ (1984): Proc Natl Acad Sci USA 81:4622–4626.
Blinkowa AL, Walker JR (1990): Nucl Acids Res 18:1725–1729.
Bramhill D, Kornberg A (1988): Cell 52:743–755.
Braun RE, Wright A (1986): Mol Gen Genet 202:246–250.
Braun RE, O'Day K, Wright A (1985): Cell 40:159–169.
Burgers PMJ, Kornberg A, Sakakibara Y (1981): Proc Natl Acad Sci USA 78:5391–5395.
Chiaramello AE, Zyskind JW (1989): J Bacteriol 171:4272–4280.
Cousin D, Belaich JP (1966): Comptes Rendus Acad Sci 263:886–888.
Crowell DN, Reznikoff WS, Raetz CRH (1987): J Bacteriol 169:5727–5734.
Das A, Yanofsky C (1984): Nucleic Acids Res 12:4757–4768.
Flower AM, McHenry CS (1986): Nucleic Acids Res 14:8091–8101.
Fuller RS, Funnell BE, Kornberg A (1984): Cell 38:889–900.
Georgopoulos C (1989): Trends Genet 5:319–321.
Georgopoulos C, Tilly K, Drahos D, Hendrix R (1982): J Bacteriol 149:1175–1177.
Gottesman MM, Hicks ML, Gellert M (1973): J Mol Biol 77:531–547.
Hansen EB, Hansen FG, von Meyenbrug K (1982): Nucleic Acids Res 10:(A) 7373–7385.
Hansen FG, Hansen EB, Atlung T (1982): EMBO J 1:1043–1048.
Hansen FG, Koefoed S, Sorensen L, Atlung T (1987): EMBO J 6:255–258.
Hawker JR, McHenry CS (1987): J Biol Chem 262:12722–12727.
Hershey HV, Taylor MW (1986): Gene 43:287–293.
Hill TM, Tecklenburg ML, Pelletier AJ, Kuempel PL (1989): Proc Natl Acad Sci USA 86:1593–1597.
Horii Z, Clark AJ (1973): J Mol Biol 80:327–344.
Hirose S, Hiraga S, Okazaki T (1983): Mol Gen Genet 189:422–431.
Johanson KO, McHenry CS (1984): J Biol Chem 259:4589–4595.
Khatri GS, MacAllister T, Sista PR, Bastia D (1989): Cell 59:667–674.
Kohara Y, Akiyama K, Isono K (1987): Cell 50:495–508.
Konrad EB, Lehman IR (1974): Proc Natl Acad Sci USA 71:2048–2051.
Kowalski D, Eddy MJ (1989): EMBO J 8:4335–4344.
Kucherer C, Lother H, Kolling R, Schauzu M, Messer W (1986): Mol Gen Genet 205:115–121.
Lasken RS, Kornberg A (1988): J Biol Chem 263:5512–5518.
LeBowitz JH, McMacken R (1986): J Biol Chem 261:4738–4748.
Lee EH, Kornberg A, Hidaka M, Kobayashi T, Horiuchi T (1989): PNAS 86:9104–9108.
Lee MS, Marians KJ (1987): Proc Natl Acad Sci USA 84:8345–8349.
Lee MS, Marians KJ (1989): J Biol Chem 264:14531–14542.
Lee S-H, Walker JR (1987): Proc Natl Acad Sci USA 84:2713–2717.
Lupski JR, Godson GN (1984): Cell 39:251–252.

Mahdi AA, Lloyd RG (1989): Nucl Acids Res 17:6781–6794.
Maki H, Maki S, Kornberg A (1988): J Biol Chem 263:6570–6578.
Maki S, Kornberg A (1988a): J Biol Chem 263:6561–6569.
Maki S, Kornberg A (1988b): J Biol Chem 263:6547–6554.
Nagata T, Horiuchi T (1974): J Mol Biol 87:369–373.
Nesin M, Lupski JR, Godson GN (1988): J Bacteriol 170:5759–5764.
Nishijima M, Bulawa CE, Raetz CRH (1981): J Bacteriol 145:113–121.
Ohmori H, Kimura M, Nagata T, Sakakibara Y (1984): Gene 28:159–170.
Olivera BM, Bonhoeffer F (1974): Nature (London) 250:513–514.
Quiñones A, Messer W (1988): Mol Gen Genet 213:118–124.
Rokeach LA, Zyskind JW (1986): Cell 46:763–771.
Sakakibara Y, Tsukano H, Sako T (1981): Gene 13:47–55.
Sako T, Sakakibara Y (1980): Mol Gen Genet 179:521–526.
Samitt CE, Hansen FG, Miller JF, Schaechter M (1989): EMBO J 8:989–993.
Schaefer C, Messer W (1989): EMBO J 8:1609–1613.
Scheurmann R, Tam S, Burgers PMJ, Lu C, Echols H (1983): Proc Natl Acad Sci USA 80:7085–7089.
Sekimizu K, Bramhill D, Kornberg A (1987): Cell 50:259–265.
Sekimizu K, Bramhill D, Kornberg A (1988): J Biol Chem 263:7124–7130.
Seufert W, Messer W (1986): EMBO J 5:3401–3406.
Seufert W, Dobrinski B, Lurz R, Messer W (1988): J Biol Chem 263:2719–2723.
Shlomai J, Kornberg A (1980): Proc Natl Acad Sci USA 77:799–803.
Sinha NK, Morris CF, Alberts BM (1980): J Biol Chem 255:4290–4303.
Studwell PS, O'Donnell M (1990): J Biol Chem 265:1171–1178.
Stuitje AR, de Wind N, van der Spek JC, Pors TH, Meijer M (1986): Nucleic Acids Res 14:2333–2344.
Tabata S, Oka A, Sugimoto K, Takanami M, Yasuda S, Hirota Y (1983): Nucleic Acids Res 11:2617–2626.
Tomasiewicz HG, McHenry CS (1987): J Bacteriol 169:5735–5744.
van den Berg EA, Geerse RH, Memelink J, Bovenberg RAL, Magnée FA, van de Putte P (1985): Nucleic Acids Res 13:1829–1840.
Wang Q, Kaguni JM (1987): Mol Gen Genet 209:518–525.
Wang Q, Kaguni JM (1989): J Biol Chem 264:7338–7344.
Welch MM, McHenry CS (1982): J Bacteriol 152:351–356.
Yamagishi J, Yoshida H, Yamayoshi M, Makamura S (1986): Mol Gen Genet 204:367–373.
Yeung T, Mullin DA, Chen K-S, Craig EA, Bardwell JCA, Walker JR (1990): J Bacteriol 172:6042–6047.
Yin K-C, Blinkowa A, Walker JR (1986): Nucleic Acids Res 14:6541–6549.
Zipursky SL, Marians KJ (1981): Proc Natl Acad Sci USA 78:6111–6115.

**Perspectives on Cellular Regulation:
From Bacteria to Cancer, pages 39–49**

Termination of Replication of the Bacterial Chromosome

Peter Kuempel

Department of Molecular, Cellular and Developmental Biology, University of Colorado, Boulder, Colorado 80309

I joined Art's lab just after he moved from Berkeley to Princeton, and I was the first of his Ph.D. students from that University. The early 1960s was an exciting time to be a graduate student in the new field of molecular biology, and I picked up a number of enduring interests and memories from working in Art's lab. I won't mention all of the now well-known individuals who worked with Art at that time either as graduate students or postdocs, but it was an amazing mix of people. Two things that I recall in particular were the lab meetings at which we all took turns presenting journal articles that told of the rapid cracking of the genetic code, and John Gerhart's kinetic studies that led to his and Art's discovery of what are now called allosteric proteins. The most important influences that I have retained were my wife, Sheryl, who was Art's technician at Princeton, which was then 99% male, and a lifelong fascination with the bacterial cell cycle. I also established valuable and long-term relationships with lab-mates Millie Masters and Willie Donachie, who also continue to study the bacterial cell cycle.

The title of my Ph.D. thesis was "Enzyme Synthesis in Synchronous Cultures of *Escherichia coli,*" and my research continues to be oriented toward bacterial growth and division. For a number of years my lab has been studying the termination of replication in *E. coli,* and that area of research has recently shown considerable activity. The intention of this article is to show why we have found this project so interesting and how the understanding of termination and the methods used to study it have steadily evolved. These studies have progressed to the point that the sequences and DNA-binding proteins that arrest DNA replication have been identified in several bacterial chromosomes and plasmids, and the mechanism by which DNA replication is inhibited is beginning to be worked out. Much still remains to be elucidated, however, about this and other aspects of the termination of the replication cycle.

Terminator Sites and Sequences

The notion of a terminus region, separate and distinct from the origin region, first became apparent with the marker-frequency experiments of Millie Masters and Paul Broda (1971). Those data indicated that replication in *E. coli* was bidirectional and was completed somewhere in the region between minutes 25 and 40 on the genetic map, opposite the origin of replication. To provide a more direct test of bidirectional replication, David Prescott and I used an autoradiographic approach. We obtained numerous examples that confirmed that initiation was bidirectional, and Figure 1A shows our favorite autoradiograph of the replication forks moving away from the origin (Kuempel et al., 1973b). Autoradiography was also used to provide a direct demonstration that the replication forks approach and meet on the circular chromosome in the region that is opposite the origin (Fig. 1, band C; Kuempel et al., 1973a). This type of approach was limited, however, since the location(s) at which the forks were meeting could not be accurately identified.

The genetic analysis of the terminus region of *E. coli* began in 1977, when it was demonstrated by our lab as well as Jean-Michel Louarn's group in Toulouse, France, that replication forks were inhibited somewhere in the region between *trp* (27 minutes) and *manA* (36 minutes; see Fig. 2). Both of our groups used an approach that involved placing replication origins near the terminus region. This was done to overcome the usual symmetry of replication, which

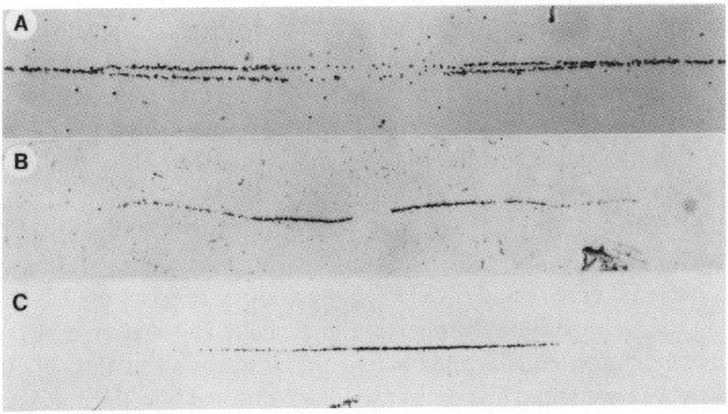

Fig. 1. Bidirectional initiation and termination of replication in *E. coli*. **(A)** Initiation of replication; the autoradiograph shows a pair of replication forks moving away from the origin of replication. The specific activity was increased after initiation, and the origin consequently has a lower grain density. The length of the grain track from fork to fork is 370 μm (Kuempel et al., 1973b). **(B, C)** Termination of replication; the autoradiographs show approaching forks **(B)** and those that have met **(C)**. The grain density is highest at the replication forks due to the increasing specific activity of precursor. The distance between the replication forks in **(B)** is 25 μm (Kuempel et al., 1973a).

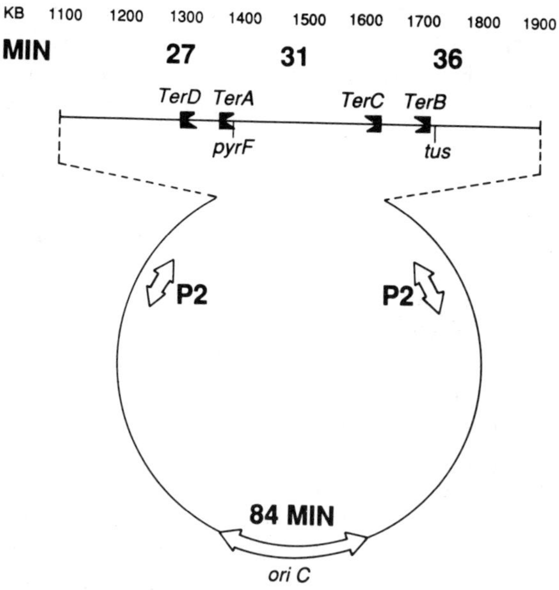

Fig. 2. Genetic map showing the terminus region and other relevant loci. *TerA, TerB, TerC,* and *TerD* are terminator sites that arrest replication forks, and Tus protein, the product of the *tus* gene, is required for function of these sites. Clockwise or counterclockwise replication of the terminus region was initiated from the appropriate P2*sig*5 prophage. The normal origin of replication is *oriC*. Terminator sites (>) inhibit only replication forks that have the appropriate (→) orientation.

makes it difficult to determine whether replication stops simply because the two forks collide with each other, or whether there are also sites that actively inhibit replication. By using asymmetrically placed origins, the fate of forks traveling in a particular direction could be studied as they passed through the terminus region, well in advance of forks arriving from the other direction.

To initiate replication from sites near the terminus region, our group used prophage P2*sig*5 (Fig. 2). The prophage was temperature inducible for initiation of replication and could not excise itself from the chromosome. Once induced, replication forks intiated at P2*sig*5 proceeded until they were blocked by a termination site. Separate insertion sites for P2*sig*5 were used to study clockwise and counterclockwise traveling replication forks. DNA–DNA hybridization to cloned fragments from various positions around the chromosome demonstrated that replication forks traveling in either direction were inhibited in the terminus region.

A more accurate understanding of where replication was actually arrested was possibly only when the terminus region had been more extensively characterized. The region contained very few loci,and its size, approximately 7 minutes, had only been estimated by bacterial conjugation. We constructed a genetic map that spanned this region, using a variety of transposon insertions and P1 transduction (Bitner and Kuempel, 1981). This confirmed the size of the region and has permitted precise location of a number of previously unmapped loci. The real impetus for more refined analysis of the termination sites came, however, from the restriction map of the terminus region constructed by Jean-Pierre Bouche (1982). This made it possible to use DNA probes from throughout the terminus region to determine where replication was inhibited. The group in Toulouse and our lab both demonstrated that inhibition did not occur at a single site in the center of the terminus region, as might have been expected. Instead, replication was inhibited in the regions that flanked the terminus, and the inhibition was polar (Hill et al., 1987; de Massy et al., 1987).

In spite of the importance of specific DNA probes in characterizing the inhibition of replication, it should be added that deletions have played an important role in locating the sites of arrest. Both Tn*10*-derived and site-directed deletions have been used. The DNA–DNA hybridization assays did not give precise locations for the arrest sites, and deletion analysis defined the locations further. Deletion analysis was possible since most of the terminus region is not essential. Surprisingly, one terminus region deletion we isolated removed 350 kb, or 7% of the chromosome. The function of most of the DNA in the terminus region is still unknown.

The location of the termination sites and their polar nature led to the concept that the sites function as a replication fork trap. Counterclockwise traveling replication forks were inhibited near *pyrF* as they exited the terminus region, and clockwise traveling replication forks were inhibited near *manA* as they exited the terminus (Fig. 2). These sites had no effect on forks as they entered the terminus. This indicates that the replication cycle is frequently completed when the bidirectional forks collide somewhere in the terminus region itself. However, if one replication fork arrives in the terminus region well ahead of the other, that fork would be inhibited as it exits the terminus, and the forks would meet at the arrest site.

A more recent discovery was that there are multiple terminator sites that flank the terminus region. To date, four arrest sites have been identified, and their locations on the genetic map are as follows (Fig. 2): *TerA,* 28.2 minutes; *TerB,* 35.6 minutes; *TerC,* 34 minutes; *TerD,* 27 minutes. A variety of procedures have been used to locate these sites and identify the relevant sequences, and it is informative to consider the widely different approaches that have been used. The procedure that we used to locate *TerA* and *TerB* within ±50 bp (Pelletier et al., 1988) was similar to that initially used by Weiss and Wake in their studies of the terminus region of *Bacillus subtilis*. A restriction fragment that contains a blocked replication fork will have a Y- shaped structure, and it consequently will have two unique properties that can be studied by Southern hybridizations. First, restriction fragments containing the blocked fork will travel slowly in agarose gels. Second, the arms of the Y can be released by shearing or digestion with the appropriate nuclease, and the length of released arm is the distance from the site of arrest to the upstream end of the restriction fragment.

The identification of *TerA* and *TerB* was facilitated by subcloning the relevant DNA fragments into ColE1 type plasmids, in which they arrested replication. This demonstrated that inhibition did not depend on a property unique to the bacterial chromosome, and it greatly simplified further manipulation of the sites. Replication was blocked, and it not only occurred at the same point as in the bacterial chromosome, it was also polar (Hill et al., 1988b; Pelletier et al., 1989). (Inhibition also required the DNA-binding Tus protein; see below). We identified the terminator sequence by sequencing the region that contained *TerB* and comparing it with the region that contained *TerA*. Fortunately, the sequence of the latter region was already available from studies of the *pyrF* gene. Computer analysis identified a 23 bp sequence, present at both sites with a difference in a single base pair (Fig. 3). The definitive test was to synthesize an oligonucleotide containing the putative 23 bp sequence, and test its effect on plasmid replication. Replication was inhibited, which confirmed the identity of the terminator sequence (Hill et al., 1988b).

Jean Michel Louarn's group located the terminator sites by inducing replication from prophage λ and determining where replication was inhibited (Francois et al., 1989). λ::Tn*10 cI857* was integrated into selected Tn*10* insertions in and near the terminus region, and replication was initiated by inactivation of the temperature-sensitive *cI857* repressor. Inhibition at termination sites was detected by DNA–DNA hybridization. This approach confirmed the location of *TerA,* and it demonstrated that the so-called T2 block sites previously studied by our two groups were actually different. T2-Boulder (Hill et

TERMINATOR SEQUENCES

TerA	**5'**	**A A T T A G T A T G T T G T A A C T A A A G T**	**3'**
Ter B		**A A T A A G T A T G T T G T A A C T A A A G T**	
TerC		**A T A T A G G A T G T T G T A A C T A A T A T**	
TerD		**C A T T A G T A T G T T G T A A C T A A A T G**	
Consensus		**A A T T A G T A T G T T G T A A C T A A A G T**	

Fig. 3. Sequences of the terminator sites in *E. coli*. The sequence of *TerA* is the consensus sequence. Differences are underlined.

al., 1987) is now called *TerB,* and T2-Toulouse (de Massy et al., 1987) is now called *TerC*. Louarn and co-workers also identified an additional site near min 27 that is now called *TerD*. It is interesting to note that *TerD* has recently been precisely located and sequenced by a procedure that is now commonly used in molecular biology research, namely computer analysis. *TerD* is in the *nar* operon at 27 minutes, and the terminator site was recognized by several labs once the sequence had been entered in GenBank.

Takashi Horiuchi's group in Fukuoka, Japan, began their studies of termination sites by studying the inhibition of replication in plasmid R6K, which also contains sites that arrest replication (Horiuchi and Hidaka, 1988). As part of those studies they developed a simple "Ter assay" for detecting whether DNA fragments contained sites that inhibited replication. Fragments were cloned into a ColE1 type plasmid, and if replication was inhibited, the restriction fragment containing the plasmid's origin of replication would have a slowly migrating, Y-shaped structure. This approach was extended to the terminus region of *E. coli,* and they cloned a number of fragments into plasmids and tested whether replication was inhibited (Hidaka et al., 1988). This procedure independently identified *TerC,* and it confirmed the locations of *TerA* and *TerB*.

The DNA sequences for the terminators are shown in Figure 3. The consensus sequence is the same as the *TerA* sequence, and *TerB* differs from the consensus by 1 base change. *TerC* deviates the most from the consensus, differing at five positions. Although the terminator sites are arranged around the terminus region in a symmetric fashion (Fig. 2), their function is not truly symmetric. We have observed that the *TerD* site in the chromosome is rarely used, since it is downstream from the strong *TerA* site. This site is used, however, if *TerA* is deleted (A. Pelletier, unpublished experiments). The situation is different on the other side of the terminus. Although forks are sometimes inhibited at *TerC* in the chromosome, many pass through this site and are then inhibited at *TerB*.

Tus, a Unique DNA-Binding Protein

The other important component required for inhibition of replication is the Tus protein, the product of the *tus* (terminator utilization substance) gene. Tus is a DNA-binding protein that binds to the terminator sequences with a K_D of 10^{-12}, and, without exception, Tus is required for the arrest of replication at termination sites in both chromosomes and plasmids. Tus is consequently a unique type of protein. By binding to specific DNA sequences, it arrests in a polar fashion replication forks that are actively moving along the chromosome. The presence of this additional component was first detected as the result of deletion mapping of the terminus region, in which we observed that deletion of the *TerB* region inactivated inhibition at the *TerA* site, located 350 kb away on the other side of the terminus region. The region required for inhibition was identified by a series of overlapping deletions, and it was demonstrated that the function could be supplied in *trans* on a plasmid (Hill et al., 1988a). This in turn simplified localization of the relevant gene by further subcloning and insertional inactivation, its identification by sequencing the region, and the demonstration that the gene product was a DNA-binding protein (Hill et al., 1988b, 1989).

The sequence of the *tus* gene has provided some very useful information about this region and its regulation. The *TerB* site is immediately upstream from *tus,* and it overlaps a weak promotor sequence. Since Tus binds to *TerB,* this suggested that *tus* was autoregulated. Further experiments have verified this: *tus* is transcribed from the weak promoter, and inactivation of *tus* leads to increased transcription (B. Roecklein, unpublished experiments). The messenger RNA as well as the protein are present in very low amounts per cell, and the DNA sequence has facilitated construction of plasmids with improved promoters and ribosome binding sites for *tus,* which produce high amounts of Tus. Several laboratories have purified Tus and are now studying its properties (Sista et al., 1989; Hidaka et al., 1989; Hill and Marians, 1990).

The availability of purified Tus led to the latest phase in studies on the termination of replication: elucidation of the mechanism by which inhibition occurs. Comparable experiments have been conducted by Deepak Bastia's lab (Khatri et al., 1989) and a collaboration between Takashi Horiuchi's and Arthur Kornberg's groups (Lee et al., 1989). Using a simple assay for helicase activity, they have observed that *dnaB* helicase is inhibited at terminator sequences, provided that the sequence is present in the correct orientation and that Tus protein is bound to the sequence. Lee et al. (1989) also demonstrated that *rep* and *uvrD* helicases were inhibited. Since the *dnaB* and *rep* proteins are an important part of the replication apparatus, this provides a simple explanation for why replication forks are inhibited in a polar fashion when they encounter a terminator site. Further studies will of course elucidate the important parts of the terminator sequence and how Tus interacts with it and with the helicases.

What Is the Role of Terminator Sites in Prokaryotic Chromosomes?

Although the details of the inhibition process are now being elucidated, the interesting question remains, what advantage is provided to cells by the arrest sites? We have noticed no growth disadvantage when the termination system is inactivated (S. Hoover, unpublished experiments). This was tested by using a *tus::kan* insertion and comparing the growth rate of such a *tus* mutant with that of an otherwise isogenic strain. The relative growth rates can be readily determined by replica plating a mixed culture after growth in various media, medium shifts, anaerobic growth, etc. This is a sensitive approach for detecting growth differences and none was observed in the conditions that have been tested.

In spite of no obvious phenotype for *tus* mutants, the presence of termination sites in other chromosomes indicates that there must be a selective pressure that maintains the system. As one example of this, we tested whether *Salmonella typhimurium* also contains a termination system. Although *E. coli* and *S. typhimurium* are often regarded as being very similar, these species diverged 120 to 160 million years ago and have about 40–50% DNA homology. We probed *S. typhimurium* DNA with an oligonucleotide that contained the *TerB* sequence, and identified three bands (A. Pelletier, unpublished experiments). The hybridization was done in conditions of low stringency, and in a control hybridization with *E. coli* DNA only three bands were identified. Consequently, this indicates that *S. typhimurium* contains a minimum of three termination sites. A *tus*-containing fragment was also identified, but more important, we also observed that the *E. coli* and *S. typhimurium* termination systems are functionally compatible. A plasmid containing the *E. coli TerA* site was transferred into *S. typhimurium* and replication was inhibited at the termination site, and this only occurred in the functional orientation. Thus, in spite of any obvious phenotype for *tus* mutants in laboratory conditions, the selective advantage conferred by a termination system has maintained similar systems in *E. coli* and *S. typhimurium*.

Terminator sites are also present in plasmid R6K, for which *E. coli* is a natural host. R6K is normally replicated in a bidirectional fashion, and it contains two terminator sites that have a sequence similar to those in *E. coli* (Horiuchi and Hidaka, 1988; Hill et al., 1988b). They are arranged as an inverted repeat, in the form of a replication fork trap, in which the inverted repeats are separated by only 73 bp. The sites require Tus protein to function (Pelletier et al., 1989; Kobayashi et al., 1989) and since R6K does not contain a *tus* gene itself, it has apparently evolved to use the Tus protein of its host. Once again, however, the advantage supplied by the terminators is not obvious. R6K is equally stable in tus^+ and tus^- strains of *E. coli,* and it will probably be necessary to inactivate other plasmid stability factors to determine the effect of the terminators on plasmid stability.

Termination sites are also present in plasmids of the RepFIIA incompatability group, and the *E. coli* Tus protein functions at these sites (Hill et al., 1988b; Hidaka et al., 1988). These plasmids have not been tested to determine if they provide their own Tus protein. These plasmids are replicated in a unidirectional fashion, and the termination sites are adjacent to the origin of replication. Consequently, the sites do not provide a replication fork trap of the type present in *E. coli* and R6K. A likely possibility is that the sites function to limit these plasmids to unidirectional replication.

Perhaps the most intriguing aspect of termination sites is that a system comparable to that of *E. coli* is also present in *B. subtilis*. This organism is a Gram-positive spore-forming bacterium, so there are considerable differences between it and *E. coli*. The work on this organism has been done in Gerry

Wake's lab in Sydney, Australia, and the *E. coli* and *B. subtilis* studies have followed parallel paths. The terminus region of *B. subtilis* contains inverted repeats (IRI and IRII), each 47 bp long, which are separated by 59 bp (Carrigan et al., 1987). These sites apparently function as a replication fork trap, similar to those in *E. coli* and R6K. Surprisingly, over a 23 bp region in which the IRI and IRII show 83% similarity, they show 57% similarity to the *E. coli* terminator sequence. Adjacent to one of the repeats is the *rtp* gene (replication termination protein), which encodes a protein that binds to the IR sequences. This protein is required for function of the termination sites (Lewis et al., 1989). The DNA sequence also suggests that expression of *rtp* is autoregulated by binding to the IR sites. The protein sequence shows no obvious similarity, however, to Tus. It is considerably smaller, and it possibly functions as a dimer. As in *E. coli,* the termination system of *B. subtilis* can be inactivated by deletion of the IR sites or inactivation of the *rtp* gene, and no phenotypic difference has been reported for these mutants.

The arrangement of the *tus* and *rtp* genes with respect to their termination sites suggests that these genes are not only autoregulated, they are also expressed in a cell cycle–specific fashion. In vitro, the Tus and *TerB* have an association rate of 3×10^8 M^{-1} sec^{-1}, and the dissociation half-life of the complex is about 200 min (M. Tecklenburg, unpublished experiments). If the rates are comparable in vivo, this suggests that the normal time of synthesis of Tus would be when the gene was replicated. If the *TerB–tus* region is replicated by counterclockwise traveling forks, replication would not be inhibited and the bound Tus protein would be transiently removed. The *tus* gene could then be transcribed. Even if a clockwise-traveling replication fork was blocked by Tus bound at *TerB,* transcription would ultimately occur: the arrival of the counterclockwise-traveling fork and subsequent completion of replication would release Tus. Although it is interesting to speculate that Tus is synthesized in a cell cycle-specific fashion, whether this affects other aspects of the cell cycle is completely unknown.

The question remains, what is the function of a termination system, and why has it evolved and been maintained? At least part of the answer might have to do with relationships between transcription and replication. It has been noted that the major transcription units in each half of the *E. coli* chromosome are oriented in the same direction as replication (Brewer, 1988), and this pattern apparently confers some selective advantage. Wu et al. (1988) have observed that a common orientation for transcription units aids in the control of supercoil density, since positive and negative supercoils generated in front of and behind transcription can compensate. Aligning replication in the same direction would have a similar effect. The arrest sites would prohibit replication forks from exiting the terminus region and replicating part of the chromosome in the backward direction.

In addition to the orientation of transcription units, there are probably other, presently unrecognized properties of the bacterial chromosome that are used optimally only if replication proceeds in a particular direction. An indication that there are other aspects to chromosome structure comes from the observation that certain chromosome inversions in *E. coli* and *S. typhimurium* cannot be constructed (Segall et al., 1988; Rebello et al., 1988).

In conclusion, work on termination has really just begun, even though one aspect of termination is now being clarified. Termination refers to events at the end of the replication cycle, and the word now has at least two meanings. It designates the arrest of replication that has been described above. But termination also includes other events, such as the meeting of replication forks and decatenation and partitioning of daughter chromosomes. The elusive signal that links replication and cell division is also thought to involve some aspect of termination. Most of us who study termination entered this field with the broader meaning of the word in mind, and now that better tools are available, it is time to pursue these other termination events as well as the arrest of replication.

References

Bitner RM, Kuempel PL (1981): P1 transduction map spanning the replication terminus of *Escherichia coli* K12. Mol Gen Genet 184:208–212.

Bouche JP (1982): Physical map of a 470 $\times$ 10^3 base-pair region flanking the terminus of DNA replication in the *Escherichia coli* K12 genome. J Mol Biol 154:1–20.

Brewer BJ (1988): When polymerases collide: Replication and the transcriptional organization of the *E. coli* chromosome. Cell 53:679–686.

Carrigan CM, Haarsma JA, Smith MT, Wake RG (1987): Sequence features of the replication terminus of the *Bacillus subtilis* chromosome. Nucleic Acids Res 15:8501–8509.

De Massy B, Bejar S, Louarn J, Louarn JM, Bouche JP (1987): Inhibition of replication forks exiting the terminus region of the *Escherichia coli* chromosome occurs at two loci separated by 5 min. Proc Natl Acad Sci USA 84:1759–1763.

Francois V, Louarn J, Louarn JM (1989): The terminus of the *Escherichia coli* chromosome is flanked by several polar replication pause sites. Mol Micro 3:995–1002.

Hidaka M, Akiyama M, Horiuchi T (1988): A consensus sequence of three DNA replication terminus sites on the *E. coli* chromosome is highly homologous to the *terR* sites of the R6K plasmid. Cell 55:467–475.

Hidaka M, Kobayashi T, Takenaka S, Takeya H, Horiuchi T (1989): Purification of a DNA replication terminus (*ter*) site-binding protein in *Escherichia coli* and identification of the structural gene. J Biol Chem 264:21031–21037.

Hill TM, Marians KJ (1990): *Escherichia coli* TUS protein acts to arrest the progression of DNA replication forks in vitro. Proc Natl Acad Sci USA 87:2481–2485.

Hill TM, Henson JM, Kuempel PL (1987): The terminus region of the *Escherichia coli* chromosome contains two separate loci that exhibit polar inhibition of replication. Proc Natl Acad Sci USA 84:1754–1758.

Hill TM, Kopp BJ, Kuempel PL (1988a): Termination of DNA replication in *Escherichia coli* requires a trans-acting factor. J Bacteriol 170:662–668.

Hill TM, Pelletier AJ, Tecklenburg ML, Kuempel PL (1988b): Identification of the DNA sequence from the *Escherichia coli* terminus region that halts replication forks. Cell 55:459–466.

Hill TM, Tecklenburg ML, Pelletier AJ, Kuempel PL (1989): Tus, the trans-acting factor required for termination of DNA replication in *Escherichia coli,* is a DNA-binding protein. Proc Natl Acad Sci USA 86:1593–1597.

Horiuchi T, Hidaka M (1988): Core sequence of two separable terminus sites of the R6K plasmid that exhibit polar inhibition of replication is a 20 bp inverted repeat. Cell 54:515–523.

Khatri GS, MacAllister T, Sista PR, Bastia D (1989): The replication terminator protein of *E. coli* is a DNA sequence-specific contra-helicase. Cell 59:667–674.

Kobayashi T, Hidaka M, Horiuchi T (1989): Evidence of a ter specific binding protein essential for the termination reaction of DNA replication in *Escherichia coli*. EMBO J 8:2435–2441.

Kuempel PL, Maglothin PD, Prescott DM (1973a): Bidirectional termination of chromosome replication in *Escherichia coli*. Mol Gen Genet 125:1–8.

Kuempel PL, Prescott DM, Maglothin PD (1973b): Autoradiographic demonstration of bidirectional replication in *Escherichia coli.* In R Wells, and R Inman (eds): DNA Synthesis in Vitro. New York: Academic Press, pp 463–472.

Lee EH, Kornberg A, Hidaka M, Kobayashi T, Horiuchi T (1989): *Escherichia coli* replication termination protein impedes the action of helicases. Proc Natl Acad Sci USA 86:9104–9108.

Lewis PJ, Smith MT, Wake RG (1989): A protein involved in termination of chromosome replication in *Bacillus subtilis* binds specifically to the *terC* site. J Bacteriol 171:3564–3567.

Masters M, Broda P (1971): Evidence for the bidirectional replication of the *Escherichia coli* chromosome. Nature (London) 232:137–140.

Pelletier AJ, Hill TM, Kuempel PL (1988): Location of sites that inhibit progression of replication forks in the terminus region of *Escherichia coli*. J Bacteriol 170:4293–4298.

PelletierAJ, Hill TM, Kuempel PL (1989): The termination sites from the *Escherichia coli* chromosome (T1 and T2) inhibit DNA replication in ColE1-derived plasmids. J Bacteriol 171:1739–1741.

Rebello J-E, Francois V, Louarn J-M (1988): Detection and possible role of two large nondivisible zones on the *Escherichia coli* chromosome. Proc Natl Acad Sci USA 85:9391–9395.

Segall A, Mahan MJ, Roth JR (1988): Rearrangement of the bacterial chromosome: Forbidden inversions. Science 241:1314–1318.

Sista PR, Mukherjee S, Patel P, Khatri GS, Bastia D (1989): A host-encoded DNA-binding protein promotes termination of plasmid replication at a sequence-specific replication terminus. Proc Natl Acad Sci USA 86:3026–3030.

Wu HY, Shyy S, Wang JC, Liu LF (1988): Transcription generates positively and negatively supercoiled domains in the template. Cell 53:433–440.

Perspectives on Cellular Regulation: From Bacteria to Cancer, pages 51–64

The Cell Cycle of *Escherichia coli*: A Growing Mystery

William D. Donachie

Institute of Cell and Molecular Biology, University of Edinburgh, Edinburgh EH9 3JR, Scotland

In January 1962, when I came to Princeton as a new post-Doc. in Art Pardee's group, research on bacterial genetics and regulatory mechanisms was at its height but very little was known about how the thousands of separate components were organized in time and space to form the adaptable and self-replicating entity that we recognize as a cell. It was typical of Art's forward-looking approach, however, that as well as working on operon structure and feedback regulatory mechanisms, his group had already started to work on temporal organization. I and my then novel *eu*karyote cultures of *Neurospora crassa* were put in a lab with two graduate students, Peter Kuempel and Millie Masters, who were trying to find ways to synchronize bacterial populations, so as to be able to follow the sequence of biochemical events that were expected to take place during the cell cycle. As a student in Edinburgh, I had been well steeped in discussions of the Problems of Development by Waddington and of the Cell Cycle by Swann and Mitchison and therefore this work greatly interested me. In addition, soon after I arrived in Princeton, Nobura Sueoka and Hiroshi Yoshikawa, our neighbours in the Moffat Lab, carried out their classical genetic experiments that showed that the bacterial chromosome is replicated in a fixed sequence, over a period of time which is comparable to that of the cell cycle itself (Yoshikawa and Sueoka, 1963). The elegance of this work, and of their subsequent demonstration of overlapping rounds of replication in fast-growing cells, made me determined to work with bacteria if I could. (After all, I had started to work with *Neurospora* because I had wanted to find an organism in which the genes were more accessible and the life cycle simpler and shorter than in the *Drosophila* with which I had worked as an undergraduate. It seemed to me then, as it does now, that what I really wanted to work with was *Escherichia coli*.)

Twenty-eight years later, after these pioneering explorations of the bacterial cell cycle by Pardee's and Sueoka's groups, can we now say how the prokaryotic cell cycle works? We cannot; but we can tell a good story.

A Beginning: The Initiation of Chromosome Replication

Work by Pardee and Louise Prestidge (1956), together with many studies from Maaløe's lab (Maaløe and Kjeldgaard, 1966) had showed that de novo protein synthesis is required before chromosomal DNA replication will begin. This is perhaps the single most important fact about the cell cycle, not only of prokaryotes but of eukaryotes as well, because it defines the event that can be considered to be the beginning of the processes that lead to genome replication and cell division. Just how well controlled this initial event is became clearer later when, after the beautiful studies of Cooper and Helmstetter (1968) had shown that initiation of chromosome replication takes place at different times (relative to division) in cells growing at different rates, I was able to calculate that initiation always took place when the ratio of cell mass to number of chromosome origins reached a constant critical value: the "Initiation Mass" (Donachie, 1968). (Incidentally, I now know what it feels like to say "Eureka!")

Sadly (or perhaps "irritatingly" would be more appropriate) we still do not know what it is that changes when cells reach initiation mass. Much recent work has implicated the DnaA protein, required specifically for initiation of DNA synthesis at the origin (*oriC*) in the timing of initiation but it is still not clear how the onset of its action is linked to cell mass. Possibly this is because many of us have been thinking along the wrong lines about initiation timing. Instead of some specific protein being synthesized at a particular stage, or achieving a critical amount or concentration, perhaps the cell itself reaches a "critical state," related to its size, in which its properties (e.g., permeability to particular ions) change suddenly ("catastrophically") with the resultant activation of dormant proteins or genes. If we were to take such a possibility seriously, perhaps we should now pay a little less attention to the effects of over- or underexpressing individual proteins, and more to those experimental treatments that appear to be able to synchronize cells or accelerate the onset of DNA replication in ways that abolish the normal correlation between the attainment of initiation mass and the onset of chromosome replication. Examples of such treatments, which synchronize populations of cells without simply selecting those of the same size or stage, are amino acid starvation, heat and cold cycles, periodic phosphate starvation, and osmotic shock. Once again, Art Pardee made some of the earliest experiments of this kind (Smith and Pardee, 1970) and showed that heat shock or osmotic shock could synchronize division in previously asynchronous populations of cells. Since that time, this kind of work has been pursued mainly by Eliora Ron, Nili Grossman, and their collaborators (Grossman and Ron, 1989). They have shown convincingly that periods of amino acid starvation can cause "premature" initiation of both chromosome replication and cell division. This shows that even small cells, well below initiation mass, have the capacity to initiate chromosome replication. Presumably this means that DnaA protein is activated by such treatments, although we still do not know whether it is the activity of pre-existing proteins or the rapid synthesis of new protein molecules that is stimulated by the period of amino acid starvation. We should certainly find out. This may then provide an essential clue as to what "change of state" takes place when unperturbed growing cells reach their initiation mass and set in train the processes leading to cell duplication.

A Next Step: Chromosome Duplication

Apart from the demonstration of a "First Prime Mover" in the cell cycle (the Initiation Mass), elucidation of subsequent events in chromosome replication has made much progress since 1962 (my egocentrically chosen reference date). Largely because of the work of Arthur Kornberg and his collaborators (Kornberg, 1974), practically all the steps in chromosome replication are understood (except, of course, what starts it). Masters (1970), Masters and Broda (1971), Bird et al. (1972), Prescott and Kuempel (1972), and others since, showed that a pair of replication forks move away from *oriC* in opposite directions until they meet on the opposite side of the circular chromosome. Peter Kuempel has shown that there are unidirectional termination sites, under the control of a *trans*-acting protein, which prevent replication forks from overrunning the terminus region (although it is not obvious why they are necessary). The systematic studies of Cooper and Helmstetter have confirmed the original idea of Maaløe and his collaborators that the rate of DNA synthesis at each replication fork, and therefore the time taken to replicate a chromosome, is largely independent of the growth rate of the cell (at constant temperature) so providing a minimum period of time between the attainment of initiation mass and subsequent cell division. Since this period is long relative to the doubling time of *E. coli* cells in most laboratory culture media, the termination of successive rounds of chromosome replication acts as a timing signal for division and accounts (at least in part, see below) for the dramatic changes in mean cell size which are seen when bacteria change their rate of growth. The fixed rate of DNA synthesis per replication fork also accounts for the overlapping rounds of replication seen by Yoshikawa and Sueoka in *B. subtilis* at high growth rates (Oishi et al., 1964) and by Cooper and Helmstetter (1968), and Masters (1970), in fast-growing *E. coli*.

A Mystery: Genome Partition

Replication of the genome and a doubling in cell mass are not enough to allow duplication of the cell. Before that can happen the duplicated copies of the chromosome must be separated and relocated within the cell, such that subsequent cell division will ensure that each sister cell has a copy. This spatial separation and relocation of sister chromosomes, the prokaryotic equivalent of mitosis, is one of the many unsolved mysteries of the bacterial cell.

In actively growing *E. coli* cells, DNA is found in the form of ''nucleoids,'' bodies of often complicated shapes that occupy a large part of the cell volume. The number and location of these bodies are therefore often hard to determine. However, once again some work by Pardee's group has shown a way to do this. Zusman et al. (1973) showed that inhibition of protein synthesis caused the DNA in nucleoids to condense into tightly packed discrete bodies that can be easily counted and their positions measured. We have used this method to show that the number of separate nucleoids in cells corresponds exactly to the number of completely replicated chromosomes (a number that varies with the growth rate of the cells). We can conclude therefore that partition of sister chromosomes takes place almost immediately after the replication of the terminus (Donachie and Begg, 1989b). Because cells with one nucleoid usually have this located close to the cell center, while those with two have them at about one-quarter and three-quarters of the cell length, we can also conclude that partition is a rapid process that takes very much less than a cell cycle to complete; not more than a minute or two at most.

We and Hiraga et al. (1990) have also shown that partition requires posttermination protein synthesis. If we could find what is the nature of this brief period of protein synthesis, we would know much more about the mechanism of partition.

DNA gyrase is required to decatenate replicated sister chromosomes (Steck and Drlica 1984) and mutants that are defective in gyrase activity cannot partition their chromosomes. However, Hiraga et al. (1990) have shown that the gyrase-dependent step is complete before the period of posttermination protein synthesis that is required for partition, because this latter step is insensitive to gyrase inhibitors.

Many mutants have now been reported in which partition is defective. However, all of these ''*par*'' mutants (*parA, B, C,* and *D*: the last of which we named in honor of Art!) have now been shown to have defects, either in the DNA gyrase A or B subunits, or in some enzyme involved in DNA synthesis per se. We (Ken Begg and I, unpublished) have shown that cells in which DNA replication has been interrupted, or slowed down relative to cell growth are unable to partition their DNA, which we think accounts for the Par$^-$ phenotype of all of those mutants. A different type of partition mutant has recently been described (Hiraga et al., 1989). In these ''*muk*'' (named for the Japanese word for partition) mutants, a certain proportion of anucleate cells is produced as the result of the occassional copartition of pairs of sister chromosomes into a single sister cell. Hiraga and his co-workers have suggested that this may be the result of random location of nucleoids within the *muk*$^-$ mutant cells, implying that the normal partition mechanism is absent in these mutants. However, to my mind, the data presented do not fully support this interpretation because, when the relatively greater length of *muk*$^-$ cells is taken into account, there seems to be no difference in the accuracy of localization of nucleoids in *mukA*$^-$ and *mukA*$^+$ cells. The *mukA* mutation has been shown to lie in the *tolC* gene, which codes for a membrane protein (involved in colicin tolerance). Ever since the first speculations on partition by Jacob et al. (1963) attachment to the cell envelope has been the favored idea for the positioning of chromosomes within bacterial cells, which makes the TolC protein particularly exciting. Its exact role in partition is however, still unknown.

We have noted an entirely different requirement for genome partition, which is geometrical (Donachie and Begg, 1989a,b). First we noted that cells growing at different rates, which have different *average* volumes and lengths, always terminate chromosome replication and undergo partition at the same length. Together with earlier observations (Donachie et al., 1976) that showed that attainment of this particular length was a prerequisite for cell division, this suggested the possibility that a minimum separation distance might also be a prerequisite for partition. This notion is supported by the kinetics of partition in asynchronous populations of cells that have been caused to complete all rounds of chromosome replication in the absence of cell growth and then allowed to resume growth in the absence of further DNA synthesis (Donachie and Begg, 1989b). Partition of the prereplicated chromosomes in such a population takes place in different cells at different times after the resumption of growth, such that the time taken for partition to have occurred in every cell is the same as that required to double the average length of the cells. This is what would be expected if partition required the attainment of a critical cell length. However a real test of the hypothesis was provided by mutants in which cell shape was altered (Donachie and Begg, 1989a). These mutants *(rodA*.ts or *pbpA*.ts) have lost the capacity to grow as rod-shaped cells at 42°C and instead grow as spherical cells. In rich media these grow at comparable rates to their rod-shaped parent strain but their average cell volume is greatly increased. In keeping with this increase in cell mass, the amount of DNA per cell is increased proportionately. Thus, the relation between initiation mass and chromosome replication is maintained but the number of cell divisions per chromosome is greatly reduced. Measurement of the dimensions of the rod-shaped and spherical cells showed what we predicted, that the change in cell mass and DNA content was the consequence of a change in cell shape without any change in cell length (i.e., the average diameter of the spherical cells was the same as the average length of the rods). The requirement of a minimum cell ''length'' (distance between cell poles) for division had therefore been maintained, but what had happened to partition of chromosomes? The spherical cells contained about six times as much DNA as the rods and therefore, ceteris paribus, should have contained six times as many nucleoids. However, because of geometric constraints, the average distance between these nucleoids would have been much less than in the rod-shaped cells. When we looked at the nucleoids in the spherical mutant cells, we found (to our great pleasure) that there was the same average number of nucleoids per cell as in normal rods. The minimum separation distance between nucleoid centers was therefore unchanged. The nucleoids themselves, however, were much larger than in the rods, in keeping with their calculated content of around six chromosomes each. (Remembering my beginning work with *Drosophila,* I call these "polytene" nucleoids.)

The discovery that chromosome partition obeys a geometric rule is in some way satisfying, inasmuch as it has predictive value, but it in no way provides a mechanistic explanation of partition. Indeed, the mystery deepens; what exactly is it that changes when a cell reaches its critical length? At the moment it seems that the answer must lie in some changes in the structure of the cell envelope (e.g., creation or duplication of sites for chromosome attachment) but looking for such answers with present technology (and experimental attitudes) is a daunting prospect.

And Finally: Cell Division

Even casual observation of *E. coli* cells growing and dividing (Fig. 1) quickly suggests a number of questions about cell division. What is the signal that causes division to begin? How is the location of the division site determined? What limits the number of divisions that takes place? What biochemical events are associated with the switch from cell elongation to septation?

Signals

In unperturbed growing cells, commitment to division (defined as the ability to initiate and complete septation in the absence of further RNA or protein synthesis) takes place almost immediately after completion of each round of chromosome replication and a short period of posttermination protein synthesis (Clark, 1968; Pierucci and Helmstetter, 1969; Jones and Donachie, 1973). It also depends on the achievement of a critical cell length at this time (Donachie et al., 1976) and the completion of partition (see above). All of these events are separately required for division at the normal time and, if any of them are prevented or delayed, division is also delayed.

Inhibition of DNA synthesis in normal cells causes the induction of the SOS response and the production of a cell division inhibitor, the SulA protein. If production of SulA is prevented, inhibition of DNA synthesis still delays cell division but this can resume, albeit at a low rate, if cell growth can continue for long enough. Division of such cells of course produces only anucleate, or "N^-", cells. This was first shown by Hirota et al. (1968) using *dnaA* mutants and by Masayori Inouye, in Art's lab, using *recA* mutants (1971). The SOS response was not induced in these cells. This suggests that failure to complete chromosome replication and partition may act as a temporary block to division, but that the real "signal" for division may only be growth to a certain size.

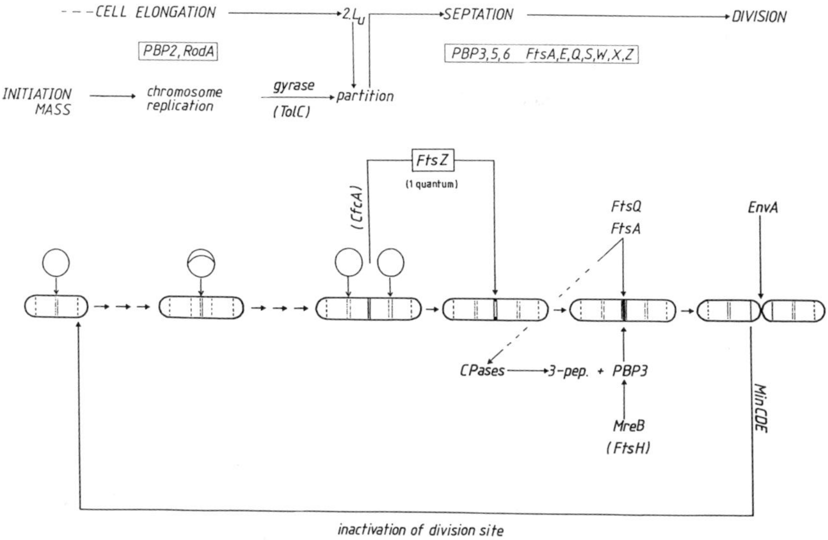

Fig. 1. An outline of the cell cycle of *Escherichia coli*. A schematic *E. coli* cell (a cylinder with hemispherical poles) is shown growing and dividing. The transverse lines represent chromosome attachment sites that later become potential division sites (dotted sites are nascent or have been inactivated). The chromosomes are shown as circles above their points of attachment to the cell envelope (and are not drawn for the septation phase). The points of action of the different morphogenetic proteins are shown (see text). CPases, carboxypeptidases I and II; 3-pep, tripeptide side chains in the peptidoglycan; 2.L_u, two unit lengths, the minimum cell length required for partition and initiation of septation.

Localization

Septa normally form between pairs of sister nucleoids. However, in cells that are growing without replicating DNA (as in the *dnaA* and *recA* mutants described above) and not producing SulA, septa form between the central nucleoid and the cell pole. This is even more dramatically shown in some of the Par mutants (such as *parD*: Hussain et al., 1987) in which DNA replication continues in the absence of partition. Such cells continue to grow and divide freely but septa form only between the central nucleoid mass (consisting of many catenated chromosomes) and the cell poles. Such observations raise the question of the relation between septal position and chromosome location (Donachie et al., 1984). Two possible suggestions are (1) that septa form at any location that is more than a certain minimum distance from a nucleoid (Woldringh et al., 1985; Hussain et al., 1987) or (2) that septa form at specific sites that arise at regular intervals during the growth of the cell, provided such sites are not already occupied by DNA. According to this second model, each envelope site would act first as a site for DNA replication and then, after postreplication release of the sister chromosomes and their relocation to newly formed sites, as a site for septation (Donachie, 1969; Donachie and Begg, 1970; Jones and Donachie, 1974).

Present evidence does not distinguish clearly between these alternative models. In some cases N^- cells are formed that are of fairly uniform size (e.g., in the *dnaA* mutant described by Hirota) while in others they are of every possible size, as if they were formed at random locations (e.g., in the *parD* mutant). What seems clear is that chromosome partition and septum localization are intimately connected, so that answering a question about one is also asking a question about the other.

Number

In normal, rod-shaped cells, one septum is formed for each pair of sister chromosomes, or for each unit of cell length. On the hypothesis that potential division sites/chromosome attachment sites arise at regular intervals of cell length, or that septa form only in spaces free of nucleoids (see above) there would therefore seem to be no problem in understanding the number of septa that form in each cycle. However, the behaviour of certain kinds of mutants suggests the existence of a limit to the number of septa that cells are capable of forming. The existence of this limit is most clear in minicell- producing mutants (*minB*). These mutants have the ability to produce small, anucleate minicells from the cell poles (Adler et al., 1967), in addition to dividing at normal, internucleoid sites. Counting the numbers of septa in such a mutant, however, showed that, despite the existence of "extra" potential division sites at the cell poles, the number of septa formed per generation was exactly the same as in min^+ cells (Teather et al., 1974). In other words, the formation of a polar septum used up some limiting amount of division capacity that would otherwise have been used in forming an internucleoid septum. Since that study, many more *minB* mutants have been isolated, one of which has a reduced capacity to make minicells, but all have the same capacity for division. This capacity may be said to be the formation of one septum per pair of sister chromosomes, or per pair of units of cell length. Another mutant that reveals this same limited division capacity is the *parD* (*gyrA*) strain. Cells of this strain at 42°C continue to grow, replicate their DNA, and divide, all at normal rates. However, probably because of the inability of the cells to decatenate replicated sister chromosomes, partition into separate nucleoids is blocked so that the cells accumulate an ever-increasing mass of DNA as a single enormous nucleoid. All septa form at apparently random locations between this mass and the cell poles, to produce only N^- cells of a wide variety of sizes (Hussain et al., 1987). Although there would appear to be plenty of room to form many more septa, this mutant also forms only one septum per pair of sister chromosomes, or per pair of unit lengths.

We have more demonstrations of this limited division capacity of cells (unpublished) but the two examples given are sufficient to describe the phenomenon. It seems to me that this is one of the most interesting aspects of the cell cycle; the quantitative control of key processes, i.e., initiation of DNA replication and formation of septa. A more subtle and even more surprising feature of "division capacity" is also revealed by minicell mutants. This is the apparently "quantal" nature of this capacity. If a binucleate cell produces a minicell, then it has used up its division capacity for that generation and cannot divide again until it has become quadrinucleate. However, at that time it will behave as if it had the ability to make two septa and each of these will form, unpredictably, at either a polar or internucleoid site. The cell therefore behaves as if it had produced two, separate packages or quanta of division capacity, each of which is used up completely in forming a septum. As we will see below, we now have a good idea of what this limited division capacity is, but not why it behaves like this.

Genes, Proteins, Enzymes, and Biochemistry

Progress in understanding the mechanism of division has come mainly from the identification of proteins that are exclusively required for division. Identification of these proteins has come either from the identification of the mutated genes in mutants that were affected only in their capacity to divide, or from the identification of the target protein of specific inhibitors of division. The identification of PBP3, the target for β-lactam inhibitors of division, was made in Pardee's lab by Brian Spratt (1975). The gene for this protein, *ftsI* (or *sep,* or *pbpB*), lies in a large cluster of genes, of which seven are involved only in cell division. (Six others code for enzymes of peptidoglycan synthesis, one is concerned with protein secretion and two are of unknown function.) The division genes in this cluster are *fts*36, *ftsI, ftsW, ftsQ, ftsA, ftsZ,* and *envA*. (Most division genes in *E. coli* are labeled "*fts*" because the nondividing mutants grow in length to produce "filamentous" cells.) Outside of this cluster, there are three genes (*ftsE, ftsX, ftsS*) that together form a cluster with *ftsY* (a gene coding for a protein that strongly resembles eukaryotic signal recognition and docking proteins), the three genes of the *minB* operon and a few separate genes that may code for proteins with regulatory action on other *fts* genes (see below). I will not attempt to deal with each of these genes and proteins in detail but I will try to pick out some of their interesting properties, which together provide some idea of the molecular events that take place when cells switch into, and out of, division.

FtsZ. Joe Lutkenhaus, in this lab, first identified the *ftsZ* locus, and has since gone on to show that this is one of the key genes in cell division (Lutkenhaus et al., 1980; Lutkenhaus, 1990). The FtsZ protein seems to be required for the earliest known step in cell division, as well as being involved in subsequent steps (Walker et al., 1975; Begg and Donachie 1985). Lutkenhaus and his collaborators have shown that the FtsZ protein is the target for a number of endogenous division inhibitor proteins, including the SOS inhibitor SulA.

(Other inhibitors will be discussed below.) They have shown that this protein is highly conserved in evolution and that recognizable homologues are produced by all bacterial species so far tested, including both Gram-negatives and Gram-positives, rods, and cocci. However, perhaps its most interesting property, in view of our previous discussion about division regulation, is that this protein alone appears to be "rate limiting" for division. Increasing levels of production of FtsZ protein have been shown to increase the division capacity of *minB* mutants so that, at a sufficiently high level of FtsZ production, the mutant cells are able to use *all* of their potential division sites, both those at the cell poles and all of the internucleoid sites. The limited "division potential" of normal cells therefore seems to be due to limited production of FtsZ. The questions now are: How is production of FtsZ regulated? What exactly does it do? How are its properties related to the apparently quantal behavior of division potential?

The role of other division proteins: FtsQ, FtsA, PBP3, and EnvA. FtsQ, FtsA, and PBP3 together with FtsZ carry out the formation of the septum; a covalently bonded pair of peptidoglycan sheets formed across the width of the cell. FtsQ and PBP3 are membrane proteins (present in very low numbers of molecules) while FtsA, although predominantly soluble, is enriched in a membrane fraction of intermediate density that is supposed to be associated with septum formation. Of these, only PBP3 has a known enzymatic activity, which is that of a septation-specific peptidoglycan transpeptidase (and, perhaps in association with FtsW, of a transglycosylase as well). EnvA appears to be required for the final splitting of the double septal layer, allowing the outer cell membrane to invaginate and cell division to be completed.

Switching off: the Min proteins. The discovery of minicell mutants showed that polar sites, which can be considered to be the remains of "old" septation sites, need to be inactivated if they are not to compete for the cell's limited supply of FtsZ protein. The recent dissection of the *minB* locus (deBoer et al., 1989) has shown us more about how this takes place. The *minB* operon consists of three genes (*minC, minD,* and *minE*). The MinC and MinD proteins together form an inhibitor of cell division, which can potentially block all division sites but, in the presence of MinE, is directed solely to polar sites. Most interestingly, Joe Lutkenhaus has shown that the MinCD block can be overcome by overproduction of FtsZ. Thus, in cells in which all division has been inhibited by the production of MinC and MinD in the absence of MinE, the capacity to divide is restored by overproduction of FtsZ. In normal cells, in which the action of the MinCD inhibitor is confined to the poles, overproduction of FtsZ can overcome this block also, so that minicells are produced freely, in addition to normal numbers of internal divisions (Ward and Lutkenhaus, 1985).

Regulation of division genes and proteins. In the large cluster of genes that includes *fts*36, *ftsI, ftsW, ftsQ, ftsA, ftsZ* and *envA,* there are no strong transcription terminators known between the division genes and they can therefore be considered to be part of an operon. However, there are also many internal promoters (often located within the genes themselves) that offer the possibility of separate regulation of blocks of genes within the operon. The presence of these internal promoters probably accounts in part for the quite different levels of production of the individual proteins, although differences in translational efficiency of different parts of the common mRNA have also been shown to be important (Mukherjee and Donachie, 1990). There is also evidence of variable regulation of transcription, linked to the cell's requirements for division proteins, for promoters in *ftsQ* and *ftsA*, immediately upstream of *ftsZ* (Dewar et al., 1989). The promoter in the *ftsQ* gene appears to be subject to regulation by upstream sequences, also lying within the *ftsQ* and *ddl* coding sequences. The properties of these regulatory sequences suggest the possibility of DNA looping in association with a regulatory protein (Dewar et al., 1990). Such behavior implies the existence of regulatory proteins and several candidates are known. MreB protein appears to act as a repressor of *ftsI* (Wachi and Matsuhashi, 1989) while FtsA appears to exert a negative effect on transcription from the promoters in *ftsQ* and *ftsA* (Dewar et al., 1989). Interestingly, the MreB and FtsA proteins show considerable sequence homology (Doi et al., 1988). The FtsH protein (Ferreira et al., 1987) regulates insertion of PBP3 into the cell membrane (Ogura et al., 1991), and the CfcA protein appears to act as a specific negative regulator of FtsZ production (Nishimura, 1989). The problem now is to determine how these regulators are themselves regulated in relation to the growing cell's periodic needs for division proteins. The only suggestion so far made is that FtsA is itself used up during septation, so that its periodic depletion could result in increased transcription from the promoters that it appears to control. This, if correct, may be considered to be a secondary level of control; producing increased amounts of division proteins once septation is already in progress.

The cell cycle of a rod-shaped bacterium can be considered to consist of an elongation phase (in which the maintenance of the cylindrical shape depends on the low abundance membrane proteins, RodA and PBP2: Spratt, 1975) followed by septation. We have recently made some observations that suggest how one part of this switch may operate (Begg et al., 1990). Briefly, we have found that cells that are unable to form septa because of very low (indeed almost undetectable) levels of PBP3 can be rescued and caused to divide by treatments which increase the proportion of peptidoglycan chains with tripeptide side chains (L-Ala:D-Glu:mDAP). We have shown that this division is carried out by the residual PBP3 molecules, thus supporting the suggestion of Botta and Park (1981) that tripeptide side chains are the preferred acceptor substrates for transpeptidation carried out by PBP3. Beck and Park (1976) have reported that the activity of carboxypeptidase II, the enzyme that produces tripeptides by removing the terminal D-alanine from tetrapeptide side chains, fluctuates during the cell cycle and reaches a maximum just before septation. They also showed that high activity of this enzyme was dependent on the activity of FtsA. This may imply either that the activation of carboxypeptidase II is an FtsA-dependent event, or that FtsA protein is itself carboxypeptidase II. Whichever, if either, is true, these observations suggest that a periodic change in the relative abundance of different kinds of acceptor side chains may be one of the events that switches cells from elongation into division.

Such changes, essential as they are for succesful cell division, are likely to be secondary events that follow on from the primary change that initiates septation. We do not know what this primary event is, but the earliest step we know depends on the activity of FtsZ protein. Finding the way in which the cell brings this protein into action should therefore be a major advance in understanding what causes cells to divide. Study of the *cfcA* gene, the activity of which appears to control *ftsZ* specifically, may therefore be very rewarding.

References

Adler HI, Fisher WD, Cohen A, Hardigree AA (1967): Miniature *Escherichia coli* cells deficient in DNA. Proc Natl Acad Sci USA 57:321–326.

Beck BD, Park JT (1976): Activity of three murein hydrolases during the cell division cycle of *Escherichia coli* K-12 as measured in toluene-treated cells. J Bacteriol 126:1250–1260.

Beck BD, Park JT (1977): Basis for the observed fluctuation of carboxypeptidase II activity during the cell cycle of BUG6, a temperature-sensitive mutant of *Escherichia coli*. J Bacteriol 130:1292–1302.

Begg KJ, Donachie WD (1985): Cell shape and division in *Escherichia coli*: experiments with shape and division mutants. J Bacteriol 163:693–703.

Begg KJ, Takasuga A, Edwards DH, Dewar SJ, Spratt BG, Adachi H, Ohta T, Matsuzawa H, Donachie WD (1990): The balance between different peptidoglycan precursors determines whether *E. coli* cells will elongate or divide. J Bacteriol 172:6697–6703.

Bird RE, Louarn J, Martuscelli J, Caro L (1972): Origin and sequence of chromosome replication in *Escherichia coli*. J Mol Biol 70:549–566.

Botta GA, Park JT (1981): Evidence for involvement of penicillin-binding protein 3 in murein synthesis during septation but not during cell elongation. J Bacteriol 145:333–340.

Clark DJ (1968): The regulation of DNA replication and cell division in *Escherichia coli* B/r. Cold Spring Harbor Symp Quant Biol 33:823–838.

Cooper S, Helmstetter CE (1968): Chromosome replication and the division cycle of *Escherichia coli* B/r. J Mol Biol 31:519–540.

de Boer PAJ, Crossley RE, Rothfield LI (1989): A division inhibitor and a topological specificity factor coded for by the minicell locus determines proper placement of the division septum in *E. coli*. Cell 56:641–649.

Dewar SJ, Begg KJ, Donachie WD (1990): Regulation of expression of the *ftsA* cell division gene by sequences in upstream genes. J Bacteriol 172:6611–6614.

Dewar SJ, Kagan-Zur V, Begg KJ, Donachie WD (1989): Transcriptional regulation of cell division genes in *Escherichia coli*. Mol Microbiol 3:1371–1377.

Doi M, Wachi M, Ishino F, Tomioka S, Ho M, Matsuhashi M (1988): Determination of the gene products of the *mre* region that functions in formation of the rod-shape of *Escherichia coli* cells and of the sequence of the *mreB* gene. J Bacteriol 170:4619–4624.

Donachie WD (1968): Relationship between cell size and time of initiation of DNA replication. Nature (London) 219:1077–1079.

Donachie WD (1969): Control of cell division in *Escherichia coli*: experiments with thymine starvation. J Bacteriol 100:260–268.

Donachie WD, Begg KJ (1970): Growth of the bacterial cell. Nature (London) 227:1220–1224.

Donachie WD, Begg KJ (1989a): Cell length, nucleoid separation and cell division of rod-shaped and spherical cells of *Escherichia coli*. J Bacteriol 171:4633–4639.

Donachie WD, Begg KJ (1989b): Chromosome partition in *Escherichia coli* requires post-replication protein synthesis. J Bacteriol 171:5405–5409.

Donachie WD, Begg KJ, Sullivan NF (1984): Morphogenes of *Escherichia coli*. In R. Losick and L. Shapiro (eds): Microbial Development. New York: Cold Spring Harbor Laboratories, pp 27–62.

Donachie WD, Begg KJ, Vicente M (1976): Cell length, cell growth and cell division. Nature (London) 264:328–333.

Ferreira LCS, Keck W, Betzner A, Schwarz U (1987): *In vivo* cell division gene product interactions in *Escherichia coli* K-12. J Bacteriol 169:5776–5781.

Grossman N, Ron EZ (1989): Apparent minimal size required for cell division in *Escherichia coli*. J Bacteriol 171:80–82.

Hill TM, Pelletier AJ, Tecklenburg ML, Kuempel PL (1988): Identification of the DNA sequence from the *E. coli* terminus region that halts replication forks. Cell 55:459–466.

Hirota Y, Jacob F, Ryter A, Buttin G, Nakai T (1968): On the process of cellular division in *Escherichia coli*. I. Asymmetrical division and production of deoxyribonucleic acid-less bacteria. J Mol Biol 35:175–192.

Hiraga S, Niki H, Ogura T, Ichinose C, Mori H, Ezaki B, Jaffé A (1989): Chromosome partitioning in *Escherichia coli*: novel mutants producing anucleate cells. J Bacteriol *171*:1496–1505.

Hiraga S, Ogura T, Niki H, Ichinose C, Mori H (1990): Positioning of replicated chromosomes in *Escherichia coli*. J Bacteriol 172:31–39.

Hussain K, Begg KJ, Salmond GPC, Donachie WD (1987): *ParD*: a new gene coding for a protein required for chromosome partitioning and septum localisation in *Escherichia coli*. Mol Microbiol 1:73–81.

Inouye M (1971): Pleiotropic of the *recA* gene of *Escherichia coli*: uncoupling of cell division from deoxyribonucleic acid replication. J Bacteriol 106:539–542.

Jacob F, Brenner S, Cuzin F (1963): On the regulation of DNA replication in bacteria. Cold Spring Harbor Symp Quant Biol 28:329–348.

Jones NC, Donachie WD (1973): Chromosome replication, transcription and control of cell division in *Escherichia coli*. Nature New Biol 243:100–103.

Jones, NC, Donachie WD (1974): Protein synthesis and the release of the replicated chromosome from the cell membrane. Nature (London) 251:252–254.

Kornberg A (1974): DNA Replication. San Francisco: W.H. Freeman.

Lutkenhaus JF (1990) Regulation of cell division in *E. coli*. Trends Genet 6:22–25.

Lutkenhaus JF, Wolf-Watz H, Donachie WD (1980): Organization of genes in the *ftsA-envA* region of the *E. coli* genetic map and identification of a new *fts* locus: *ftsZ*. J Bacteriol 142:615–620.

Maaløe O, Kjeldgaard NO (1966): The Control of Macromolecular Synthesis. New York and Amsterdam: W.A. Benjamin.

Masters M (1970): Origin and direction of replication of the chromosome in *Escherichia coli*. Proc Natl Acad Sci USA 65:601–608.

Masters M, Broda P (1971): Evidence for the bidirectional replication of the *E. coli* chromosome. Nature New Biol 232:137–140.

Mukherjee A, Donachie WD (1990): Differential translation of cell division proteins. J Bacteriol 172:6106–6111.

Nishimura A (1989): A new gene controlling the frequency of cell division per round of DNA replication in *Escherichia coli*. Mol Gen Genet 215:286–293.

Oishi M, Yoshikawa H, Sueoka N (1964): Synchronous and dichotomous replication of the *Bacillus subtilis* chromosome during spore germination. Nature (London) 204:1069–1073.

Ogura T, Tomoyasu T, Yuki T, Morimura S, Begg KJ, Donachie WD, Mori H, Niki H, Hiraga S (1991): Structure and function of the *ftsH* gene in *Escherichia coli*. Res Microbiol, in press.

Pardee AB, Prestidge LS (1956): The dependence of nucleic acid synthesis on the presence of amino acids in *E. coli*. J Bacteriol 71:677–683.

Pierucci O, Helmstetter CE (1969): Chromosome replication protein synthesis and cell division in *Escherichia coli*. Fed Proc 28:1755–1760.

Prescott DM, Kuempel PL (1972): Bidirectional replication of the chromosome in *Escherichia coli*. Proc Natl Acad Sci USA 69:2842–2845.

Smith HS, Pardee AB (1970): Accumulation of a protein required for division during the cell cycle of *Escherichia coli*. J Bacteriol 101:901–909.

Spratt BG (1975): Distinct penicillin-binding proteins involved in the division, elongation and shape of *E. coli* K-12. Proc Natl Acad Sci USA 72:2999–3001.

Steck TR, Drlica K (1984): Bacterial chromosome segregation: evidence for DNA gyrase involvement in decatenation. Cell 36:1081–1088.

Teather RM, Collins, JF, Donachie WD (1974): Quantal behaviour of a diffusible factor which initiates septum formation at potential division sites in *Escherichia coli*. J Bacteriol 118:407–413.

Wachi M, Matsuhashi M (1989): Negative control of cell division by *mreB*, a gene that functions in determining the rod-shape of *Escherichia coli* cells. J Bacteriol 171:3123–3127.

Walker J, Kovarik A, Allan J, Gustafson (1975): Regulation of bacterial cell division: temperature-sensitive mutants of *Escherichia coli* that are defective in cell division. J Bacteriol 123:693–703.

Ward JE, Jr, Lutkenhaus JF (1985): Overproduction of FtsZ induces minicell formation in *E. coli*. Cell 42:941–949.

Woldringh CL, Valkenburg JAC, Pas E, Taschner PEM, Huls P, Weintjes FB (1985): Physiological and geometrical conditions for cell division in *Escherichia coli*. Ann Inst Pasteur/Microbiol 136A131–138.

Yoshikawa H, Sueoka N (1963): Sequential replication of the *Bacillus subtilis* chromosome I. Comparison of marker frequencies in exponential and stationary growth phases. Proc Natl Acad Sci USA 49:559–566.

Zusman DR, Carbonell A, Haga JI (1973): Nucleoid condensation and cell division in *Escherichia coli* MX7472ts52 after inhibition of protein synthesis. J Bacteriol 115:1167–1178.

**Perspectives on Cellular Regulation:
From Bacteria to Cancer, pages 65–72**

Cellular Aggregation During Fruiting Body Formation in *Myxococcus xanthus*

David R. Zusman

Department of Molecular and Cell Biology, University of California, Berkeley, California 94720

From Studying Cell Division in *Escherichia coli* to Fruiting Body Formation in *Myxococcus*

I spent two years as a postdoctoral fellow with Art Pardee, from 1970 to 1972, working on the behavior of cell division mutants of *Escherichia coli* in the presence of various inhibitors. The experience taught me a great deal about humility. Not only did *E. coli* cell division prove much more complex than we expected, but we were forever having difficulties finding and then holding onto a handle on the problem. I shared a tiny laboratory with Masayori Inouye, who at that time already had 45 publications, a graduate student Po Chi Wu, who could sight read articles in the Journal of Biological Chemistry and play concert piano, and a dishwasher who was a recognized authority on life. I was particularly influenced at this time by a lecture Pardee gave to his class on how to go about becoming a great scientist with little or no experience. He explained that the first and most critical step was to pick a research area that interests you. He warned his students not to join a bandwagon, because other labs will undoubtedly do a better job. But instead to choose a small area and become a world authority on the problem. He cautioned that it was important to choose a problem that can go somewhere in a reasonable period of time, certainly within our research lifetime.

When I arrived in Berkeley early in 1973, I was faced with all of the difficulties of setting up a laboratory in a department that was so poor we recycled chipped pipets. I decided to heed Pardee's advice and revived some old strains of *Myxococcus xanthus*, an organism I had studied as a graduate student at UCLA (with Eugene Rosenberg) and that seemed to fill the requirements. I was particularly interested in how the individual bacteria communicated with each other to aggregate to form fruiting bodies, multicellular structures that contain spores (Zusman, 1984; Shimkets, 1990). The fruiting bodies of one

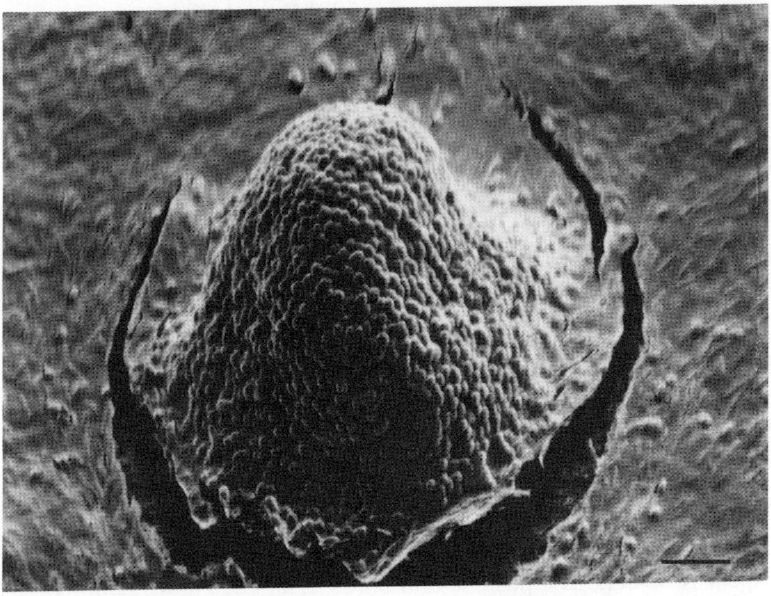

Fig. 1. Scanning electron micrograph of a fruiting body of *Myxococcus xanthus* strain DZF1. The bar is 0.01 mm. Courtesy of K.A. O'Connor.

myxobacterial species are shown in Figure 1. They are brightly colored, usually yellow, orange, or red and visible with the naked eye (about 0.1 to 0.5 mm in height). The fruiting bodies serve two functions: they allow cells to survive periods of starvation stress in their soil environment and they serve as a means of spore dispersal, since the fruiting bodies are often ingested by insects and birds and the spores dispersed in distant locations. But before discussing fruiting bodies, we must first consider the vegetative life cycle of the myxobacteria. Our discussion here will be limited to *M. xanthus*, the most studied of the myxobacteria.

Vegetative Life-Style of *Myxococcus*

M. xanthus is a Gram-negative, rod-shaped bacterium, typically about 5 μm in length and 0.5 μm in diameter. In nature, the bacteria inhabit the same ecological niche as the eukaryotic cellular slime molds, damp soils rich in decaying organic matter (Kaiser, 1986). Like the cellular slime molds, they are predators that kill and digest both Gram-positive and Gram-negative bacteria, as well as some larger eukaryotes such as yeasts. They can also be grown on defined or semidefined laboratory media. The myxobacteria cannot engulf other microorganisms but rely instead on extracellular antibiotics that kill or immobilize their prey, and on lytic enzymes and extracellular proteases, nucleases, lipases, phosphatases, and various polysaccharide-degrading enzymes. The reliance on extracellular bacteriolytic and degradative activities may be responsible for the strong social interactions exhibited by these organisms: they tend to aggregate into large "hunting groups" that may contain several million individuals. Presumably, this ensures that enough antibiotics and lytic activities are present at one location to defeat the hardiest of microorganisms. However, if food is not available, the bacteria trigger their developmental program.

Developmental Program of *M. xanthus*

Two Cell Types Appear During Development

The developmental program of *M. xanthus* has been studied morphologically, and by using biochemical markers and transcriptional fusion probes (Kaiser et al., 1985). Perhaps the most interesting recent finding is that heterogeneity soon develops within the starving population (O'Connor and Zusman, 1989). While the large majority of cells (80–90%) aggregate to form raised mounds in which the cells all sporulate, the remaining cells follow a different developmental fate. These cells, called peripheral rods, remain as rod-shaped arrested cells that act as sentinels around and between fruiting bodies. These cells do not normally aggregate or sporulate unless they are harvested and resuspended at high concentration on a fresh substrate (O'Connor and Zusman, J Bacteriol, in press). The peripheral rods express many biochemical markers that distinguish them from both vegetative cells and sporulating cells. If a limited or transient food source becomes available, these cells have been shown to respond immediately, unlike spores that require several hours and a relatively high threshold of food to germinate.

Patterns of Cellular Interaction During Fruiting Body Formation

The aggregating cells have been followed morphologically by scanning electron microscopy and time-lapse video microscopy (O'Connor and Zusman, 1989). The first sign of aggregation is the formation of large spiral patterns of cells in stacked monolayers parallel to the substratum. The cells in spirals are tightly coherent; they remain associated even after being scraped from the agar surface. Terraces form through the stacking of aggregation centers in adjacent monolayers. The formation and maintenance of terraced monolayers suggest that cells participate in side-to-side and pole-to-pole interactions that differ from dorsal-to-ventral interactions. As cells continue to accumulate, the terraces become mounds. The mounds become hemispherical in shape as development proceeds. Sporulation can be concurrent with the morphogenesis of mounds. Eventually at least 98% of the cells in the aggregates become spores.

Biochemical Markers of Development

During the process of cellular aggregation and sporulation several interesting biochemical markers were identified. Protein S was the first and perhaps the most studied of these markers (Inouye et al., 1979). This protein is synthesized early in development and soon becomes the most abundant developmentally regulated protein in the cell. At later times it is transported to the outer surface of the cell where it self-assembles into an outer surface coat on the spores. An almost identical protein, protein S1, is produced at a later time and accumulates inside of spores (Downard and Zusman, 1985; Teintze et al., 1985). Another developmentally regulated protein, myxobacterial hemagglutinin, is also produced at early times (Cumsky and Zusman, 1979). It is most abundant when cells aggregate and appears to be localized at the cell poles (Nelson et al., 1981). The protein was shown to be a galactose-binding lectin (Cumsky and Zusman, 1979). Mutants that are missing this protein show some defects in cellular aggregation (Romeo and Zusman, 1987). Interestingly, these mutants seem impaired at forming end-to-end associations, which are characteristic of wild-type aggregating cells. Alkaline phosphatase was recently shown to be a useful marker for late development since it is spore-specific, accumulating in maturing spores (Weinberg and Zusman, 1990).

Aggregation-Defective Mutants of *M. xanthus*

***tag* mutants are temperature-dependent for aggregation.** A large number of nonfruiting missense mutants were also isolated in an attempt to analyze the process of aggregation (Morrison and Zusman, 1979). One particularly interesting class, called *tag*, were temperature-dependent for aggregation. These mutants aggregated and fruited normally at 28°C but at 34°C failed to form fruiting bodies. The cells aggregated into a flat monolayer of cells at 34°C but were unable to display the vertical interactions that appear to be necessary for mound formation. Nine *tag* complementation groups spanning 8.5 kilobase pairs (kbp) of DNA were identified through the mapping of 28 independent Tn*5* insertions (O'Connor and Zusman, 1990). Surprisingly, all of the insertion and deletion mutants had the same phenotype as the missense mutants: they were temperature sensitive for mound formation. These results suggest that *M. xanthus* has at least two sets of genes for developmental aggregation. The *tag* genes constitute one set of these genes; they are required for normal development at 34°C but are not required for normal development at 28°C.

"Frizzy" mutants are defective in cellular aggregation. Another interesting class of nonaggregating mutants was called "frizzy" (*frz*) (Zusman, 1982; Zusman et al., 1990). These mutants failed to aggregate into discrete mounds but rather aggregated into "frizzy" filaments that contain myxospores (Fig. 2). Thirty-six mutants exhibiting this phenotype were all found to be linked to the same Tn*5* insertion site. Using the kanamycin resistance drug marker in Tn*5* as

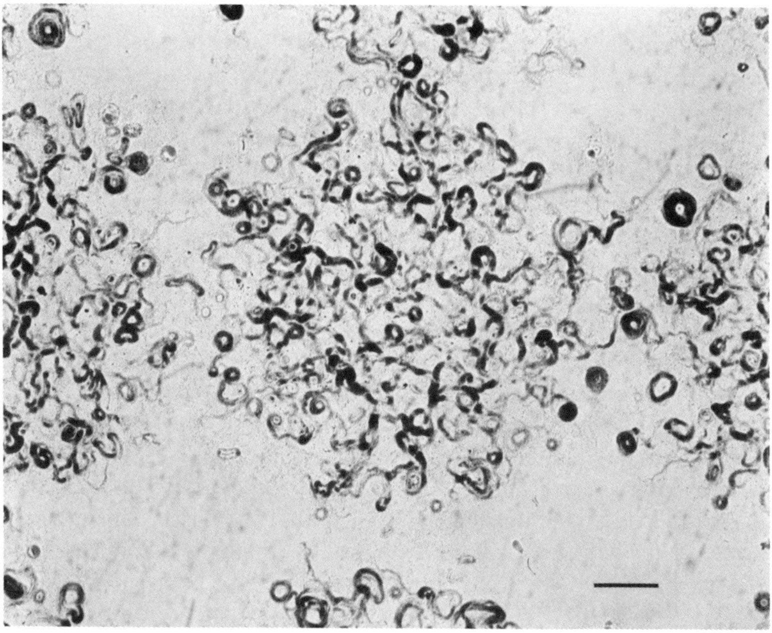

Fig. 2. Frizzy colony morphology. A colony of the *frzE* mutant DZF1536 was grown on fruiting agar for 7 days and photographed. The bar is 1 mm.

a selectable marker, the *frz* genes were cloned and mapped. Genetic analysis revealed five *frz* complementation groups on the cloned DNA (*frzA, frzB, frzC, frzE,* and *frzF*) (Blackhart and Zusman, 1985a). Strains with mutations in any one of these genes had the Frz phenotype. Another complementation group, *frzD*, was defined by two Tn*5* mutations mapping in the C-terminus of *frzCD*. These mutants formed nonspreading colonies of motile cells. Tn*5* insertions in the one kbp region between *frzE* and *frzF* defined *frzG*. These mutants showed minor defects in aggregation and formed irregular fruiting bodies.

The "frizzy" genes are similar to enteric chemotaxis genes. The first clue to the function of the *frz* genes came from observations of the motility pattern of the *frz* mutants. *M. xanthus* does not contain flagella. It moves by gliding motility, a motility system that involves smooth movement of the bacteria in the direction of the long axis on a solid surface. Cells generally move within slime trails, but they are capable of movement, albeit somewhat slower, in the absence of these trails. The mechanism of gliding motility is unknown. *M. xanthus* generally moves about 2–4 μm/min, about 1000-fold slower than enteric bacteria. Time-lapse video microscopy shows that the *frz* genes are involved in determining the frequency at which cells reverse their direction of gliding (Blackhart and Zusman, 1985b). When plated at low cell density on motility agar, wild-type cells reverse their direction of gliding approximately every 7–8 min; net movement occurs since the interval between reversals can vary widely. *frzA*, *-B*, *-C*, *-E*, and *-F* cells rarely reverse direction, approximately every 1–2 hr; *frzD* and *frzG* cells reverse more frequently than wild type, every 1–2 and 4–5 min, respectively. These behavior patterns are reminiscent of chemotaxis mutants of enteric bacteria that swim smoothly and rarely tumble or tumble and rarely swim smoothly.

The similarity of the *frz* genes to chemotaxis genes was supported by DNA sequence analysis of the *frz* region (McBride et al., 1989). The deduced gene products, with the exception of FrzB, showed strong homology to the chemotaxis genes of *Salmonella typhimurium* and *E. coli*. The FrzA protein exhibited 28% amino acid identity with the CheW protein. The FrzCD protein contains a region of about 250 amino acids that is similar to the C-terminal portion of the methyl-accepting chemotaxis receptor proteins (MCPs). The regions of strongest similarity are those that are also highly conserved among the enteric MCPs. For example, FrzCD and *S. typhimurium* Tar exhibit 40% amino acid identity over a 141-residue region. FrzCD also contains a region with potentially significant similarity to the DNA-binding region of a *Bacillus subtilis* sigma factor. The FrzE protein was found to be homologous to both CheA and CheY (McCleary and Zusman, 1990), which are members of a family of "two component response regulators." FrzE also contains an unusual 68 amino acid region that is 38% alanine and 34% proline. Similar alanine- and proline-rich regions are also seen in peptides of the pyruvate dehydrogenase multienzyme complex of *E. coli* and have been shown to make the protein very flexible at this region. Since CheA has been shown to autophosphorylate and then transfer its phosphate to CheY, the alanine- and proline-rich region of FrzE might constitute a flexible hinge that would facilitate phosphate transfer within different domains of the same protein. FrzE is thus likely to be a second messenger that relays information between the signaling protein, FrzCD, and the gliding motor.

Several lines of evidence show that FrzCD is modified by methylation and that FrzF (which is homologous to CheR) is a methyl transferase (McCleary et al., 1990: (1) FrzCD contains regions similar in sequence to the methylatable sites of the MCPs; (2) FrzCD migrates as a ladder of bands between 45 and 41 kDa; (3) glutamate methyl esters are base labile; treatment of cell extracts with alkali prior to electrophoresis chased the ladder of bands into a more slowly migrating pattern; (4) a ladder of radiolabeled FrzCD bands was detected when developing *M. xanthus* cells were labeled with the methyl donor [^{3}H]*S*-adenosyl methionine; and (5) mutants of the putative methyltransferase, *frzF*, failed to produce the modified forms of FrzCD.

The last *frz* gene product identified was FrzG. It was found to be homologous to CheB, the methylesterase. Thus, although *M. xanthus* is unflagellated, it appears to have a sensory transduction system that is similar in many of its components to those found in flagellated bacteria.

The "frizzy" genes are developmentally regulated. The expression of the first four genes of the *frz* gene cluster (*frzA, frzB, frzCD*, and *frzE*) was studied by using Tn*5*–*lac* transcriptional fusions as reporters of gene expression (Weinberg and Zusman, 1989). These *frz* genes were found to be developmentally regulated with their transcription peaking about the time of early mound formation. Northern blot hybridization analysis suggested that the *frz* genes are arranged as an operon. To test this hypothesis, double mutants were constructed that contained Tn*5*-*132* (a modified form of Tn*5* that has the tetracycline resistance element rather than the kanamycin resistance element) either upstream or downstream of the reporter Tn*5*–*lac*. The expression of the *frz* genes in the double mutants was consistent with the hypothesis that the first four genes are organized as an operon with an internal promoter. Insertion mutations in *frzCD* lowered gene expression whether they were upstream or downstream of the reporter Tn*5*–*lac*, suggesting that the FrzCD proteins regulate transcription of the entire operon from a promoter upstream of *frzA*. The expression of the *frz* genes during development is important for the regulation of cell communication during the aggregation process. The methylation of the putative receptor protein, FrzCD, may provide a biochemical assay for the signals that are exchanged between cells during aggregation.

Acknowledgments

The research in my laboratory is supported by Public Health Service Grant GM20509 from the National Institutes of Health and by National Science Foundation Grant DMB-8820799.

References

Blackhart BD, Zusman DR (1985a): Cloning and complementation analyses of the "frizzy" genes of *Myxococcus xanthus*. Mol Gen Genet 198:243–254.

Blackhart BD, Zusman DR (1985b): "Frizzy" genes of *Myxococcus xanthus* are involved in the control of rate of reversal of gliding motility. Proc Natl Acad Sci USA 82:8767–8770.

Cumsky M, Zusman DR (1979): Myxobacterial hemagglutinin: A developmentally-specific lectin of *Myxococcus xanthus*. Proc Natl Acad Sci USA 76:5505–5509.

Downard JS, Zusman DR (1985): Differential expression of protein S genes during *Myxococcus xanthus* development. J Bacteriol 161:1146–1155.

Inouye M, Inouye S, Zusman DR (1979): Biosynthesis and self-assembly of protein S, a development-specific protein of *Myxococcus xanthus*. Proc Natl Acad Sci USA 76:209–213.

Kaiser D (1986): Control of multicellular development: *Dictyostelium* and *Myxococcus*. Annu Rev Genet 20:539–566.

Kaiser D, Kroos L, Kuspa A (1985): Cell interactions govern the temporal pattern of *Myxococcus* development. Cold Spring Harbor Symp Quant Biol 50:823–830.

McBride MJ, Weinberg RA, Zusman DR (1989): "Frizzy" aggregation genes of the gliding bacterium *Myxococcus xanthus* show sequence similarities to the chemotaxis genes of enteric bacteria. Proc Natl Acad Sci USA 86:424–428.

McCleary WR, McBride MJ, Zusman DR (1990): Developmental sensory transduction in *Myxococcus xanthus* involves methylation and demethylation of FrzCD. J Bacteriol 172:4877–4887.

McCleary WR, Zusman DR (1990): FrzE of *Myxococcus xanthus* is homologous to both cheA and cheY of *Salmonella typhimurium*. Proc Natl Acad Sci USA 87:5898–5902.

Morrison CE, Zusman DR (1979): *Myxococcus xanthus* mutants with temperature-sensitive, stage-specific defects: Evidence for independent pathways in development. J Bacteriol 140:1036–1042.

Nelson DR, Cumsky MG, Zusman DR (1981): Localization of myxobacterial hemagglutinin in the periplasmic space and on the cell surface of *Myxococcus xanthus* during developmental aggregation. J Biol Chem 256:12589–12595.

O'Connor KA, Zusman DR (1989): Patterns of cellular interactions during fruiting body formation in *Myxococcus xanthus*. J Bacteriol 171:6013–6024.

O'Connor KA, Zusman DR (1990): Genetic analysis of *tag* mutants of *Myxococcus xanthus* provides evidence for two developmental aggregation systems. J Bacteriol 172:3868–3878.

Romeo JM, Zusman DR (1987): Cloning of the gene for myxobacterial hemagglutinin and the isolation and analysis of structural gene mutations. J Bacteriol 169:3801–3808.

Shimkets LJ (1990): Social and developmental biology of the myxobacteria. Microbiol Rev 54:473–501.

Teintze M, Thomas R, Furuichi T, Inouye M, Inouye S (1985): Two homologous genes coding for spore-specific proteins are expressed at different times during development of *Myxococcus xanthus*. J Bacteriol 163:121–125.

Weinberg RA, Zusman DR (1989): Evidence that the *frz* genes of *Myxococcus xanthus* are developmentally regulated. J Bacteriol 171:6174–6186.

Weinberg RA, Zusman DR (1990): Alkaline, acid, and neutral phosphatase activities are induced during development in *Myxococcus xanthus*. J Bacteriol 172:2294–2302.

Zusman DR (1982): Frizzy mutants: a new class of aggregation-defective developmental mutants of *Myxococcus xanthus*. J Bacteriol 150:1430–1437.

Zusman DR (1984): Cell-cell interactions and development in *Myxococcus xanthus*. Q Rev Biol 59:119–138.

Zusman DR, McBride MJ, McCleary WR, O'Connor KA (1990): Control of directed motility in *Myxococcus xanthus*. Symp Soc Gen Microbiol 46:199–218.

Perspectives on Cellular Regulation:
From Bacteria to Cancer, pages 73–83

Mosaic Genes, Hybrid Penicillin-Binding Proteins, and the Origins of Penicillin Resistance in *Neisseria meningitidis* and *Streptococcus pneumoniae*

Brian G. Spratt, Christopher G. Dowson, Qian-yun Zhang, Lucas D. Bowler, James A. Brannigan, and Agnes Hutchison

Microbial Genetics Group, School of Biological Sciences, University of Sussex, Falmer, Brighton BN1 9QG, United Kingdom

Introduction

At the time I joined Art Pardee's laboratory at Princeton in 1973 it was becoming apparent that the mechanism of action of penicillin was considerably more complicated than had been envisaged a few years earlier. In particular it had become clear that bacteria possessed multiple penicillin-sensitive enzymes, although nothing was then known of their roles in the killing action of β-lactam antibiotics. The development in Pardee's laboratory of a convenient method for studying penicillin-sensitive enzymes as penicillin-binding proteins (PBPs; Spratt and Pardee, 1975) opened the way to a genetic dissection of the role of each of these enzymes in the killing action of β-lactam antibiotics, and led to an understanding of the key functions that these proteins play in cell division, cell elongation, and the determination of cell shape in *Escherichia coli*. The model for the mechanism of action of penicillin that resulted from this work that, with characteristic generosity, Art allowed me to publish on my own (Spratt, 1975), has stood the test of time.

In this article we will very briefly outline the general properties of PBPs, and will then describe recent results that show that the evolution of altered PBPs in penicillin-resistant clinical isolates of Neisserial and Streptococcal species has occurred by a quite unexpected mechanism, that involves the creation of mosaic genes, which encode hybrid PBPs with reduced affinity for penicillin, by recombinational events that recruit parts of the homologous genes from closely related bacterial species.

More extensive reviews of the properties of PBPs, and on PBP-mediated resistance to β-lactam antibiotics, can be found in Spratt and Cromie (1988) and Spratt (1989).

Basic Properties of High-Molecular-Weight PBPs

β-Lactam antibiotics kill bacteria by inactivating a set of penicillin-sensitive enzymes that catalyze the final transpeptidation steps of peptidoglycan synthesis. These enzymes can be detected as PBPs since their inactivation by penicillin involves the formation of an essentially irreversible penicilloyl–enzyme complex that is analogous to the acyl–enzyme intermediate formed with the normal transpeptidase substrate. In most bacteria, the higher molecular weight (M_r) PBPs are essential for cell growth, whereas the lower M_r PBPs are nonessential. Bacteria possess multiple high M_r PBPs since the synthesis of different types of peptidoglycan during the cell cycle (e.g., during cell division versus cell elongation) is mediated by distinct PBPs (Spratt, 1975).

High M_r PBPs are embedded in the cytoplasmic membrane at their amino-termini and have their catalytic domain(s) extending into the periplasm where they have access to their peptidoglycan substrate (Spratt and Cromie, 1988). Although the evidence is largely equivocal, some of these PBPs appear to be bifunctional, catalyzing both a penicillin-insensitive transglycosylation reaction and a penicillin-sensitive transpeptidation reaction. A considerable body of evidence supports the view that the penicillin-sensitive transpeptidase domain is located toward the carboxy-terminus of high M_r PBPs (Spratt and Cromie, 1988). The function of the amino-terminal region is unclear, but in those PBPs that are bifunctional (e.g., PBP 1B of *E. coli*), it may form the transglycosylase domain.

PBP-Mediated Resistance to Penicillin

In most bacterial species resistance to penicillin, and other members of the β-lactam group of antibiotics, is due to the production of a β-lactamase that destroys the antibiotic. In a few species resistance has emerged by the modification of the physiologically important high M_r PBPs that are the killing targets for β-lactam antibiotics. In Gram-negative bacteria, resistance to penicillin can also be obtained by alterations of the outer membrane that reduce the access of the antibiotic to the PBPs, although this mechanism of resistance is usually not very significant unless accompanied by the production of a β-lactamase or alterations of PBPs (Spratt, 1989).

PBP-mediated resistance to penicillin appears to be difficult to achieve for two reasons. First, penicillin and other β-lactam antibiotics are structural analogs of the normal peptide substrate of high M_r PBPs. The development of a PBP that is resistant to inhibition by penicillin therefore requires a subtle remodeling of its active center, so that it can exclude the antibiotic, without impairing its ability to process the substrate. Laboratory studies have shown that, at least for PBP 3 of *E. coli,* it is not possible to discriminate greatly between most β-lactams and substrate by single amino acid substitutions (Hedge and Spratt, 1985). It is therefore likely that in most cases multiple amino acid substitutions need to be introduced into the transpeptidase domain to produce large reductions in the affinities of a PBP for β-lactam antibiotics.

Second, β-lactam antibiotics kill bacteria by inactivating multiple killing targets. For example, the binding of a β-lactam to either PBP 1, or PBP 2, will result in the killing of *Neisseria gonorrhoeae*. At the concentrations achieved in vivo, penicillin will inactivate each of these PBPs, and killing will occur by two independent routes. The emergence of resistance to penicillin has therefore necessitated the development of penicillin-resistant forms of both of these killing targets (Dougherty et al., 1980). Similarly, in *Streptococcus pneumoniae* there are five high M_r PBPs each of which appear to be killing targets, and the development of resistance to penicillin has required large reductions in the affinities of each of these five enzymes (Zighelboim and Tomasz, 1980).

The combined effects of the need to accumulate multiple amino acid substitutions, in multiple killing targets, has presumably slowed the emergence of PBP-mediated resistance to penicillin. This type of resistance has, however, arisen in several clinically important pathogens (Spratt, 1989). Perhaps the most dramatic example of PBP-mediated resistance to penicillin is provided by *S. pneumoniae* (the pneumococcus). Penicillin resistance was unknown in pneumococci until the late 1960s when isolates with moderate levels of resistance were encountered in Australia and Papua New Guinea. Subsequently, isolates with increased levels of penicillin resistance, and with resistance to other groups of antibiotics, appeared in South Africa in the late 1970s (reviewed by Klugman and Koornhof, 1988). Whereas typical penicillin-sensitive pneumococci have minimal inhibitory concentrations (MICs) of about 0.006 μg benzylpenicillin/ml, the resistant strains have MICs as high as 16 μg/ml. Resistance to penicillin in these strains is entirely due to the production of altered forms of each of the high M_r PBPs with greatly decreased affinity for penicillin (Zighelboim and Tomasz, 1980). As pneumococci have not acquired the ability to produce β-lactamases, and are Gram-positive and thus unable to reduce the access of penicillin to the PBPs, the development of resistance by the production of altered PBPs appears to be their only route to resistance.

PBP-mediated resistance to penicillin has also been well documented in *N. gonorrhoeae* (the gonococcus). The majority of penicillin-resistant gonococci are β-lactamase producers and have high levels of resistance. Non-β-lactamase-producing strains that have clinically significant levels of resistance to penicillin are also widely encountered. The latter strains, which have MICs as high as 4 μg benzylpenicillin/ml, in contrast to 0.004 μg/ml for sensitive strains, produce altered forms of each of the two high M_r PBPs (PBP 1 and PBP 2) and also have alterations in the permeability of the gonococcal outer membrane (Dougherty et al. 1980; Faruki and Sparling, 1986).

Other species in which resistance to penicillin mediated by alterations of PBPs has been demonstrated include *Staphylococcus aureus, Staphylococcus epidermidis, Neisseria meningitidis,* and *Haemophilus influenzae* (reviewed in Spratt, 1989).

Mosaic PBP 2 Genes in Penicillin-Resistant Neisseriae

We will start by describing the origins of the altered forms of PBP 2 that exist in penicillin-resistant clinical isolates of pathogenic and commensal *Neisseria* species. *Neisseria* species possess two physiologically-important high M_r PBPs that are the killing targets for β-lactam antibiotics. As described above, penicillin resistance in *N. gonorrhoeae* has occurred by the production of altered forms of both of these PBPs (PBP 1 and PBP 2), together with alterations of the outer membrane. In recent years penicillin-resistant clinical isolates of the other pathogenic member of the genus, *N. meningitidis* (the meningococcus), have appeared particularly in Spain and the United Kingdom. These strains have only moderate levels of resistance (MICs of 0.32–1.28 μg benzylpenicillin/ml compared to 0.02 μg/ml for sensitive strains) and appear to still respond to penicillin therapy. Any further increase in the level of resistance is, however, likely to result in treatment failure since high concentrations of benzylpenicillin cannot be achieved in the cerebrospinal fluid.

Resistance to penicillin in meningococci is due, at least in part, to the production of altered forms of PBP 2 that have decreased affinity for penicillin (Mendelman et al., 1988). Resistance to penicillin has also emerged in the commensal species, *N. lactamica* and *N. polysaccharea*. Although the mechanism of resistance is less characterized than in the pathogenic species, it also involves alterations of PBP 2.

The sequences of the PBP 2 gene of four penicillin-sensitive and six penicillin-resistant clinical isolates of *N. meningitidis* have been compared (Spratt et al., 1989 and unpublished). The sequences of the gene from the penicillin-sensitive strains, which include isolates of each of the three main meningococcal serogroups, are very uniform (<0.3% divergence between any pair of strains). In contrast, the gene from each of the penicillin-resistant strains (Fig. 1) consists of regions that are essentially identical to the corresponding regions of the gene from the sensitive strains (''sensitive blocks''), alternating with regions that show 14–23% divergence (''resistant blocks'').

The simplest interpretation of these data is that the mosaic structures of the PBP 2 genes have arisen by recombinational events that have replaced parts of the meningococcal gene with regions from the PBP 2 gene of other species

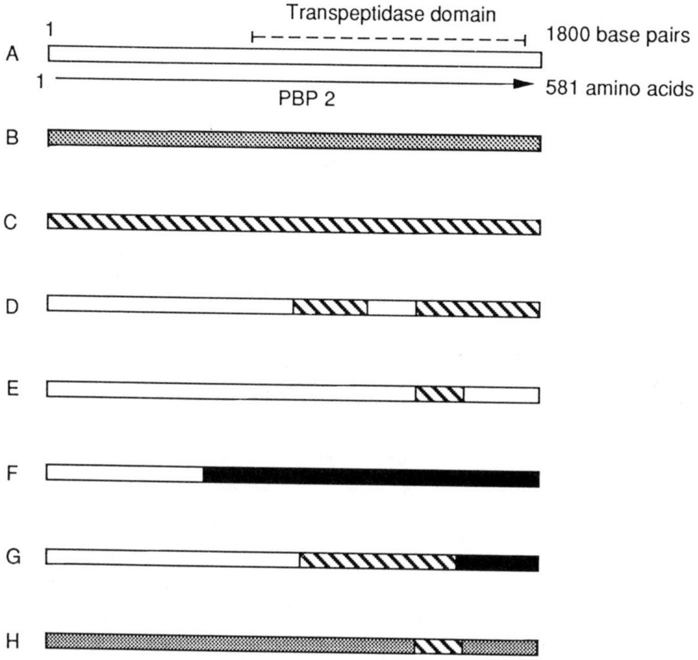

Fig. 1. Mosaic PBP 2 genes of penicillin-resistant *N. meningitidis* and *N. lactamica*. The PBP 2 gene of a penicillin-sensitive strain of *N. meningitidis* is represented in **(A).** The thin line terminating in an arrow illustrates the region encoding PBP 2. The PBP 2 genes from a penicillin-sensitive *N. lactamica* and a penicillin-sensitive *N. flavescens* are represented in **(B)** and **(C)**, respectively. **(D)–(H)** represent the PBP 2 genes of penicillin-resistant *Neiseria* isolates. **(D)** *N. meningitidis* S738; **(E)** *N. meningitidis* 74-JC; **(F)** *N. meningitidis* K196; **(G)** *N. meningitidis* 32-IC; **(H)** *N. lactamica* NL2535. The mosaic structures of the PBP 2 genes from the resistant strains are shown, together with the origins of the "sensitive" and "resistant" blocks (see text). = *N. meningitidis* DNA, = *N. lactamica* DNA, = *N. flavescens* DNA, = DNA from an unidentified species.

that show 14–23% sequence divergence from the meningococcus. Since meningococci are naturally transformable, it is very likely that these recombinational events have occurred by this process. The formation of a mosaic gene results in a hybrid PBP that has decreased affinity for penicillin (see below). As expected, the recombinational events have occurred almost exclusively within the region encoding the penicillin-sensitive transpeptidase domain (≈codons 260–581).

Confirmation that the resistant blocks have been introduced into the meningococcus from other Neisserial species has been obtained by using oligonucleotides that correspond to regions from the resistant blocks as probes to identify the donor species among a strain collection that includes each of the described Neisserial species. These experiments indicated that *N. flavescens* was the likely source of some of the resistant blocks. Sequencing of the PBP 2 gene of *N. flavescens* has confirmed that the majority of penicillin-resistant meningococci contain resistant blocks that have been introduced from *N. flavescens* (Spratt et al., 1989 and unpublished).

Figure 1 shows a diagrammatic representation of the origins of different regions of the mosaic PBP 2 genes found in some of the penicillin-resistant strains of *N. meningitidis* that we have studied. Strain S738 contains two blocks of DNA that are ≈22% diverged from those found in the PBP 2 gene of penicillin-sensitive strains of *N. meningitidis* but are essentially identical to the corresponding region of the PBP 2 gene of *N. flavescens*. The PBP 2 gene of a second penicillin-resistant strain (74-JC) contains a single block derived from *N. flavescens*. The resistant strain K196 contains a block from a so far unidentified species whereas strain 32-IC contains blocks from both *N. flavescens* and the unidentified donor species.

Penicillin resistance in commensal *Neisseria* species, e.g., *N. lactamica* and *N. polysaccharea*, as well as in *N. gonorrhoeae* (Spratt, 1988), has also occurred by the creation of mosaic PBP 2 genes that contain regions derived from other *Neisseria* species. Figure 1 shows the PBP 2 gene of a penicillin-resistant strain of *N. lactamica* (MIC of 0.4 μg benzylpenicillin/ml) that differs from that of penicillin-sensitive strains (MICs of 0.02 μg/ml) only by the presence of a block of DNA that is identical to that found in the PBP 2 gene of *N. flavescens* (R. Lujan, J.A. Sáez-Nieto, Q-Y. Zhang, and B.G. Spratt, unpublished).

Comparisons of the sequences of the PBP 2 genes of the six penicillin-resistant meningococci, and high resolution fingerprinting of the PBP 2 genes from further strains isolated in the UK and Spain, have shown that mosaic PBP 2 genes have arisen on more than one occasion (Zhang et al., 1990). At present it is unclear whether the mosaic PBP 2 genes have arisen independently in meningococci, gonococci, and commensal *Neisseria* species, or whether they arose in one species and have been spread horizontally into the other species.

Mosaic PBP 2B Genes in Penicillin-Resistant Streptococci

Penicillin-resistant pneumococci appeared in the late 1960s and are now found at varying frequencies in most countries (Klugman and Koornhof, 1988). Almost nothing is known about the origins of these strains. Thus it is not known whether they arose only once, and have subsequently spread around the world, or, alternatively, whether they have multiple origins. Pneumococci are naturally transformable, and the role of clonal spread in the dissemination of penicillin-resistant strains, versus the formation of novel resistant strains by the horizontal transfer of each of the altered PBP genes from resistant strains into sensitive strains, has not been addressed.

To investigate these epidemiological aspects of penicillin-resistant pneumococci, we chose to compare the sequences of the PBP 2B genes of a collection of resistant strains (and control penicillin-sensitive strains) isolated from various countries over the last 20 years. The PBP 2B gene was chosen since there is good evidence that it is an important killing target for penicillin and since large decreases in its affinity are found in penicillin-resistant strains.

The complete PBP 2B gene was cloned and sequenced from the penicillin-sensitive laboratory strain R6, and the region encoding the transpeptidase domain has been sequenced from a further 5 sensitive strains and from 17 penicillin-resistant strains (Dowson et al., 1989a,b, and unpublished). The sequences of the PBP 2B gene from the penicillin-sensitive strains were very uniform showing a maximum sequence diversity of 1% between strains. A similar uniformity was found within the amylomaltase gene from both penicillin-sensitive and penicillin-resistant strains. In sharp contrast, the PBP 2B genes from each of the resistant strains (Fig. 2) were very different to those of the sensitive strains and had a mosaic structure, consisting of blocks of sequence that were very similar to those in the gene from sensitive strains (sensitive

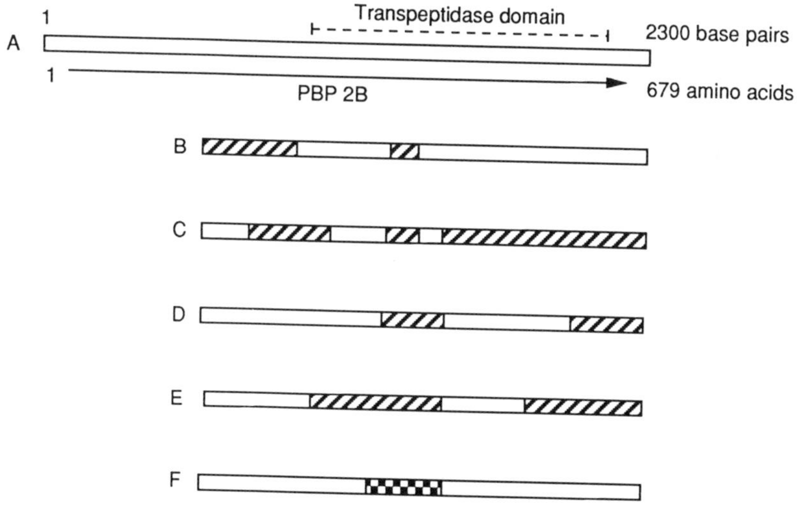

Fig. 2. Mosaic PBP 2B genes of penicillin-resistant *S. pneumoniae*. The PBP 2B gene of a penicillin-sensitive strain of *S. pneumoniae* is represented in **(A)**. The thin line terminating in an arrow shows the region encoding PBP 2B. The mosaic structures of the Class A PBP $2B^R$ genes from the penicillin-resistant strains 29044 **(B)**, 53139/72 **(C)**, 64147 **(D)**, VA1 **(E)**, and of the Class B PBP $2B^R$ gene from the resistant strain DN87/577 **(F)**, are shown (the proximal part of the PBP 2B gene of the penicillin-resistant strains has not been sequenced). The sequences of the unhatched regions are almost identical to the corresponding regions from penicillin-sensitive strains, whereas those of the hatched regions are approximately 14–21% diverged. The two types of hatching in the Class A and B PBP $2B^R$ genes indicate that these regions are believed to have been derived from two separate sources (see text and Dowson et al., 1989b for further details).

blocks), alternating with blocks that showed ≈14–21% sequence divergence (resistant blocks).

The presence of blocks of sequence that are 14–21% diverged contrasts sharply with the uniformity of the sequences of the PBP 2B genes from sensitive strains, and of the amylomaltase genes from both sensitive and resistant strains, and indicates that these resistant blocks have been introduced, presumably by transformation, from some other source. The sources of the resistant blocks have not so far been identified, but are presumed to be streptococcal species that are about 14–21% diverged from *S. pneumoniae*.

The penicillin-resistant strains fall into two major groups with respect to the sequences of their PBP 2B genes. One group of strains possess mosaic PBP 2B genes that are identical in sequence (the class B PBP $2B^R$ gene). These strains, which were isolated in Spain and the UK, are all serotype 23, are also resistant to both chloramphenicol and tetracyline and are probably members of the same resistant clone (Dowson et al., 1989b).

The majority of the resistant strains, which included isolates from Papua New Guinea, South Africa, the United States, and Europe, possess mosaic PBP 2B genes (class A PBP $2B^R$ genes) that had an independent origin from the class B PBP $2B^R$ genes. Altered PBP $2B^R$ genes have therefore arisen on at least two occasions, and have involved the replacement of parts of the pneumococcal gene with the corresponding regions from two different species, to produce the class A and class B PBP $2B^R$ genes.

The class A PBP $2B^R$ genes of different resistant pneumococci vary both in their block structures (i.e., the patterns of alternating sensitive and resistant blocks), and by slight differences in the sequences of the sensitive and resistant blocks. At present it is difficult to decide whether the class A genes had a single origin, and have subsequently diverged, or whether they have multiple origins, and their similarities are due to the introduction of sequences from different members of the same genetically diverse streptococcal species (see Dowson et al., 1989b).

Penicillin-resistant isolates of viridans group streptococci that produce altered forms of high M_r PBPs have also been isolated in recent years. At least in the isolates of *S. sanguis* and *S. oralis* that we have examined, resistance to penicillin has emerged by the transfer of mosaic PBP genes from penicillin-resistant pneumococci into these viridans streptococci (Dowson et al., unpublished). The situation is particularly clear in *S. sanguis,* where nucleotide sequencing has shown that the *S. sanguis* homolog of PBP 2B has been replaced in resistant strains with sequences that are identical to the class B PBP $2B^R$ gene found in penicillin-resistant pneumococci. The PBP 2B gene of penicillin-resistant strains of *S. sanguis* therefore contains regions of DNA derived both from penicillin-sensitive pneumococci (i.e., the sensitive blocks of the class B PBP $2B^R$ gene) and from the unknown streptococcal species that was the source of the resistant blocks in the class B PBP $2B^R$ gene.

Production of Hybrid Enzymes by Horizontal Gene Transfer

The results we have described show that the altered PBPs found in penicillin-resistant isolates of both Neisserial and Streptococcal species have arisen by interspecies recombinational events, probably mediated by transformation, which replace parts of the PBP genes with the corresponding regions from the homologous genes of closely related species.

Genetic transformation allows a bacterium to sample the genetic variation that occurs, both within its own species, and in other species that are sufficiently similar for homologous recombination to occur (probably ≥70% nucleotide sequence similarity). Species that differ by 20–30% in nucleotide sequence produce variant forms of the same PBP that differ by perhaps 10–15% in amino acid sequence. Closely related species, or different members of a genetically diverse species, are therefore likely to produce variant forms of the PBP, which, by virtue of differences in their amino acid sequences, possess differing kinetic properties, e.g., affinities for β-lactam antibiotics. Genetic transformation allows a bacterial species that is very sensitive to penicillin, because it produces PBPs with very high affinity, to become more resistant to penicillin by the recruitment of homologous PBP genes (or the relevant regions of them) from any related species that happens to produce lower affinity forms of these enzymes.

In most penicillin-resistant *N. meningitidis* strains recruitment of parts of the PBP 2 gene from *N. flavescens* has occurred. It might therefore be expected that the latter species produces a PBP 2 that has a lower affinity for penicillin, and is more resistant to penicillin, than other *Neisseria* species. This appears to be the case as *N. flavescens* isolates, including those obtained in the preantibiotic era, have MICs of 0.2–0.4 μg benzylpenicillin/ml, which is approximately 10-fold higher than most other members of this genus.

There are a few other reports of situations in which genetic diversity is produced by transformation. A well-documented example is antigenic variation of the *N. gonorrhoeae* pilin protein (Seifert et al., 1988). Gonococci possess only a single intact pilin gene (*pilE*) but contain multiple (nonexpressed) variant partial pilin sequences (*pilS* genes). During infection, bacteria that produce antigenically variant pilins arise by transformational events that replace parts of the expressed *pilE* gene with the corresponding regions from *pilS* genes released by the autolysis of neighboring cells. The presence of multiple *pilS* genes and a single *pilE* gene allows antigenic variation to occur in an infection caused by a single gonococcal strain.

Transformation can also generate diversity where there is only a single copy of a gene, by the formation of mosaic genes through transformational events between genetically diverse members of the same species or, as in the case of PBP genes, by utilizing the diversity between related species. The production of variant forms of the IgA protease of *N. gonorrhoeae*, that differ in their immunological properties and their proteolytic specificity, appears to result from the production of mosaic genes (Halter et al., 1989). Doubtless other examples of this type of phenomenon remain to be discovered in bacteria that are naturally transformable, especially among genes that encode cell surface antigens or virulence factors.

Concluding Remarks

Penicillin-resistant forms of high M_r PBPs in penicillin-resistant clinical isolates of meningococci, gonococci, commensal *Neisseria* species, pneumococci, and viridans streptococci have unexpectedly all occurred by the formation of mosaic genes, rather than by the accumulation of amino acid substitutions that gradually decrease the affinity of the PBPs for penicillin. It is very likely that the formation of mosaic genes has occurred by genetic transformation and it is interesting that, with one exception, all of the best characterized examples of PBP-mediated resistance have occurred in bacterial species that are naturally transformable. The one exception is *Staphylococcus aureus* where resistance to methicillin (a penicillin) has arisen by the acquisition, by an illegitimate recombinational event, of a penicillin-resistant PBP from an unknown source that has no homolog in penicillin-sensitive strains (Song et al., 1987).

Acknowledgments

The work from our laboratory was funded by the Medical Research Council and the Wellcome Trust. B.G.S. is a Wellcome Trust Principal Research Fellow. Q-Y.Z was supported by a grant from the Society for General Microbiology. We are grateful to John Maynard Smith for stimulating discussions and to our clinical colleagues for providing interesting organisms.

This paper is dedicated to Dr. Arthur B. Pardee on his 70th birthday.

References

Dougherty TJ, Koller AE, Tomasz A (1980) Penicillin-binding proteins of penicillin-susceptible and intrinsically resistant *Neisseria gonorrhoeae*. Antimicrob Agents Chemother 18:730–737.

Dowson CG, Hutchison A, Spratt BG (1989a): Extensive remodelling of the transpeptidase domain of penicillin-binding protein 2B of a penicillin- resistant South African isolate of *Streptococcus pneumoniae*. Mol Microbiol 3:95–102.

Dowson CG, Hutchison A, Brannigan JA, George RC, Hansman D, Liñares J, Tomasz A, Maynard Smith J, Spratt BG (1989b): Horizontal transfer of penicillin-binding protein genes in penicillin-resistant clinical isolates of *Streptococcus pneumoniae*. Proc Natl Acad Sci USA 86:8842–8846.

Faruki H, Sparling PF (1986): Genetics of resistance in a non-β-lactamase-producing gonococcus with relatively high-level penicillin resistance. Antimicrob Agents Chemother 30:856–860.

Halter R, Pohlner J, Meyer TF (1989): Mosaic-like organization of IgA protease genes in *Neisseria gonorrhoeae* generated by horizontal genetic exchange *in vivo*. EMBO J 8:2737–2744.

Hedge PJ, Spratt BG (1985): Resistance to β-lactam antibiotics by re-modelling the active site of an *E. coli* penicillin-binding protein. Nature (London) 318:478–480.

Klugman KP, Koornhof HJ (1988): Drug resistance patterns and serogroups or serotypes of pneumococcal isolates from cerebrospinal fluid or blood. J Infect Dis 158:956–964.

Mendelman PM, Campos J, Chaffin DO, Serfass DA, Smith AL, Sáez-Nieto JA (1988): Relative penicillin G resistance in *Neisseria meningitidis* and reduced affinity of penicillin-binding protein 3. Antimicrob Agents Chemother 32:706–709.

Seifert HS, Ajioka RS, Marchal C, Sparling PF, So M (1988): DNA transformation leads to pilin antigenic variation in *Neisseria gonorrhoea*. Nature (London) 336:392–395.

Song MD, Wachi M, Doi M, Ishino F, Matsuhashi M (1987): Evolution of an inducible penicillin-target protein in methicillin-resistance *Staphylococcus aureus* by gene fusion. FEBS Lett 221:167–171.

Spratt BG (1975): Distinct penicillin-binding proteins involved in the division, elongation, and shape of *Escherichia coli* K12. Proc Natl Acad Sci USA 72:2999–3003.

Spratt BG (1988): Hybrid penicillin-binding proteins in penicillin-resistant strains of *Neisseria gonorrhoeae*. Nature (London) 332:173–176.

Spratt BG (1989): Resistance to β-lactam antibiotics mediated by alterations of penicillin-binding proteins. In L.E. Bryan (ed): Microbial Resistance to Drugs. Berlin: Springer-Verlag, pp. 77–100.

Spratt BG, Cromie KD (1988): Penicillin-binding proteins of Gram-negative bacteria. Rev Infect Dis 10:699–711.

Spratt BG, Pardee AB (1975): Penicillin-binding proteins and cell shape in *E. coli*. Nature (London) 254:516–517.

Spratt BG, Zhang Q-Y, Jones DM, Hutchison A, Brannigan JA, Dowson CG (1989): Recruitment of a penicillin-binding protein gene from *Neisseria flavescens* during the emergence of penicillin resistance in *Neisseria meningitidis*. Proc Natl Acad Sci USA 86:8988–8992.

Zhang Q-Y, Jones DM, Sáez-Nieto JA, Pérez Trallero E, Spratt BG (1990): Genetic diversity of penicillin-binding protein 2 genes of penicillin-resistant strains of *Neisseria meningitidis* revealed by fingerprinting of amplified DNA. Antimicrob Agents Chemother 34:1523–1528.

Zighelboim S, Tomasz A (1980): Penicillin-binding proteins of multiply antibiotic-resistant South African strains of *Streptococcus pneumoniae*. Antimicrob Agents Chemother 17:434–442.

Perspectives on Cellular Regulation:
From Bacteria to Cancer, pages 85–94

The All-Purpose Gene Fusion

Jon Beckwith

Department of Microbiology and Molecular Genetics, Harvard Medical School, Boston, Massachusetts 02115

I left Harvard for Berkeley to begin my postdoctoral work in Art Pardee's lab in 1961. This was to be my first experience with *Escherichia coli* and its genetics. Shortly after my arrival, Art announced that he was moving to Princeton and gave me the choice of continuing with him or finding someone else at Berkeley. I couldn't find anyone there whose work was of great interest to me, so I agreed to go to Princeton. It was the right choice for me scientifically. First, the period I spent in Art's lab was formative for me scientifically in a number of ways. Second, it was much easier to get work done in Princeton than in Berkeley. There were not the distractions of Berkeley and environs. On the other hand, I always regretted having to leave Berkeley after only six months. Life in the Bay Area in 1961 was a sharp change from Cambridge. A revolution in life style and politics was just beginning.

I learned some important lessons about science from Art. First, I was struck with his approach to science. He had the ability to smell out areas that were off the beaten track, yet had the potential to yield novel insights. He was certainly not a camp-follower in any sense. As has been pointed out elsewhere (Judson, 1979), Art was responsible for initiating a number of important areas of molecular biology. And that brings me to the second lesson. Although Art has an impressive list of firsts to his credit, he never received as much recognition for these as he deserved. This may have been due in part to a ''fault'' which I remember him once bemoaning to me. In particular, he was pointing out how much credit Changeux and Monod got for their studies on feedback inhibition compared to those of himself and John Gerhart. He attributed this, in part, to his inability to come up with sexy (he didn't use this word) titles for his articles when compared to Monod's skill at the art. This discussion has stuck with me; I have always tried, often unsuccessfully, to use titles that would attract people to read papers from my lab. It is also a point I make with students and postdocs in the lab.

O° Mutants in Pardee's Lab

Gene fusions have been an integral part of my research for over 25 years. My interest in fusions can be traced back to a project I began in Art Pardee's lab in 1962. This project took off from something Art had already started and was, in turn, derived from the studies of Jacob, Monod and co-workers on the operator region of the *lac* operon.

Both the Pasteur group and Art were engaged in trying to define the *cis*-acting elements responsible for the expression and regulation of genes. While the repressor had been established as a molecule that shuts off gene expression and RNA polymerase as a molecule that initiates gene expression, the question of where these molecules acted on the DNA was not raised at first. It was the inspiration of Jacob (Judson, 1979) to see that there must be an operator and to devise ways of defining it with mutations. In the first paper on the operator (Jacob et al., 1960), it was assumed that the site of repression and the site of initiation of transcription were one and the same. Interestingly, there was not even a discussion of whether there was one or more sites for these functions. The reader was expected to see this assumption as a given, inherent in the concepts of regulation. Why this was so is not clear. It may have been that the most natural way to imagine the repressor acting was by direct competition for the RNA polymerase binding site. At that time, enzyme inhibitors were known to act in this way, and the general concept of allosteric regulation was yet to come.

Strengthening the identity of operator and the RNA polymerase binding site, later to be known as the promoter, was the existence of the O^o mutations. These were mutations that mapped in the same general region as O^c mutations, but were phenotypically *lacZ*$^-$ and *lacY*$^-$. Given the knowledge at the time, the most ready explanation for these mutations was that they had altered the transcription initiation site leading to failure to express the operon. The proximity of O^o and O^c mutations reinforced the proposal that the operator site served the two functions.

Art Pardee used the O^o mutants to ask a different question. Recognizing that different constitutive genes were expressed at different levels, he supposed that this could be due to differing strengths of promoters. To test this, he isolated Lac$^+$ revertants of O^o mutants and found that they varied in their level of expression of the *lac* operon. At this point, I joined the project to further characterize the revertants. Since, at that time, we assumed that the interpretation of the O^o mutants by the Pasteur group was correct, we concluded that variations in operator (operator defined in part as the transcription initiation site) strength could be responsible for variation in constitutive gene expression (Pardee and Beckwith, 1962).

Both the Pasteur definition of the operator and our explanation of O^o revertants were subsequently shown to be faulty. As discussed below, my later analysis of the O^o mutations indicated that they did not map in the same region as O^c mutations, and they were, in fact, only extreme examples of polar mutations of the *lacZ* gene.

Despite the wrong paths taken in these studies, there is still an impressive aspect to this work. Scientific breakthroughs come for many different reasons. In some cases, it is a combination of accident and the prepared mind; in others, it is the synthesis of accumulated data into an insightful hypothesis (the repressor hypothesis is one of these). However, another component of major scientific advances is asking the right questions. The question Jacob asked was "where does the repressor act?" Pardee's question was "How can constitutive genes be expressed at different levels?" Today, these questions may seem painfully obvious ones with painfully obvious answers; but, they were not in these early days of molecular biology. I often have the frustrated feeling that we would be moving faster in the research in my own lab if we could only think of the right questions.

O^o Mutants and Fusions

The first hint that something was amiss both with the understanding of the O^o mutations and our analysis of O^o revertants came when I found that many of the revertants we had studied were due to unlinked suppressor mutations (Beckwith, 1963). These suppressors, which we later came to define as ochre suppressors (Brenner and Beckwith, 1965), were acting at the translation level to restore operon expression. In addition, John Gerhart and I, then a graduate student in Art's lab, came up with the idea of looking for revertants of the O^o mutants in which *lacY* but not *lacZ* expression was restored. Such revertants could be sought by selecting on medium where melibiose was the sole carbon source. Art and Louise Prestidge had recently shown that melibiose (at 42°) requires the *lacY* permease but not β-galactosidase for its utilization. When I characterized these revertants in some detail, I obtained further evidence against the proposal that O^o mutations mapped in the operator.

Many of the revertants genetically linked to the *lac* operon were due to deletions (Beckwith, 1964). One *lacZ* deletion I obtained, M15 (now widely used in α-acceptor strains for β-galactosidase complementation), deleted all known O^o sites but retained full sensitivity to repressor *and* exhibited maximal levels of *lac* operon expression. This one deletion eliminated the concept that O^o mutations were in the operator and that the region they defined was responsible for determining the level of expression of the *lac* operon. The O^o mutations turned out to be nonsense and frame-shift mutations with strong polar effects in the early portion of the *lacZ* gene. The continuation of this project begun in Art's lab was to help carry me through subsequent postdoctoral stints in the laboratories of Bill Hayes, Sydney Brenner, and François Jacob.

Among the melibiose$^+$ revertants of the O^o mutants were *lacZ* deletions that did reduce the expression of the operon. Since these deletions also removed the entire *lacI* gene, I proposed that these were due to deletions fusing *lac* to an adjacent operon which was expressed at a lower level. These were my first fusions. Subsequently, Jacob, Ullmann, and Monod used the melibiose approach with *lacI*^{s} mutants to obtain a large collection of such deletions and then to obtain fusions of the *lac* operon to the *purE* gene. With these latter fusions, the *lac* permease was placed under the control of the *purE* promoter and regulatory elements (Jacob et al., 1965).

Genetic fusions were not unknown in molecular biology in 1964. A classic case of such a fusion was the deletion r1589, which joined the rIIA and B genes of bacteriophage T4 (Champe and Benzer, 1962). This fusion had been used in the discovery of nonsense and frame-shift mutations, discoveries that were essential for the elucidation of the nature of the genetic code. However, the *purE–lacY* fusions were the first in which the regulation of expression of one gene could be shown to be newly subjected to the regulation of another. This regulatory feature of these fusions opened up a general use of fusions that is an important component of biological research today.

Lac Fusions as a General Tool in Molecular Biology (references for most of this work can be found in Silhavy and Beckwith, 1985)

One of the projects I pursued while a postdoctoral fellow in François Jacob's laboratory was the effect on *lac* operon expression of its transposition to different chromosomal sites. These studies were made possible by the work of Cuzin and Jacob (1964), who had isolated a number of strains in which an F′-*lac* was inserted at various chromosomal positions. I set out to obtain a larger collection of these transpositions. One of these strains, by chance, contained the *lac* genes inserted quite near the tryptophan (*trp*) operon. My labmate in Paris was Ethan Signer, who had been working with φ80 transducing phages. The attachment site for φ80 was also close to the operon. He suggested that we lysogenize the transposition strain with φ80 and see whether we could obtain φ80*lac* transducing phages. The experiment was tried and worked (Beckwith et al., 1966). The φ80*lac* phage became an important tool both in genetic and in vitro studies on gene regulation. We also developed a way to demand insertion of the F′-*lac* near *att80*. By selecting for the inactivation of the *tonB* gene at the same time that we selected for *lac* insertion, we required that the episome insert in the *tonB* gene. Insertion in this site again placed *lac* near *att80,* allowing isolation of φ80*lac* phages.

These studies and subsequent elaboration of them represented a form of cloning of bacterial genes. When I set up my laboratory at Harvard Medical School in 1965, one of the projects we undertook was to generalize this approach to cloning bacterial genes. We extended the techniques used for the isolation of φ80*lac* phages to obtain a φ80*ara* transducing phage (Gottesman and Beckwith, 1969) and a λ *lac* phage (Ippen et al., 1971).

The ϕ80*lac* phages also provided a means of generating new classes of fusions of the *lac* operon. The proximity of ϕ80*lac* to *trp* when inserted at *att80* permitted selection of deletions that fused the two operons. In addition to fusions similar to those obtained by Jacob and co-workers, in which the *lacY* gene was put under the control of tryptophan, we also obtained fusions in which the entire *lac* operon (and, thus, lactose metabolism) was dependent on the tryptophan concentration in the medium. These *trp–lac* fusions were useful for studies of both the *lac* and *trp* operons. Their analysis helped elucidate the function of the *lac* promoter and operator and defined the transcription termination signal of the *trp* operon.

Seeing the utility of *lac* fusions for studying regulation, a graduate student of mine at Harvard, Malcolm Casadaban, set out to devise an even more generalizable strategy for obtaining such fusions. He developed an approach in which *lac* could be inserted in any gene and fusions obtained in a subsequent selection. The steps involved inserting bacteriophage Mu into the gene in question, using a specially constructed phage carrying *lac* and a portion of Mu to insert *lac* into the gene, and then selecting for Lac^+ derivatives. Over the years, Malcolm developed easier to use systems for generating such fusions including the widely used Mud*lac* phages.

A new variation on the fusion approach was made possible by a discovery in the laboratory of Benno Müller-Hill. He and Joseph Kania found that they could delete a portion of the *lacZ* gene corresponding to the amino-terminus of the protein, replace that amino-terminus with varying lengths of the *lac* repressor, and still express β-galactosidase activity. This meant that not only could operon fusions be obtained, but gene fusions could be easily detected in which β-galactosidase was fused to another protein. This finding came at a time when I was switching my interests from problems of gene regulation to those of protein secretion. One of the approaches I was considering was that of fusing a cytoplasmic protein to a secreted protein in an effort to define the signals in the latter protein which were responsible for its export. The ability to fuse β-galactosidase in such a way as to retain its activity meant that the properties of such fusion strains would be much easier to follow. Malcolm Casadaban developed a new λ *lac* phage that allowed ready selection of such fusions.

We immediately began to use this approach to study the incorporation into the cell envelope of alkaline phosphatase, LamB protein, maltose binding protein, and the MalF protein. Unexpected properties of such gene fusions have provided the basis for much of the genetic work on the study of both protein secretion and membrane protein insertion. We initiated our studies in the expectation that by attaching an export signal to β-galactosidase, we would be able to transport the enzyme across the cytoplasmic membrane into the periplasm or outer membrane. Instead, we found that an export signal would initiate the translocation process, but that β-galactosidase was incompetent to be fully transferred through the membrane. The hybrid protein was imbedded in the cytoplasmic membrane.

Strains producing the "stuck" fusion proteins exhibited two interesting properties. First, synthesis of high amounts of these proteins blocked the cell's secretion machinery and caused cell death. Selection for mutations resistant to the lethal effects of the hybrid protein yielded signal sequence mutants that prevented the membrane incorporation of the protein. In Tom Silhavy's and my lab, we showed that these mutations were in the hydrophobic core of the signal sequence of the exported protein. These studies demonstrated directly that signal sequences were essential to the export process.

A second property of these fusion strains is that the localization of the hybrid proteins to the membrane prevents the proper assembly of β-galactosidase into active enzyme. Very low enzymatic activities are observed and, as a result, the strains exhibit a Lac^- phenotype. A Lac^+ phenotype can be restored by a mutation that interferes with the export process. That is, localization of the hybrid protein to the cytoplasm restores β-galactosidase activity. Because of this feature of strains carrying fusions between β-galactosidase and an exported protein, the system provides a direct genetic selection for internalization of an exported protein. Selection for Lac^+ revertants yields both signal sequence mutants and mutants defective in components of the cell's secretory apparatus. These studies have allowed the definition of several *sec* genes coding for protein components of the cell that are required for translocation of proteins across the cytoplasmic membrane.

Development of the Alkaline Phosphatase Fusion Approach

(references for most of this work can be found in Manoil et al., 1990)

One of the set of signal sequence mutations we obtained from the Lac^+ selections was in the *phoA* gene, which codes for the periplasmic enzyme, alkaline phosphatase. These mutations resulted in the bulk of the alkaline phosphatase being internalized to the cytoplasm in the form of unprocessed precursor. We noted that the amount of remaining alkaline phosphatase enzymatic activity detected in these strains corresponded to the small amount of protein that was exported to the periplasm. These and subsequent results led us to conclude that when alkaline phosphatase was restricted to the cytoplasm, it could not assemble into active enzyme. We now know that part of the reason for this lack of activity is the failure to form essential disulfide bonds in the reducing environment of the cytoplasm (Derman and Beckwith, unpublished results). This may well be a general way the cell has evolved for avoiding the deleterious effects of the internalization of degradative proteins normally localized to the periplasm.

Andrew Wright, who was doing a sabbatical in my lab, decided to use this feature of alkaline phosphatase to develop a fusion system that would employ alkaline phosphatase as a probe for protein export signals. He constructed plasmids that contained a *phoA* gene missing the DNA corresponding to the signal sequence, and, along with his graduate student, Charlie Hoffman, used them to show that a functional signal sequence from another protein would restore alkaline phosphatase activity. Colin Manoil, a postdoc in the lab, then extended these results to permit an in vivo approach to obtaining fusions of alkaline phosphatase to other proteins. He began with a *phoA* fragment obtained from a Wright–Hoffman plasmid, which contained DNA for most of the structural information for the protein but was missing its promoter, translation initiation signals, and the DNA corresponding to the signal sequence. This fragment was incorporated into the transposon Tn*5*. The resulting transposon, Tn*phoA*, is capable of inserting into genes in such a way as to generate gene fusions. These gene fusions exhibit alkaline phosphatase activity only if the gene to which *phoA* is fused codes for some kind of protein export signal.

Alkaline phosphatase fusions provide a new kind of reporter gene. Fusions of the *lac* operon and other analogous fusion systems have been generally used for monitoring the *expression* of genes to which *lac* is fused. The *phoA* gene, on the other hand, can be used to monitor the *location* of the gene product to which alkaline phosphatase is attached. This sensor quality of the *phoA* gene has resulted in a variety of new uses for the gene fusion approach. It allows one to distinguish genes for cell envelope proteins from those for cytoplasmic proteins. Since only the former class will code for protein export signals, only fusions to these genes will produce hybrid proteins with alkaline phosphatase activity. This property of *phoA* has permitted the detection of plasmid-carried or chromosomal genes for cell envelope proteins. Chromosomes can be scanned for such genes that are subject to a common regulation or involved in a common process. This approach has proved particularly useful with pathogenic bacteria. Most bacterial virulence determinants are cell envelope or secreted proteins, including toxins, hemolysins, pili, and other cell-surface components. Wide host-range plasmids carrying Tn*phoA* have been introduced into pathogenic bacteria, including *Vibrio cholera, Bordetella pertussis,* and *Salmonella typhimurium*. The detection of active *phoA* fusions has been used to identify genes that code for virulence determinants in these pathogenic bacteria.

Another application of the sensor property of alkaline phosphatase is in the topological mapping of cytoplasmic membrane proteins. Fusions of alkaline phosphatase to periplasmic domains of such membrane proteins show high enzymatic activity, while those to cytoplasmic domains generally show low activity. When several proteins of known topology have been mapped using this technique, the correct structure was predicted. As a result, alkaline phosphatase fusions have been used to analyze the topology of a number of membrane proteins of complex structure.

Although, generally, the *lac* fusions have served as a means of studying gene regulation, they also can be used in the study of membrane protein topology. Since protein export signals, when attached to β-galactosidase, cause it to be inserted in the membrane where it has low activity, one can distinguish periplasmic from cytoplasmic domains of such proteins using *lacZ* fusions (Froshauer et al., 1988).

I point out that some caution must be used in employing fusions to analyze aspects of protein structure. The way the alkaline phosphatase fusions are currently constructed, they result in the removal of a carboxy-terminal portion of the protein. If topology is determined by the interaction between amino-terminal and carboxy-terminal portions of membrane proteins, then hybrid proteins missing portions of the protein could assume an inappropriate topology. Further, the alkaline phosphatase itself may interfere in some way with the assembly process. So far, this has not been found to be the case with proteins the structure of which was already known. Furthermore, the current biochemical and biophysical techniques for determining the arrangement of proteins in membranes are quite limited. Thus, the alkaline phosphatase fusion approach adds a powerful tool to the array of strategies available for studying this question.

Alkaline phosphatase fusions also have allowed us to analyze the signals (topogenic sequences) within membrane proteins that determine their topology. These studies have revealed, for instance, that positively charged amino acids in the cytoplasmic domains of membrane proteins play an important role in maintaining the correct topology (Boyd and Beckwith, 1989).

A recent improvement on the alkaline phosphatase fusion approach involves the construction of a *phoA* cassette, which allows the insertion of alkaline phosphatase into cell envelope proteins (M. Ehrmann and J. Beckwith, unpublished results). In such insertions, the alkaline phosphatase is tethered both at its amino- and carboxy-terminus to the other protein. In this way, we have eliminated the problem of the missing carboxy-terminus associated with fusions. The doubly tethered alkaline phosphatase is still enzymatically active, and, thus, can still be used as an assay for the location of the domain to which it is fused. Our results show that properties of these "sandwich" fusions give a clearer picture of membrane protein topology than the standard fusions.

Conclusions

Genetic fusions have provided a powerful tool for studying an array of biological problems. They are one of the common approaches used today in molecular biology. The range of questions that can be studied using fusions has expanded with the development of reporter genes that act as sensors of subcellular location.

I would like to speculate that, in the future, the kinds of problems for which fusions are used will expand even further. I am thinking particularly of questions related to protein folding and structure. I have already described how alkaline phosphatase fusions can be used to determine features of the structure of cytoplasmic membrane proteins. There is a recent example of gene fusions being used to study protein–protein interactions. Fields and Song (1989) employed portions of the GAL4 DNA binding protein of *Saccharomyces cerevisiae* as sensors for proteins that can form complexes. The GAL4 protein is composed of two domains, one for DNA binding and the other for transcriptional activation of an adjacent gene. The DNA binding domain was fused to one protein and the activation domain to a second protein, the two proteins being ones which are known to form a complex. When constructs producing these two hybrid proteins were introduced into the same cell, the *GAL* genes were activated. The authors propose that this system could be used in general to assay for protein–protein interaction.

A second recent example of gene fusions employed for protein structure questions involves a study on protein folding. Luzzago and Cesareni (1989) found that when the α-subunit of β-galactosidase was fused to human ferritin, the ferritin enveloped the α preventing it from being available for complementation with another fragment of β-galactosidase. Mutations that allowed α complementation were ones disrupting the folding of the ferritin thus freeing the α-peptide for complementation. The collection of mutations obtained in this way could be used to study the folding pathway of ferritin. We have suggested another potential approach to the genetics of the protein folding problem using alkaline phosphatase fusions (Lee et al., 1989).

To be even more speculative, it may be possible to use fusions to analyze aspects of protein secondary structure. Can one develop a fusion system which can sense a preceding α-helix or β-sheet? Some basis for optimism on this score comes from recent studies which demonstrate the propagation of an α-helix into an attached polypeptide in what could be considered a gene fusion (Bansal et al., 1990). While these speculations may seem a little extreme at this point, it would have been hard to imagine a few years ago that fusions could give so much information on such problems as membrane protein structure.

References

Bansal A, Stradley SJ, Gierasch L (1990): Conformational studies of peptides corresponding to the LDL receptor cytoplasmic tail and transmembrane domain. In J.J. Villafranca (ed): Current Research in Protein Chemistry. New York: Academic Press , pp 331–338.

Beckwith J (1963): Restoration of operon activity by suppressors. Biochim Biophys Acta 76:162–164.

Beckwith JR (1964): A deletion analysis of the *lac* operator region in *E. coli*. J Mol Biol 8:427–430.

Beckwith JR, Signer ER, Epstein W (1966): Transposition of the *lac* region of *E. coli*. Cold Spring Harbor Symp Quant Biol 31:393–401.

Boyd D, Beckwith J (1989): Positively charged amino acid residues can act as topogenic determinants in membrane proteins. Proc Natl Acad Sci USA 86:9446–9450.

Brenner S, Beckwith JR (1965): Ochre mutants, a new class of suppressible nonsense mutants. J Mol Biol 3:629–637.

Champe SP, Benzer S (1962): An active cistron fragment. J Mol Biol 4:288–292.

Cuzin F, Jacob F (1964): Délétions chromosomiques et intégration d'un episome sexuel F-*lac*$^{+}$ chez *Escherichia coli* K12. CR Acad Sci 258:1350–1352.

Fields S, Song O (1989): A novel genetic system to detect protein-protein interactions. Nature (London) 340:245–246.

Froshauer S, Green GN, Boyd D, McGovern K, Beckwith J (1988): Genetic analysis of the membrane insertion and topology of MalF, a cytoplasmic membrane protein of *Escherichia coli*. J Mol Biol 200:501–555.

Gottesman S, Beckwith JR (1969): Directed transposition of the *ara* operon: A technique for the isolation of specialized transducing phages for any *E. coli* gene. J Mol Biol 44:117–127.

Ippen K, Shapiro JA, Beckwith J (1971): Transposition of the *lac* region to the *gal* region of the *Escherichia coli* chromosome: Isolation of *lac* transducing phages. J Bacteriol 108:5–9.

Jacob F, Perrin D, Sanchez C, Monod J (1960): L'opéron: groupe de gènes a expression coordonée par un opérateur. CR Acad Sci 250:1717–1729.

Jacob F, Ullman A, Monod J (1965): Délétions fusionnant l'opéron lactose et un opéron purine chez *Escherichia coli*. J Mol Biol 42:511–520.

Judson HF (1979): The Eighth Day of Creation. New York: Simon and Schuster.

Lee C, Li P, Inouye H, Beckwith J (1989): Genetic studies on the inability of β-galactosidase to be translocated across the *E. coli* cytoplasmic membrane. J Bacteriol 171:4609–4616.

Luzzago A, Cesareni G (1989): Isolation of point mutations that affect the folding of the H chain of human ferritin in *E. coli*. EMBO J 8:569–576.

Manoil C, Mekalanos JJ, Beckwith J (1990): Alkaline phosphatase fusions: Sensors of subcellular location. J Bacteriol 172:515–518.

Pardee AB, Beckwith JR (1962): Genetic determination of constitutive enzyme levels. Biochim Bipohys Acta 60:452–454.

Silhavy TJ, Beckwith J (1985): Uses of *lac* fusions for the study of biological problems. Microbiol Rev 49:398–418.

**Perspectives on Cellular Regulation:
From Bacteria to Cancer, pages 95–104**

Minor Codon, Cold Shock Protein, and Protein Folding

Masayori Inouye

Department of Biochemistry, Robert Wood Johnson Medical School, University of Medicine and Dentistry of New Jersey, Piscataway, New Jersey 08854

Without a Refrigerated Centrifuge

My first "culture shock," after coming to the United States from Japan in 1968, was to find that there was no refrigerated centrifuge in Art Pardee's laboratory, at least not the type that I had become accustomed to in Japan. Art's version of the refrigerated centrifuge was actually just a naked tabletop centrifuge in a small cold room. The cold room had a thick, wooden door, giving the impression that one was about to enter an ancient dungeon. The centrifuge itself was housed in an iron cage that looked as though it was designed to provide lodging for somebody's parakeets. The top of the cage swung open on a hinge so that one could load samples. There was no speed meter, instead a transformer with a sliding knob on top was connected to the centrifuge so that the speed could be controlled manually. The only sophisticated equipment I could find in Art's laboratory was an old Beckman spectrophotometer. Of course, there was no ultracentrifuge.

In those days, Japan was still far behind the United States in the fields of molecular biology and biochemistry. Japanese laboratories were poorly equipped and research budgets were very limited. Nevertheless, most of the major laboratories in biochemistry were already equipped with a refrigerated centrifuge. I came to Art's laboratory because it was relatively common for young Japanese molecular biologists to spend a few years as a postdoctoral fellow in the United States, the Mecca of molecular biology. Upon returning to Japan, many of these pilgrims would recount tales of cutting edge equipment, expansive laboratory space, and huge research budgets. After working for five years in Dr. Tsugita's laboratory at Osaka University on T4 phage lysozyme, I felt that it was time that I made my own pilgrimage. For my second postdoctoral research, I had chosen Art's laboratory, which was clearly at the forefront of molecular biology. There was no doubt in my mind that I was going to a prime laboratory in the United States. Therefore, when I found that Tsugita's laboratory was better equipped than Art's, it was a serious blow to my expectations. It took some time for me to realize that great science does not necessarily come from richly equipped laboratories, rather it comes from a richly equipped mind.

Among the many significant things that I learned during my three and a half years in Art's laboratory, I can put "Pardeeism" at the top of the list. It is the concept that the idea is paramount, and that if the idea is sound, everything else will follow. Who needs a refrigerated centrifuge? Better to have a creative idea and use the tabletop centrifuge, even if it is a little frightening. To this day, I can still clearly remember the fear that I felt each time I ran the centrifuge, praying that it would not be the last time before it exploded. Once I started it, I would quickly run out of the room, slamming the heavy, wooden door behind me. More terrifying, however, was entering the cold room to retrieve something while the centrifuge was running. Have you ever opened a Sorvall centrifuge running at 15,000 rpms?

I would like to dedicate this chapter to "Pardeeism," selecting three current topics from my laboratory. Although I must admit that I have four Sorvall centrifuges in my laboratory, I am a strong believer in "Pardeeism." It has been one of my most important philosophies in pursuing experiments over the last twenty years and I continue to strive to make my research "state of the Art."

Minor Codon Modulator Hypothesis

Since I worked on genetic codons used in T4 phage lysozyme during my first postdoctoral research in Japan, I have been interested in codon usages in various organisms and, in particular, nonrandom usage of synonymous codons in various organisms (Ikemura 1981a,b; Maruyama et al., 1986). In *Escherichia coli,* codon usage in abundant proteins such as ribosomal proteins and major membrane proteins is extremely biased, and such nonrandom usage was speculated to be due to preferential usage of major isoaccepting species of tRNA (Post et al., 1979; Nomura et al., 1980). From the analysis of codon usage in various *E. coli* proteins and relative quantities of tRNAs, a strong positive correlation between tRNA content and the occurrence of respective codons has been demonstrated, and it was proposed that *E. coli* genes encoding abundant protein species selectively use the "optimal codon" or major codons as determined by the abundance of isoaccepting tRNA (Ikemura, 1981a).

Because of these observations it is generally believed that protein production from a gene containing minor codons or nonoptimal codons is less efficient than from a gene containing no minor codons. However, contrary to this notion, some researchers have failed to demonstrate the negative effect of minor codons on gene expression (Holm, 1986; Sharp and Li, 1986). Further, it was recently demonstrated that the difference in overall translation time between the genes with and without minor codons was small, although there is a difference in translation rate between common codons and minor codons (Sorensen et al., 1989; Kurland, 1987). Therefore, it was concluded that the minor codon usage does not contribute to significant effects on gene expression at the level of translation.

On the other hand, in *Saccharomyces cerevisiae* it has been demonstrated that replacing an increasing number of major codons with synonymous minor codons at the 5′-end of the coding sequence of the gene for phosphoglycerate kinase caused a dramatic decrease of the gene expression (Hoekema et al., 1987). In this work decreasing efficiency of mRNA translation was attributed to mRNA destabilization.

What is the actual effect of minor codons on gene expression? Are there any general roles of minor codons in gene expression? To answer these questions, Ms. Giafen Chen and I first focused our attention on AGA/AGG codons for arginine (Chen and Inouye, 1990). Among codons used in *E. coli* AGA and AGG are the least used codons, and in most of the cases arginine is coded by CGU (49%), CGC (38%), CGG (6%), or CGA (4%). The usages of AGA and AGG are only 1.7 and 0.8%, respectively. It is also known that AGA and AGG codons are recognized by a very rare tRNA, and interestingly the gene for this tRNA has been identified to be *dnaY* (Garcia et al., 1986). When we analyzed 678 polypeptides available in GenBank, 452 proteins lack both AGA and AGG codons (Chen and Inouye, 1990). Among the remaining 226 proteins, 132 have either a single AGA or AGG codon, and to our surprise it became very clear that in these proteins either the AGA or AGG codon is preferentially used within the first 25 codons. Although it is less significant, the same tendency was observed for the proteins containing more than one AGA/AGG codon (Chen and Inouye, 1990). It is important to note that approximately 40% of the genes containing a single AGA or AGG codon have the minor codon within the first 25 codons. This preferential usage of minor codons within the first 25 codons was observed not only for AGA/AGG codons but also for other minor codons such as UCA plus AGU for serine, AUA for isoleucine, ACA for threonine, CCC for proline, and GGG for glycine (Chen and Inouye, 1990).

The analysis described above raised an intriguing possibility that gene expression in *E. coli* may be modulated by AGA/AGG and other minor codons by their replacement near the initiation codon. Thus we propose the following hypothesis designated minor codon modulator hypothesis or AGA/AGG modulator hypothesis: The minor codons cause ribosomal pause on a mRNA if the availability of their tRNAs becomes limited under certain growth conditions. The closer the pause occurs to the initiation site, the more effective the inhibitory effect of the paused ribosome is on translational initiation because of the formation of a queue of ribosomes from the pause site toward the initiation codon (the details, see Inouye and Chen, 1990). Not only AGA/AGG but in addition most of the other minor codons are preferentially located within the first 25 codons of the initiation codon in *E. coli* mRNAs. This strongly suggests that *E. coli* mRNAs have evolved to most effectively attenuate translation when cell growth becomes limited; a paused ribosome binding to mRNA near the translational initiation site prevents the entry of another ribosome, thus working as a repressor at the level of translation. On the basis of this hypothesis, we examined the effects of AGG codons on gene expression by inserting one to five AGG codons after the tenth codon from the initiation codon of the *lacZ* gene (Chen and Inouye, 1990). We found that the production of β-galactosidase decreased as more AGG codons were inserted. With five AGG codons, β-galactosidase production completely ceased after a mid-log phase of cell growth. After 22 hr of the *lacZ* induction, the overall production of β-galactosidase was only 11% of the control production without insertion of any arginine codons. In contrast, when five CGU codons, the major arginine codon were inserted instead of AGG, the production of β-galactosidase continued even after stationary phase and the overall production was 66% of the control.

Apparently there are two distinct phases for the β-galactosidase production from the *lacZ* gene containing five AGG codons during cell growth, phase I and phase II. During phase I, corresponding to early growth phase, the minor codons show no effect, while in phase II, after mid-log phase, β-galactosidase production completely stops. The dramatic negative effect of the AGG codons in phase II is likely due to the limited availability of charged tRNA for AGG. However, on the basis of the AGA/AGG modulator hypothesis, the negative effect was possibly observed because the minor codons were positioned very close to the initiation codon in the *lacZ* gene. If this is the case, one should be able to suppress the negative effect of the five AGG codons on the *lacZ* expression by increasing the distance between the initiation codon and the site of the AGG codons. Indeed, as the distance was increased by simply inserting sequences of various sizes between the initiation codon and the site of the AGG codons, β-galactosidase production increased almost linearly up to 8-fold. When the distance became 76 amino acid residues, the yield of β-galactosidase approached the level obtained with the control gene lacking any arginine codon insertions (Chen and Inouye, 1990). In other words, the negative effect of the minor codons was completely suppressed and there was no more phase II effect either, clearly supporting the AGA/AGG modulator hypothesis.

The proposed minor codon (or AGA/AGG) modulator hypothesis suggests a novel mechanism for global regulation of gene expression and cellular functions. In particular, the AGA/AGG codons being encoded by the *dnaY* product are considered to play a very important role in various cellular activities. During nutritional deprivation, the AGA/AGG codons function as a modulator to inhibit the production of key proteins in *E. coli* involved in DNA replication, protein synthesis, and cell division. Under such nutrient-limited conditions, the distance between the initiation codon and the AGA/AGG codons may have subtle effects on the synthesis of key proteins that regulate global cellular activities. It is interesting to point out that among a group of the proteins that have a single AGA or AGG codon within the first 25 codons, there are indeed various essential gene products such as (1) a protein required for DNA replication (single-strand DNA binding protein), (2) proteins associated with gene regulation (adenylate cyclase, LexA protein, *leu* operon leader peptide, and IHF), (3) proteins associated with protein synthesis (gly-tRNA synthetase α-subunit, phe-tRNA synthetase β-subunit, and ribosomal S10 protein), and (4) other essential proteins (FtsA protein, SecY protein, EnvZ protein, and ribonuclease P).

The Major Cold Shock Protein of *E. coli*

Heat shock proteins have been well documented as a group of proteins whose production is induced at high temperature. In *E. coli,* when cell cultures are transferred from 30° to 42°C, 17 "heat shock" proteins are known to play important roles in protecting cells from serious thermal damage (Neidhardt et al., 1984). Then how about cold shock proteins? Are there any proteins whose production is specifically induced at low temperature? In 1975, Dr. Simon Halegoua, then a graduate student in my laboratory, made a preliminary observation suggesting induction of a protein of a small molecular weight when cells were kept at low temperature. Since then I have been interested in this protein, but I was not successful in persuading new graduate students to further characterize this protein, until Mr. Joel Goldstein joined my laboratory.

Indeed we found that when exponentially growing *E. coli* cell cultures were transferred from 37° to 10° or 15°C, the production of a 7.4-kDa cytoplasmic protein (CS7.4) was prominently induced (Goldstein et al., 1990). The rate of CS7.4 production reached as high as 13% of total protein synthesis within 1–1.5 hr after shift to 10°C, and subsequently dropped to a lower basal level. Regulation of CS7.4 expression was very strict such that synthesis of the protein was undetectable at 37°C. We were able to purify the protein and obtain a partial amino acid sequence from which we could synthesize oligonucleotide probes to clone the gene encoding CS7.4. From the nucleotide sequence of the gene, it was found that the gene (*cspA*) encodes a hydrophilic protein of 70 amino acid residues. The protein contains seven lysine residues but no arginine residues. Interestingly, it contains a relatively large number of aromatic residues; six phenylalanine, one tyrosine, and one tryptophan. When the amino acid residues are plotted on a helical net, 12 of 16 charged residue are adjacent as oppositely charged pairs, suggesting that a large portion of CS7.4 may be in an α-helical conformation.

What is the function of the cold shock protein? It is possible that CS7.4 is involved in an adaptive process required for cells' dormancy at low temperature. Another attractive hypothesis is that CS7.4 may serve as an antifreeze protein. Antifreeze proteins are low-molecular-weight proteins commonly found at high concentration in the serum of polar dwelling marine fishes (Hew et al., 1986) and in the hemolymph of insects that winter in subfreezing climates (Duman and Horwath, 1983). The fish antifreeze proteins may be divided into four classes based on structural properties (Hew et al., 1986). These include glycoprotein, alanine-rich, and cysteine-rich classes, as well as a fourth class that is neither alanine nor cysteine rich. CS7.4 bears no similarity to any of these groups except with respect to its small size and its considerable proportion of hydrophilic residues. The nonglycoprotein fish antifreeze proteins, and to an even greater extent the insect thermal hysteresis (antifreeze) proteins, have a high proportion of hydrophilic residues that are thought to be important in hydrogen binding to the lattice of nascent ice crystals (Duman and Horwath, 1983, Yang et al., 1988). The crystallographic structure of the alanine-rich antifreeze polypeptide from the fish winter flounder has recently been determined (Yang et al., 1988). This 36 residue polypeptide appears to consist of a single α-helix extending throughout the entire molecule. The possibility that CS7.4 may be largely α-helical is significant since it may indicate homology at the level of function and secondary structure that is not represented in the primary structure.

Another interesting question is how the *cspA* gene is regulated. It is tightly regulated by temperature; no CS7.4 production was observed at 37°C, while at low temperature, its production was rapidly induced. The gene expression may be regulated by both positive and negative factors. The transitory nature of the *cspA* expression also suggests that CS7.4 itself may repress its own synthesis. It is interesting to note that yeast cells are also able to produce specific proteins at low temperature. A few genes for such yeast cold-shock proteins have been cloned (K. Kondo and M. Inouye, unpublished results) and their characterization is now in progress.

Induced Template Hypothesis

Subtilisin E is an alkaline serine protease produced by *Bacillus subtilus* (Boyer and Carton, 1968). This enzyme caught my attention because of its production in *B. subtilus* from an unusual precursor. Subtilisin E is known to be produced from preprosubtilisin consisting of the pre-sequence of 29 amino acid residues, the pro-sequence of 77 residues, and the mature protease, subtilisin E, of 275 residues (Wong and Doi, 1986). The pre-sequence has been shown to function as the signal peptide for protein secretion across the membrane (Wong and Doi, 1986). However, the role of the pro-sequence has been obscure and I was interested in the role of the pro-sequence. On the other hand, the X-ray crystallographic structure of subtilisin has been determined (Wright et al., 1969; Drench et al., 1972), and this enzyme has been extensively used as model systems for protein engineering in terms of the protein stability, substrate specificities, and physicochemical studies (see Ohta and Inouye, 1990).

To study the function of the pro-sequence, we first attempted to express the subtilisin gene in *E. coli*. For this purpose, we utilized a high expression secretion vector developed in our laboratory. The pre-sequence of preprosubtilisin was replaced with the OmpA signal peptide and the expression of the fusion gene was regulated by the *lacZ* promoter-operator and the strong *lpp* promoter (Ikemura et al., 1987). We found that in the presence of a *lac* inducer active subtilisin E was produced in the periplasmic space. To our surprise when the OmpA signal peptide was directly fused to the mature subtilisin sequence, no protease activity was detected in spite of the fact that a large amount of a protein with identical primary structure to active subtilisin was produced. Subsequently, we fused the OmpA signal peptide to the 15th or 44th residue from the amino terminal end of the pro-sequence. Again, no active subtilisin was produced. From these results, we have proposed that the pro-sequence is essential for guiding appropriate folding of the enzymatically active conformation of subtilisin E.

During the course of these experiments, we noticed that a substantial amount of prosubtilisin can be accumulated in the periplasmic space in addition to active subtilisin. We found that the processing of prosubtilisin to active subtilisin can be controlled by the inducer concentration and more importantly by the culture temperature (Takagi et al., 1988). When the cells were grown at 23°C, the products were mostly active subtilisin, while at 37°C prosubtilisin became a major product with a small amount of active subtilisin. It appeared that at 37°C prosubtilisin could not be properly folded resulting in the accumulation of a large amount of denatured prosubtilisin. This enabled us to isolate prosubtilisin and to study the mechanism of processing of prosubtilisins to active subtilisin (Ikemura and Inouye, 1988; Ohta and Inouye, 1990). We found that during the purification procedure, prosubtilisin E had to be kept in 6 M guanidine-HCl or 5 M urea. Otherwise it was readily converted into active subtilisin.

The processing mechanism was studied by dialyzing the purified prosubtilisin denatured in 5 M urea or 6 M guanidine-HCl against various buffer solutions. The processing of prosubtilisin to active subtilisin was detected by measuring the protease activity as well as by identifying the production of subtilisin by SDS-polyacrylamide gel electrophoresis. We found that the processing was very sensitive to the anions and ionic strengths used. Approximately 20% of prosubtilisin E was converted to active subtilisin E when the purified prosubtilisin was dialyzed against 0.5 M $(NH_4)_2SO_4$ and 1 mM $CaCl_2$ in 10 mM Tris-HCl buffer (pH7.0). If the concentrations of $(NH_4)_2SO_4$ were less than 0.3 M, no activity was recovered (Ohta and Inouye, 1990).

Importantly, the activation process was not inhibited by *Streptomyces* subtilisin inhibitor, a specific inhibitor from subtilisin. Furthermore, the subtilisin activity increased almost linearly with the prosubtilisin E concentration, indicating that the processing reaction is a unimolecular process (Ohta and Inouye, 1990). From these results we concluded that on renaturating denatured prosubtilisin, prosubtilisin is autoprocessed intramolecularly to yield the cleaved propeptide and active subtilisin. As expected from the in vivo experiment, we were not able to renature subtilisin E once it was denatured in 6 M guanidine-HCl.

We then wondered whether refolding of denatured subtilisin E could be complemented by the exogenously added propeptide. We started collaboration on this aspect with Dr. Frank Jordan's group, Department of Chemistry at Rutgers University. To our surprise, we found that mutant prosubtilisin was able to intermolecularly complement proper folding of denatured subtilisin (Zhu et al., 1989). The prosubtilisin mutant has an active-center mutation. As a result this mutant prosubtilisin was unable to be processed to active enzyme either in vivo or in vitro. Subsequently, we have attempted to examine the effect of the synthetic propeptide on refolding of denatured subtilisin. The 77-mer propeptide was synthesized and purified by Dr. S. Aimoto, Institute for Protein Research, Osaka University. We found that the synthetic propeptide worked almost as well as the mutant prosubtilisin for refolding of denatured subtilisin (Ohta et al., unpublished results).

How does the propeptide intermolecularly complement renaturation of denatured subtilisin? Our results clearly demonstrate that the propeptide is able to intermolecularly guide proper folding of denatured subtilisin. In one model, it appears to work as a mould for proper folding of denatured subtilisin. Since the propeptide is mixed with subtilisin denatured in 6 M guanidine-HCl, it is unlikely that denatured subtilisin can be refolded on a preformed mould or template made of the propeptide. Instead, the propeptide probably exists in random conformations at the beginning, and as the renaturation process proceeds part(s) of the propeptide start to interact with specific part(s) of denatured subtilisin. Interaction between the two polypeptides could prevent inappropriate intramolecular interactions between two different parts of the subtilisin molecule, which would lead to incorrect folding. As renaturation proceeds, additional intermolecular interactions between the two molecules will continue to develop resulting in the final conformation of active subtilisin to which the properly folded propeptide is bound. Thus, the function of the propeptide as a mould for folding of active subtilisin is induced during the renaturation process through interactions between the two molecules.

This hypothesis (induced template hypothesis) predicts the existence of folding intermediates resulting from specific interactions between the propeptide and mature portion of the subtilisin molecule. Using our system, it is possible to identify such intermediates of protein folding. The molecular mechanisms of protein folding are as yet very poorly understood, and I believe that further study of the system described here will substantially advance our understanding of protein folding. Currently we are attempting to crystallize pure prosubtilisin and also to isolate propeptide mutants that are unable to process to active subtilisin.

References

Boyer HW, Carton BC (1968): Production of two proteolytic enzymes by a transformable strain of *Bacillus subtilis*. Archs Biochem Biophys 128:442–455.

Chen G, Inouye M (1990): Suppression of the negative effect of minor arginine codons on gene expression; preferential usage of minor codons within the first 25 codons of the *Escherichia coli* genes. Nucleic Acids Res 18:1465–1473.

Drenth J, Hol WGJ, Jansonius J, Koekuek R (1972): Subtilisin novo: The three-dimensional structure and its comparison with subtilisin BPN′. Eur J Biochem 26:177–181.

Duman J, Horwath K (1983): The role of hemolymph proteins in the cold tolerance of insects. Annu Rev Physiol 45:261–270.

Garcia GM, Mar PK, Mullin DA, Walker JR, Prather NE (1986): The *E. coli* dnaY gene encodes an arginine transfer RNA. Cell 45:453–459.

Goldstein J, Pollitt S, Inouye M (1990): Major cold shock protein of *Escherichia coli*. Proc Natl Acad Sci USA 87:283–287.

Hew CL, Scott GK, Davies PL (1986): Molecular Biology of Antifreeze. New York: Elsevier, pp 117–123.

Hoekema A, Kastelein RA, Vasser M, de Boer HA (1987): Codon replacement in the pGK1 gene of *Saccharomyces cerevisiae*: Experimental approach to study the role of biased codon usage in gene expression. Mol Cell Biol 7:2914–2924.

Holm L (1986): Codon usage and gene expression. Nucleic Acids Res 14:3075–3087.

Ikemura H, Inouye M (1988): *In vitro* processing of pro-subtilisin produced in *Escherichia coli*. J Biol Chem 263:12959–12963.

Ikemura H, Takagi H, Inouye M (1987): Requirement of pro-sequence for the production of active subtilisin E in *Escherichia coli*. J Biol Chem 262:7859–7864.

Ikemura T (1981a): Correlation between the abundance of *Escherichia coli* transfer RNAs and the occurrence of the respective codons in its protein genes. J Mol Biol 146:1–21.

Ikemura T (1981b): Correlation between the abundance of *Escherichia coli* transfer RNAs and the occurrence of the respective codons in its protein genes: A proposal for a synonymous codon choice that is optimal for the *E. coli* translational system. J Mol Biol 151:389–409.

Inouye M, Chen G (1990): Regulation of gene expression by minor codons in *Escherichia coli*: Minor codon modulator hypothesis. In J.E.G. McCarthy (ed): Post-transcriptional Control of Gene Expression. NATO ASI Series, in press.

Kurland CG (1987): Strategies for efficiency and accuracy in gene expression. Trends Biochem 12:126–128.

Maruyama T, Gojobori T, Aota SI, Ikemura T (1986): Codon usage tabulated from the GenBank genetic sequence data. Nucleic Acids Res 14:r151–4197.

Neidhardt FC, VanBogelen RA, Vaughn V (1984): The genetics and regulation of heat-shock proteins. Annu Rev Genet 18:295–329.

Normura M, Yates JL, Dean D, Post LE (1980); Feedback regulation of ribosomal protein gene expression in *Escherichia coli*: Structural homology of ribosomal RNA and ribosomal protein mRNA. Proc Natl Acad Sci USA 77:7084–7088.

Ohta Y, Inouye M (1990): Pro-subtilisin E: Purification and characterization of its autoprocessing to active subtilisin E *in vitro*. Mol Micro 4:295–304.

Post LE, Stycharz GD, Nomura M, Lewis H, Dennis PP (1979): Nucleotide sequence of the ribosomal protein gene cluster adjacent to the gene for RNA polymerase subunit β in *Escherichia coli*. Proc Natl Acad Sci USA 76:1697–1701.

Sharp PM, Li WH (1986): Codon usage in regulatory gene in *Escherichia coli* does not reflect selection for ''rare'' codons. Nucleic Acids Res 14:7737–7749.

Sorensen MA, Kurland CG, Pedersen S (1989): Codon usage determines translation rate in *Escherichia coli*. J Mol Biol 207:365–377.

Takagi H, Morinaga Y, Tsuchiya M, Ikemura H, Inouye M (1988): Control of folding of proteins secreted by a high expression secretion vector, pIN-III-ompA: 16-fold increase in production of active subtilisin E in *Escherichia coli*. Bio/Technology 6:948–950.

Wong SL, Doi RH (1986): Determination of the signal peptidase cleavage site in the preprosubtilisin of *Bacillus subtilis*. J Biol Chem 261:10176–10181.

Wright CS, Alden RA, Kraut J (1989): Structure of subtilisin BPN′ at 2.5 A resolution. Nature (London) 221:235–242.

Yang DSC, Sax M, Chakrabartty A, Hew CL (1988): Crystal structure of an antifreeze polypeptide and its mechanistic implications. Nature (London) 333:232–237.

Zhu X, Ohta Y, Jordan F, Inouye M (1989): Pro-sequence of subtilisin can guide the refolding of denatured subtilisin in an intermolecular process. Nature (London) 339:483–484.

II. NORMAL AND ABNORMAL GROWTH AND DIFFERENTIATION IN EUKARYOTES

At Princeton, and subsequently at Harvard Medical School, Arthur Pardee's interests shifted to eukaryotic cell regulation, which he approached with critical and creative thought and with a panoramic view. Some of his studies in eukaryotes had clear origins in his earlier studies in prokaryotes (e.g., the role of nutrient transport and accumulation of intracellular protein in cell division control); others evolved from his ability to look at complex problems in novel ways, while reducing them to manageable hypotheses and experiments without compromising an appreciation for the complexity. His eukaryotic studies have covered, and continue to include, the role of growth factors, receptors, and signal-transducing mechanisms; nutrient transport and membrane function; cell cycle control; synthesis and repair of DNA; regulation and function of protooncogenes; and transcriptional control of cell cycle-related genes. Because of his long-standing interest in the cancer problem, Art generally found a way to explore how the normal processes under study in his lab became deranged in tumor cells, and he recently brought his years of experience in cell cycle control and DNA repair mechanisms to practical fruition through some very promising clinical trials.

The breadth of Art Pardee's interests in eukaryotic cell biology is reflected in the diversity of topics presented in this section. This section is a representative, but necessarily incomplete, account of the current research interests of some of Art's former students and postdoctoral fellows who have pursued problems in eukaryotic cell biology. These students and fellows trained with Art during his time at Princeton and Harvard. It is interesting that not all of the authors in this section worked on eukaryotic cells in Art's lab. Some worked with him on prokaryotes but later tackled problems in eukaryotic cell biology.

Perspectives on Cellular Regulation:
From Bacteria to Cancer, pages 107–119

Regulation of Expression of the Ornithine Decarboxylase Gene by Intracellular Signal Transduction Pathways

Mitchell S. Abrahamsen and David R. Morris

Department of Biochemistry, University of Washington, Seattle, Washington 98195

Introduction

There are few genes that are controlled by as complex a network of regulatory interactions as that encoding ornithine decarboxylase (ODC). ODC catalyzes a key regulated step in the synthesis of the polyamines and the activity of this enzyme is modulated by a broad array of effectors. Depending on the particular tissue or cell type studied, ODC activity can be rapidly and dramatically elevated by a variety of exogenous stimuli (Morris and Fillingame, 1974; Bachrach, 1980) that are coupled to several signal transduction pathways, including those involving protein kinase C (PKC), protein kinase A (PKA), protein tyrosine kinases, and steroid hormone receptors. It is indeed a challenge to molecular and cellular biologists to unravel the regulatory complexities of this interesting enzyme.

The polyamines (putrescine, spermidine, and spermine) comprise a family of low-molecular-weight compounds that occurs ubiquitously in nature. The biosynthesis of polyamines is tightly controlled in cells by various trophic stimuli. Although many aspects of the functions of these compounds in the intact cell have yet to be defined, it is clear that the polyamines are necessary for growth processes in both prokaryotic and eukaryotic organisms, and in particular for DNA replication (Marton and Morris, 1987). It is therefore not surprising that enhanced synthesis of the polyamines, controlled by large increases in the cellular activity of ODC, is a universal concomitant of growth stimulation in mammalian cells.

The mechanisms that modulate the activity of this enzyme in cells are varied and involve regulation of ODC mRNA level, alterations in the efficiency of translation of ODC mRNA, and posttranslational modification of the structure and stability of the ODC protein. Although all of these means of control can be highly significant under specific physiological conditions (for examples, see Holtta and Pohjanpelto, 1986; White et al., 1987a; van Daalen Wetters et al., 1989), this chapter will focus on regulation of the level of ODC mRNA. Our major emphasis will be on the potential mechanisms by which cytosolic signal transduction pathways may control expression of the ODC gene.

ODC Gene

To further explore the mechanisms that regulate ODC mRNA levels, several groups have recently cloned a functional ODC gene. Mouse, rat, and human genomes contain a family of ODC-related DNA sequences, one of which has been identified as a processed pseudogene located within the 5′-flanking region of the mouse V_K19A immunoglobulin gene (Kelley and Perry, 1987). The cloning of an active mouse ODC gene was facilitated by the use of drug-resistant cell lines that overproduce the enzyme due to amplification of one of these ODC-related sequences (Brabant et al., 1988; Katz and Kahana, 1988). Interestingly, this ODC gene was coamplified with the gene for the M2 subunit of ribonucleotide reductase during the selection of a hydoxyurea-resistant hamster cell line (Srinivasan et al., 1987). Results from in situ hybridization have indicated that both of these genes are located on the same band of human chromosome 2p.

The structure of the mouse ODC gene is shown in Figure 1A. The ODC gene is split into 12 exons and spans approximately 6.2 kb. Exons 1 and 2, along with the first 16 nucleotides of exon 3 contain the 5′ noncoding region of the ODC mRNA. The remaining part of exon 3, exons 4 through 11, and the first 142 bp of exon 12 encode the ODC protein. The rest of exon 12 corresponds to the 3′ noncoding regions of the two ODC mRNA species detected in mouse, which arise from the alternative use of two polyadenylation signals separated by 422 nucleotides. Genomic DNA sequences have also been reported for rat (van Steeg et al., 1988; Wen et al., 1989) and human (Fitzgerald and Flanagan, 1989) ODC. A comparison of the mouse, rat, and human genes reveals a striking conservation of genomic organization. The splice sites utilized are identical in the protein-coding region of all three genes, yet the introns of the human gene are generally larger than those of the mouse and rat. The nucleotide sequence in the coding region of exons is highly conserved, as well as an 82% identity within the first 148 bp of the 5′-flanking regions of the mouse, rat and human ODC genes.

Sequence immediately 5′ to the transcription start site of the mouse ODC gene contains a typical TATAA box in a GC-rich region with potential binding sites for transcription factors SP1 (Briggs et al., 1986) and AP-2 (Imagawa et al., 1987). The 5′-flanking sequence and the promoter of the ODC gene, when placed upstream of the bacterial chloramphenical acetyltransferase (CAT)

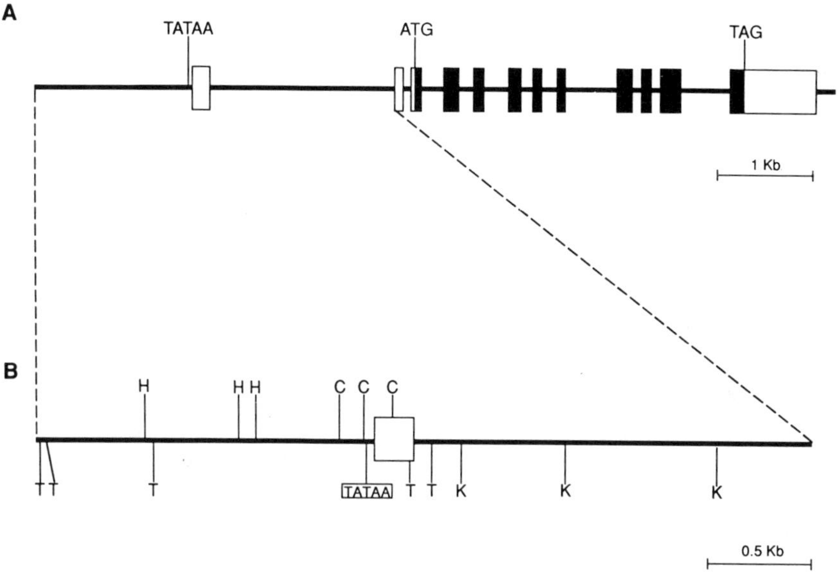

Fig. 1. Structural map and potential regulatory sequences of the mouse ornithine decarboxylase gene. **(A)** Schematic representation of the structure of the ODC gene. (Solid box) Coding region; (open box) untranslated region; (solid line) intervening and flanking genomic sequence. TATAA box, translational initiation (ATG) and stop (TAG) codons are indicated. **(B)** Potential regulatory sequences of the ODC gene. (Open box) Exon 1; (solid line) 5′ flanking sequences and intron 1. Regulatory sequences are T, TPA-responsive element; H, steroid hormone-responsive element; C, cAMP-responsive element; K, NF-KB binding site.

gene and introduced into mouse cells, directs the expression of the bacterial gene and is comparable in promoter strength to the Rous sarcoma virus long terminal repeat (Brabant et al., 1988; Katz and Kahana, 1988).

Further analysis of the 5′-flanking region, exon 1, and intron 1 of the mouse ODC gene reveals sequences with strong homology to consensus binding sites for known transcription factors, which may be responsible for regulating the expression of the ODC gene. As seen in Figure 1B, the 5′-flanking region contains two putative cAMP-responsive elements (CRE) at −48 and −175 bp, three potential 12-*O*-tetradecanoylphorbol 13-acetate (TPA)-responsive elements (TRE) at −1646, −1631, and −1081 bp, and three steroid hormone-responsive elements (HRE) at −1135, −670, and −600 bp. Exon 1 contains an additional CRE at +94 bp and an additional TRE at +167 bp. Intron 1 of the ODC gene contains three potential binding sites for the transcription factor NF-KB, which has been shown to respond to both TPA and cAMP (Shirakawa and Mizel, 1989), at +416, +929, and +1639 bp. The importance of any of these putative binding sites has yet to be determined and their presentation merely provides for a discussion of possible models for regulation of the ODC gene.

Transcriptional and Posttranscriptional Regulation of ODC mRNA

The increases in ODC mRNA levels achieved by addition of growth factors to quiescent fibroblasts or PC12 pheochromocytoma cells (Katz and Kahana, 1987; Greenberg et al., 1985), or activation of PKC or PKA-dependent signaling pathways in fibroblasts and the adrenal cell line Y1 (our unpublished results), are preceded by a corresponding increase in the transcription rate of the ODC gene. In all of these cases, there appears to be a strong correlation between the transcriptional activity of the gene and the level of ODC mRNA. Given the undetectable level of ODC transcription in quiescent fibroblastic cells, where the ODC mRNA levels are low, it seems apparent that any stimulation that results in increased ODC mRNA levels must be accompanied by increased transcription of the ODC gene.

In contrast to the cell types discussed above, ODC, along with several other genes, is regulated posttranscriptionally in mitogen-activated T lymphocytes (White et al., 1987b). Unlike fibroblasts, quiescent T cells have a high basal level of ODC transcription yet low mRNA levels. In experiments conducted with T cells in parallel with fibroblasts (Abrahamsen and Morris, 1990), we have confirmed that there is no increase in transcriptional activity of the ODC gene which would correspond to the substantial accumulation of ODC mRNA after mitogenic stimulation. Additionally, we find no evidence for transcriptional attenuation within the ODC gene as is seen in some other genes that are up-regulated during mitogenic stimulation of T cells (Lindsten et al., 1988a,b).

The high level ODC transcription in resting T cells along with the lack of message accumulation implies that ODC transcripts are very unstable in resting T cells. Furthermore these results suggest that after mitogenic activation, a posttranscriptional stabilization results in elevated ODC mRNA levels. We have determined that there is no change in the stability of the mature ODC mRNA after mitogenic stimulation, a result that implicates a posttranscriptional regulatory mechanism within the nucleus of T cells (Abrahamsen and Morris, 1990). The intranuclear event leading to elevation of ODC mRNA on activation of T cells could involve stabilization of a pre-mRNA species that is unstable in resting cells, a change in pre-mRNA processing from an unproductive pathway to one that leads to mature message, or an enhancement of the transport of processed ODC mRNA out of the nucleus.

Protein Kinase C-Regulated Pathways

Up-regulation of ODC activity was one of the first cellular events to be associated with activation of PKC. Early studies of the action on mouse skin of the potent tumor promoter TPA demonstrated that this activator of PKC induced rapid elevations of cellular levels of ODC activity (O'Brien, 1976) and mRNA (Gilmour et al., 1987). Similar to the observations with epidermis in vivo, activation of PKC with TPA is sufficient to induce expression of ODC mRNA in a variety of cell types in culture, including fibroblasts (Hovis et al., 1986) and T lymphocytes (Morris et al., 1988). In these experiments, down-regulation of PKC by pretreatment with high concentrations of TPA abolished the increase of ODC mRNA after stimulation with the phorbol ester, thus establishing the role of PKC in this process.

Activation of PKC-mediated phosphorylation has been observed to be an early response of cells to a variety of growth factors. PKC activation is a consequence of elevated intracellular diacylglycerol, resulting from the receptor-mediated stimulation of phoshatidylinositide breakdown (Nishizuka, 1986). Interestingly, down-regulation of PKC produced only a partial inhibition of ODC induction in fibroblasts in response to several growth factors (Hovis et al., 1986). These results suggest that growth stimuli activate at least two pathways (a PKC-dependent and a PKC-independent) regulating ODC expression.

Regulation of ODC mRNA level by TPA and growth factors is achieved through modulation of the transcription rate of the gene in fibroblasts and adrenal cell lines (Katz and Kahana, 1987; Greenberg et al., 1985; our unpublished results). Several PKC-regulated transcription factors are now known. AP-1, which is a complex mixture of dimers between members of the *jun* and the *fos* families of gene products, binds to the TRE and was the first of the TPA-responsive transcription factors to be recognized (Curran and Franza, 1988; Vogt and Bos, 1989). NF-KB is a factor that is expressed constitutively in B lymphocytes and is important for immunoglobulin gene expression; additionally, NF-KB-like factors are induced by treatment of other cell types with phorbol esters (Lenardo and Baltimore, 1989). Consensus binding sites for both AP-1 and NF-KB have been identified within the 5′-flanking region and transcribed noncoding regions of the ODC gene (Fig. 1B). Transcription factor AP-2 is of interest in that it seems to be regulated by both PKC- and PKA-dependent pathways (Imagawa et al., 1987). Possible AP-2 sites have been noticed 5′ of the ODC transcriptional start site (Fitzgerald and Flanagan, 1989); however, the importance of these homologies cannot be assessed given the high GC content of both the consensus AP-2 binding site and the 5′-flanking region of the ODC gene. It will only be possible to assign significance to the putative binding sites for these three factors when the appropriate mutagenesis and protein-binding studies have been performed.

As discussed in the previous section, in contrast to the transcriptional control of the ODC gene in some cell types, several genes in T lymphocytes, including ODC, show no increase in transcriptional activity in the face of substantial elevations of the corresponding mRNA levels (White et al., 1987b; Abrahamsen and Morris, 1990). In lymphocytes, TPA by itself induces ODC mRNA, and down-regulation of PKC completely abolishes the response of the ODC message to activation through the T cell receptor (Morris et al., 1988). Thus, in this cell type, it appears that stimulation of a PKC signaling pathway is both necessary and sufficient to posttranscriptionally activate ODC expression and, additionally, there seems to be no involvement of a PKC-independent pathway. The nature of the posttranscriptional processes involved in ODC expression in these cells must be understood before the details of its regulation can be further defined.

PKC-Independent Growth Factor Pathways

The growth factor-stimulated signal transduction pathways involving activation of PKC via breakdown of the phosphatidylinositides are well established (see above). However, there are clear indications in the literature that there exist other, hitherto undefined, pathways that are involved in regulation of the immediate early class of genes induced by growth factors. Insulin, EGF, and PDGF were shown to promote phosphorylation of intracellular substrates in cells severely depleted of PKC (Blackshear, 1989). A similar experimental paradigm revealed that stimulation of c-*myc* (Coughlin et al., 1985) and c-*fos* (Blackshear, 1989) expression in response to several growth factors was only partially inhibited by down-regulation of PKC, suggesting involvement of a PKC-independent pathway in the regulation of these genes.

Likewise, one component of the activation of ODC expression in response to growth factors is also mediated by a PKC-independent pathway. Down-regulation of PKC completely abolished the increase of ODC mRNA in NIH-3T3 cells after phorbol ester stimulation (Hovis et al., 1986). In contrast, this treatment left a 20–40% residual response of this gene to serum, PDGF, or FGF. Therefore, in fibroblasts, where ODC regulation is largely transcriptional (see above), ODC seems to be regulated by both PKC-dependent and -independent pathways. On the other hand, in T lymphocytes, where control of ODC mRNA level is via a posttranscriptional mechanism, regulation is mediated by a pure PKC pathway; down-regulation of PKC completely abolishes induction of ODC mRNA by TPA and by binding of lectin to the T cell receptor (Morris et al., 1988).

Thus, there exists the interesting situation in fibroblasts where a single growth factor, such as FGF, activates expression of a single gene by two, apparently independent pathways. The identity of one pathway as involving PKC seems definite. The other, PKC-independent pathway has not been identified. The lack of induction of ODC mRNA by calcium ionophores in fibroblasts (Hovis et al., 1986) or in T lymphocytes (Morris et al., 1988) argues against the direct involvement of elevated cytosolic calcium in a PKC-independent pathway. Receptor-linked protein tyrosine kinases seem likely candidates for initiating this second pathway. In this context, it is interesting to point out the central role that the product of the protooncogene c-*raf* (Raf-1) seems to have in cytosolic signal transduction pathways. Raf-1 is a protein serine kinase, which is itself phosphorylated on tyrosine by growth factor receptor kinases (Morrison et al., 1989) and on serine and threonine by treatment of cells with phorbol esters (Morrison et al., 1988). Expression of activated Raf-1 in fibroblasts activates expression from both the c-*fos* serum responsive element and from constructs containing the TRE, but not the CRE (Kaibuchi et al., 1989). Thus, a simple model for regulation of ODC transcription in response to growth factors would involve the protein tyrosine kinase and PKC pathways intersecting at Raf-1. Activated Raf-1 in turn would modulate, either directly or indirectly through phosphorylation, the activity of a transcription factor (perhaps AP-1?) involved in regulation of the ODC gene. This model would predict that both signaling pathways would regulate the ODC gene through a single *cis*-acting element. Alternatively, the PKC-dependent and PKC-independent pathways could be independent and parallel, acting through two sets of transcription factors and *cis* elements.

Steroid Receptor-Mediated Pathways

Most of the cellular actions of the steroid hormones seem to be mediated through their interactions with specific intracellular binding proteins. The specific receptors for the steroid hormones, together with those for thyroxine, vitamin D_3, dioxin, and retinoic acid, form a highly related "superfamily" of proteins that function as transcription factors (Evans, 1988). The members of the family of steroid receptors are sequence-specific DNA binding proteins, which bind to highly related palindromic sequences known as hormone-responsive elements (HREs). The proteins of this family are constructed of separable "zinc finger" DNA-binding domains, ligand-binding domains, and transcriptional activation domains (Evans, 1988). In addition to their role in transcriptional regulation, steroids have also been found to modulate, in certain systems, mRNA stability, protein processing, and protein turnover (Berger and Watson, 1989). In these latter instances, the observed physiological effects are not necessarily a primary response to steroid treatment.

Many of the ligands that interact with the members of the steroid receptor superfamily were found to strongly elevate ODC activity in responsive tissues (Morris and Fillingame, 1974; Bachrach, 1980). These ligands included glucocorticoids, estrogens, androgens, and thyroxine. In most instances the mechanisms of these hormonally regulated elevations of ODC activity have not been defined. Given the multiple sites at which steroid hormones may exert their regulatory activities and at which ODC activity potentially can be controlled, simple measurements of enzymic activity give no insight into molecular mechanisms of regulation. An additional caveat is that, in all cases studied, it is not clear whether induction of ODC activity is a direct result of steroid hormone action or an indirect corollary of growth responses of particular tissues to these hormones.

The molecular mechanisms of steroid regulation of ODC have been best defined in androgen-stimulated mouse kidney (Berger and Watson, 1989). In this tissue, there is a rapid, several hundred-fold elevation of ODC activity in response to the hormone. This increase in enzyme activity seems to result largely from a 25-fold increase in ODC mRNA level, together with increased stability of the ODC protein. It seems that elevated transcription of the ODC gene in nuclei isolated from hormone-treated kidney can largely account for the increase in mRNA content.

It should be emphasized that, even in the well-characterized mouse kidney system, the increased rate of transcription has not been shown to be a direct result of hormone treatment. However, it is intriguing to point out that there are three good matches (either 10 or 11 out of 13 base pairs) to the consensus HRE palindrome upstream of the ODC promoter (Fig. 1B). Because of the similarity in nucleotide sequence of all of the HREs, it is difficult to state with certainty which hormone receptor these HREs might interact with, but they seem most similar to estrogen-responsive elements. Other less well-conserved homologies to the HRE consensus are found within exon 1 and intron 1 (not shown in Fig. 1B). Definition of the significance of these putative HREs awaits appropriate studies of protein binding and expression of mutant constructs.

Protein Kinase A-Regulated Pathways

In a variety of systems, cAMP mediates the hormonal stimulation of a variety of mammalian genes, including ODC (Russel and Durie, 1978). The induction of ODC by these hormones is dependent on their ability to activate adenylate cyclase, resulting in increased cAMP levels and activation of PKA. Although no measurements of ODC mRNA levels were made, there is indirect evidence, assessed by studies using actinomycin D and cycloheximide, that ODC is transcriptionally regulated in these systems (Russell, 1981). We have shown that elevation of cAMP in fibroblasts or an adrenal cell line induces substantial increases in ODC mRNA levels which are preceded by a corresponding stimulation of the transcription rate of the ODC gene (unpublished results).

Generally, transcriptional induction of genes by cAMP is rapid, peaking at 30 min and declining gradually over 24 hr (Sasaki et al., 1984; Lewis et al., 1987). This burst in transcription is resistant to inhibitors of protein synthesis, suggesting that cAMP may stimulate gene expression by modulation of existing nuclear factors. Since all known cellular responses of cAMP occur via the catalytic subunit of PKA, it appears likely that this enzyme mediates the phosphorylation of specific factors that are critical for the transcriptional response. Many genes that are transcriptionally regulated by cAMP contain a conserved sequence in the 5′-flanking region that is homologous or identical to the palindromic CRE (Roesler et al., 1988; Jameson et al., 1989). A 43-kDa nuclear protein, which binds specifically to the CRE, CREB, has been isolated and characterized (Montminy and Bilezikjian, 1987; Yamamoto et al., 1988). The amino acid sequence of CREB predicts a single consensus recognition site for PKA-catalyzed phosphorylation. This site is efficiently phosphorylated both in vitro and in vivo and is critical for the ability of CREB to activate gene transcription through the CRE (Gonzalez and Montminy, 1989). Another transcription factor, ATF, is known to bind a sequence that is very similar (or identical) to the CRE (Roesler et al., 1988) and has been shown to be a member of an extensive family of DNA-binding proteins (Hai et al., 1989). Recently, a 120-kDa protein has been purified, which also displays sequence-specific binding to the CRE (Andrisani and Dixon, 1990). Unlike CREB, this protein is readily phosphorylated in vitro by PKC but not by PKA, suggesting that this protein, and the CRE, may be involved in other signal transduction pathways. Other transcription factors, including AP-2 (Imagawa et al., 1987) and NF-KB (Shirakawa and Mizel, 1989), have also been reported to be regulated cAMP.

Sequence analysis of the ODC gene reveals that it contains three possible CRE sequences that are clustered around the start site of transcription (see Fig. 1B) and are conserved between mouse, rat, and human. To explore the role these three CREs may play in the transcriptional induction of the ODC gene, we have fused these ODC sequences to the bacterial CAT gene and introduced the constructs into an adrenal cell line (unpublished results). A 620-bp fragment of the ODC promoter containing all three CREs and eliminating all of the NF-KB sites in intron 1 (see Fig. 1B) is as effective in directing the transcriptional response of PKA as a fragment containing 3 kb of ODC 5′-flanking sequence, exon 1, intron 1, and part of exon 2. A 150-bp fragment containing only the single CRE proximal to the TATAA box is still able to confer almost all of the cAMP inducibility of the ODC gene. We have also been able to demonstrate by DNase I footprinting that purified CREB, as well as factors present in extracts of the adrenal cells, binds to these sequences. Based on these results, it seems likely that these CRE sequences will turn out to be responsible for the cAMP-induced expression of the ODC gene.

In T lymphocytes and myeloid cells the situation is very different. cAMP, which in these cells inhibits proliferation, did not stimulate ODC gene expression and actually inhibited growth factor-stimulated increases in ODC mRNA levels (Farrar et al., 1988; Farrar and Harel-Bellan, 1989). This effect is not solely specific for ODC, since cAMP also inhibits mitogen-induced c-*myc* expression in mouse thymocytes (Moore et al., 1986) and interleukin-2 (IL-2) stimulation of c-*myc* mRNA in a cytotoxic T cell line (Farrar et al., 1987). It is not surprising that an antiproliferative kinase pathway, dependent on cAMP, may affect the expression of genes stimulated by growth factors; however, cAMP does not generally suppress induced gene expression. Elevation of intracellular cAMP stimulates c-*fos*, c-*myb*, and IL-2 receptor mRNA accumulation in lymphocytes (Farrar et al., 1987). The precise mechanism(s) of this selective inhibition gene expression remains to be determined, but the relationship between cAMP as a growth-suppressive signal, and the inhibition of ODC, whose expression is invariably associated with cell growth, may turn out to be particularly important.

Perspectives

For many years the literature was deluged with descriptions of the induction of ODC activity in every imaginable cell type and in response to every conceivable stimulus. Although ODC induction provided a valuable marker for cellular activation, these studies of changes in enzyme activity provided little insight into the molecular mechanisms of regulation of this fascinating gene. With the cloning of the ODC gene and use of its promoter to drive expression of reporter genes, there has been a major change in the experimental approaches available to study regulation of its expression. Putative binding sites for regulated transcription factors have been identified by scanning the available DNA sequences. However, protein binding studies and mutational analysis must be performed before the functional significance of these putative *cis*-acting elements can be ascertained within various cell types. Interestingly, based on the preliminary sequence information, it appears that considerable redundancy in regulated elements may exist. For example, there are possible binding sites for transcription factors AP-1, NF-KB, and AP-2, all of which are regulated by PKC-dependent pathways. cAMP regulates NF-KB elements and some AP-1 sites (Deutsch et al., 1988), in addition to the CRE. These redundancies provide not only for cross-talk between these two signaling pathways, but also for the possibility that the same signaling pathway might regulate ODC transcription via different factors under diverse physiological conditions or in different cell types. These aspects of ODC regulation are certain to be defined soon. Additionally, we expect timely progress to be made on elucidating the signal transduction pathways that regulate ODC expression at levels other than transcription.

Alterations in ODC expression have been associated with various complex biological phenomena including differentiation, neoplasia, and growth. The study of ODC regulation will continue to be a valuable tool for understanding the complex signal transduction pathways involved in these cellular processes. With the availability of the cloned gene, and the ability to study expression of various constructs in cells of interest and, through transgene technology, in the intact animal, we anticipate that many aspects of the biology of this intriguing gene will come under intense study.

Acknowledgments

This paper is dedicated to Dr. Arthur B. Pardee on the occasion of his seventieth birthday. One of us (D.R.M.) had the opportunity to work with Dr. Pardee as a postdoctoral fellow from 1964 to 1966. Art suggested prior to my arrival in his laboratory that, given their ubiquity at high concentration in cells, the polyamines might be interesting to look at. This suggestion has kept me busy for the past 25 years. In addition to pointing me in the direction of polyamines, Art's early interest in growth regulation of bacterial and mammalian cells, his insightful and imaginative way of looking at data, and his knack for always maintaining a biological perspective continue to influence my research to this day.

References

Abrahamsen MS, Morris DR (1990): Cell type-specific mechanisms of regulating expression of the ornithine decarboxylase gene after growth stimulation. Mol Cell Biol 10:5525–5528.

Andrisani O, Dixon JE (1990): Identification and purification of a novel 120-kDa protein that recognizes the cAMP-responsive element. J Biol Chem 265:3212–3218.

Bachrach U (1980): The induction of ornithine decarboxylase in normal and neoplastic cells. In J.M. Gaugas (ed): Polyamines in Biomedical Research. New York: John Wiley, pp. 81–107.

Berger FG, Watson G (1989): Androgen-regulated gene expression. Annu Rev Physiol 51:51–65.

Blackshear PJ (1989): Insulin-stimulated protein biosynthesis as a paradigm of protein kinase C-independent growth factor action. Clin Res 37:15–25.

Brabant M, McConlogue L, van Daalen Wetters T, Coffino P (1988): Mouse ornithine decarboxylase gene: cloning, structure, and expression. Proc Natl Acad Sci USA 85:2200–2204.

Briggs MR, Kadonaga JT, Bell SP, Tjian R (1986): Purification and biochemical characterization of the promoter-specific transcription factor, Sp1. Science 234:47–52.

Coughlin SR, Lee WMF, Williams PW, Giels GM, Williams LT (1985): c-myc gene expression is stimulated by agents that activate protein kinase c and does not account for the mitogenic effect of PDGF. Cell 43:243–251.

Curran T, Franza BRJ (1988): Fos and jun: The AP-1 connection. Cell 55:395–397.

Deutsch PJ, Hoeffler JP, Jameson JL, Habener JF (1988): Cyclic AMP and phorbol ester-stimulated transcription mediated by similar DNA elements that bind distinct proteins. Proc Natl Acad Sci USA 85:7922–7926.

Evans RM (1988): The steroid and thyroid hormone receptor superfamily. Science 240:889–895.

Farrar WL, Evans SW, Rapp UR, Cleveland JL (1987): Effects of anti-proliferative cyclic adenosine 3′,5′-monophosphate on interleukin 2-stimulated gene expression. J Immunol 139:2075–2080.

Farrar WL, Harel-Bellan A (1989): Myeloid growth factor(s) regulation of ornithine decarboxylase: Effects of antiproliferative signals interferon-gamma and cAMP. Blood 73:1468–1475.

Farrar WL, Vinocour M, Cleveland JL, Harel-Bellan A (1988): Regulation of ornithine decarboxylase activity by IL-2 and cyclic AMP. J Immunol 141:967–971.

Fitzgerald MC, Flanagan MA (1989): Characterization and sequence analysis of the human ornithine decarboxylase gene. DNA 8:623–634.

Gilmour SK, Verma AK, Madara T, O'Brien TG (1987): Regulation of ornithine decarboxylase gene expression in mouse epidermis and epidermal tumors during two-stage tumorigenesis. Cancer Res. 47:1221–1225.

Gonzalez GA, Montminy MR (1989): Cyclic AMP stimulates somatostatin gene transcription by phosphorylation of CREB at serine 133. Cell 59:675–680.

Greenberg ME, Greene LA, Ziff EB (1985): Nerve growth factor and epidermal growth factor induce rapid transient changes in proto-oncogene transcription in PC12 cells. J Biol Chem 260:14101–14110.

Hai T, Liu F, Coukos WJ, Green MR (1989): Transcription factor ATF cDNA clones: An extensive family of leucine zipper proteins able to selectively form DNA-binding heterodimers. Genes Dev 3:2083–2090.

Holtta E, Pohjanpelto P (1986): Control of ornithine decarboxylase in chinese hamster ovary cells by polyamines. Translational inhibition of synthesis and acceleration of degradation of the enzyme by putrescine, spermidine, and spermine. J Biol Chem 261:9502–9508.

Hovis JG, Stumpo DJ, Halsey DL, Blackshear PJ (1986): Effects of mitogens on ornithine decarboxylase activity and messenger RNA levels in normal and protein kinase C-deficient NIH-3T3 fibroblasts. J Biol Chem 261:10380–10386.

Imagawa M, Chiu R, Karin M (1987): Transcription factor AP-2 mediates induction by two different signal-transduction pathways: Protein kinase C and cAMP. Cell 51:251–260.

Jameson JL, Albanese C, Habener JF (1989): Distinct adjacent protein-binding domains in the glycoprotein hormone α gene interact independently with a cAMP-responsive enhancer. J Biol Chem 264:16190–16196.

Kaibuchi K, Fukumoto Y, Oku N, Hori Y, Yamamoto T, Toyoshima K, Takai Y (1989): Activation of the serum response element and 12-O-tetra-decanoylphorbol-13-acetate response element by the activated c-Raf-1 protein in a manner independent of protein kinase-C. J Biol Chem 264:20855–20858.

Katz A, Kahana C (1987): Transcriptional activation of mammalian ornithine decarboxylase during stimulated growth. Mol Cell Biol 7:2641–2643.

Katz A, Kahana C (1988): Isolation and characterization of the mouse ornithine decarboxylase gene. J Biol Chem 263:7604–7609.

Kelley DE, Perry RP (1987): Association of an ornithine decarboxylase processed pseudogene with members of a V_K immunoglobin gene family provides a useful evolutionary clock. Nucleic Acids Res 15:7199.

Lenardo MJ, Baltimore D (1989): NF-Kappa-B—a pleiotropic mediator of inducible and tissue-specific gene control. Cell 58:227–229.

Lewis EJ, Harrington CA, Chikaraishi DM (1987): Transcriptional regulation of the tyrosine hydroxylase gene by glucocorticoid and cyclic AMP. Proc Natl Acad Sci USA 84:3550–3554.

Lindsten T, June CH, Thompson CB (1988a): Multiple mechanisms regulate c-myc gene expression during normal T-cell activation. EMBO J 7:2787–2794.

Lindsten T, June CH, Thompson CB, Leiden JM (1988b): Regulation of 4F2 heavy-chain gene expression during normal T-cell activation can be mediated by multiple distinct molecular mechanisms. Mol Cell Biol 8:3820–3826.

Marton LJ, Morris DR (1987): Molecular and cellular functions of the polyamines. In P.P. McCann, A.E. Pegg, and A. Sjoerdsma (eds): Inhibition of Polyamine Metabolism. Orlando, FL: Academic Press, pp. 79–105.

Montminy MR, Bilezikjian LM (1987): Binding of a nuclear protein to the cyclic-AMP response element of the somatostatin gene. Nature (London) 328:175–178.

Moore JP, Todd JA, Hesketh TR, Metcalfe JC (1986): c-fos and c-myc gene activation, ionic signals, and DNA synthesis in thymocytes. J Biol Chem 261:8158–8161.

Morris DR, Allen ML, Rabinovitch PS, Kuepfer CA, White MW (1988): Mitogenic signaling pathways regulating expression of c-myc and ornithine decarboxylase genes in bovine T-lymphocytes. Biochemistry 27:8689–8693.

Morris DR, Fillingame RH (1974): Regulation of amino acid decarboxylation. Annu Rev Biochem 43:303–325.

Morrison DK, Kaplan DR, Escobedo JA, Rapp UR, Roberts TM, Williams LT (1989): Direct

activation of the serine threonine kinase activity of Raf-1 through phosphorylation by the PDGF beta-receptor. Cell 58:649–657.
Morrison DK, Kaplan DR, Rapp U, Roberts TM (1988): Signal transduction from membrane to cytoplasm: Growth factors and membrane-bound oncogene products increase Raf-1 phosphorylation and associated protein kinase activity. Proc Natl Acad Sci USA 85:8855–8859.
Nishizuka Y (1986): Studies and perspectives on protein kinase C. Science 233:305–312.
O'Brien TG (1976): The induction of ornithine decarboxylase as an early, possibly obligatory, event in mouse skin carcinogenesis. Cancer Res 36:2644–2653.
Roesler WJ, Vandenbark GR, Hanson RW (1988): Cyclic AMP and the induction of eukaryotic gene transcription. J Biol Chem 263:9063–9066.
Russell DH (1981): Ornithine decarboxylase: Transcriptional induction by tropic hormones via a cAMP and cAMP-dependent protein kinase pathway. In D.R. Morris and L.J. Marton (eds): Polyamines in Biology and Medicine. New York: Marcel Dekker, pp. 109–125.
Russell DH, Durie BGM (1978): Polyamines as Biochemical Markers of Normal and Malignant Growth. New York: Raven Press, ch. 5, pp. 59–88.
Sasaki K, Cripe TP, Koch SR, Andreone TL, Peterson DD, Beale EB, Granner KK (1984): Multihormonal regulation of phosphoenolpyruvate carboxykinase gene transcription. J Biol Chem 259:15242–15251.
Shirakawa F, Mizel SB (1989): In vitro activation and nuclear translocation of NF-Kappa-B catalyzed by cyclic AMP-dependent protein kinase and protein kinase-C. Mol Cell Biol 9:2424–2430.
Srinivasan PR, Tonin PN, Wensing EJ, Lewis WH (1987): The gene for ornithine decarboxylase is co-amplified in hydroxyurea-resistant hamster cells. J Biol Chem 262:12871–12878.
van Daalen Wetters T, Brabant M, Coffino P (1989): Regulation of mouse ornithine decarboxylase activity by cell growth, serum and tetradecannoyl phorbol acetate is governed primarily by sequences within the coding region of the gene. Nucleic Acids Res 17:9843–9860.
van Steeg H, van Oostrom TM, van Kranen HJ, van Kreijl CF (1988): Nucleotide sequence of the rat ornithine decarboxylase gene. Nucleic Acids Res 16:8173–8174.
Vogt PK, Bos TJ (1989): The oncogene jun and nuclear signaling. Trends Biochem Sci 14:172–175.
Wen L, Huang JK, Blackshear PJ (1989): Rat ornithine decarboxylase gene. Nucleotide sequence, potential regulatory elements, and comparison to the mouse gene. J Biol Chem 264:9016–9021.
White MW, Kameji T, Pegg AE, Morris DR (1987a): Increased efficiency of translation of ornithine decarboxylase mRNA in mitogen-activated lymphocytes. Eur J Biochem 170:87–92.
White MW, Oberhauser AK, Kuepfer CA, Morris DR (1987b): Different early-signaling pathways coupled to transcriptional and posttranscriptional regulation of gene expression during mitogenic activation of T lymphocytes. Mol Cell Biol 7:3004–3007.
Yamamoto KR, Gonzalez GA, Biggs WH, Montminy MR (1988): Phosphorylation-induced binding and transcriptional efficacy of nuclear factor CREB. Nature (London) 334:494–498.

Perspectives on Cellular Regulation:
From Bacteria to Cancer, pages 121–127

Retinoic Acid Regulation of Mammalian Gene Expression and Differentiation

Lorraine J. Gudas

Department of Biological Chemistry and Molecular Pharmacology, Harvard Medical School, and Division of Cellular and Molecular Biology, Dana-Farber Cancer Institute, Boston, Massachusetts 02115

I was a graduate student in Dr. Arthur Pardee's laboratory in the Department of Biochemical Sciences at Princeton University from 1971 through 1975. I had chosen to attend Princeton University because Dr. Pardee was a professor there; I had heard him speak about sulfate transport in bacteria at Amherst College while I was an undergraduate at Smith College, and I had decided that bacterial transport was an exciting area, an area in which I would like to do thesis work. I was fortunate to be able to be a graduate student in Art Pardee's lab, as he would accept only two graduate students at a time. The other student in the lab at that time was Keith Fournier.

I learned a tremendous amount about doing science from Art. One of the most important things he taught me was how to organize my results in a manuscript and to present the data in the clearest and most exciting fashion. He also encouraged me to develop more self-confidence and a belief in my scientific abilities. During my last two years of graduate school, Dr. Pardee, Dr. Sarah Via, a French professor at Princeton, and I played piano trios once a week. Art played the cello, I played the violin, and Sarah played the piano. These music sessions were great fun, and they allowed me to see a different part of Art's character. He was much more relaxed during these evening music sessions than he was during the day, as he was very busy as chairman of the Biochemical Sciences Department at that time. Often Ruth Sager and John Wagner listened to our music; they said that they enjoyed it, but Ruth was romantically involved with Art and John with me, so our audience was clearly biased. During graduate school and over the years since I received my Ph.D. from Princeton in 1975, Art has been a stalwart supporter of my career. I have appreciated this support very much, especially since I was one of the first female graduate students at Princeton and was initially insecure in my position as a woman scientist. I have been very lucky to have been associated with Dr. Pardee and to have had his help and guidance over the years.

My graduate work in Dr. Pardee's laboratory focused on the regulation of bacterial DNA repair. In a PNAS paper in 1975 (Gudas and Pardee, 1975) we proposed a model for the regulation of the recA/lexA-mediated "SOS" DNA repair system (Fig. 1A). This model was formulated from experiments on the genetic and metabolic regulation of the synthesis of a protein designated Protein X; we later demonstrated this protein to be the recA protein (Gudas and Mount, 1977). The model generated much excitement as it synthesized both genetic and biochemical data to form a hypothesis about the regulation of the "SOS" DNA repair pathway.

The current research in my laboratory also concerns regulatory pathways, but these are now being studied in mammalian cells. We are currently attempting to dissect the pathway by which the vitamin A derivative retinoic acid causes the differentiation of embryonic teratocarcinoma stem cells. Teratocarcinomas of mice contain tumorigenic stem cells that can proliferate indefinitely in culture; these stem cells resemble the pluripotent embryonic stem cells of the inner cell mass of the blastocyst. One particular stem cell line, F9, differentiates in monolayer culture into a homogeneous population of primitive endoderm when given physiological concentrations of the vitamin A derivative retinoic acid, and into parietal endoderm cells when given retinoic acid and dibutyryl cyclic AMP, a cyclic AMP analog. Parietal endoderm cells are extraembryonic "epithelial" cells formed from the inner cell mass of the blastocyst during early mouse embryogenesis. Thus, F9 cells can be used to study one step in mouse embryogenesis, and as a model system to study the mechanism of action of retinoic acid.

There has been enormous progress in the past few years with respect to understanding the molecular mechanisms by which retinoids (vitamin A and its natural and synthetic derivatives) regulate cell growth, cell differentiation, and pattern formation in development. Vitamin A (retinol) is required throughout life by mammals, and probably by most or all vertebrates. Humans obtain retinol through their diets; retinol is stored in the liver, transported through the serum via a serum retinol-binding protein, and presumably oxidized to the more active form of the vitamin, retinoic acid, in appropriate cell types.

Earlier studies focused primarily on the effects of retinoids on epithelial growth and differentiation, but more recent work has demonstrated that retinoids can influence the growth and differentiation of several types of cultured cells, including but not limited to melanomas, neuroblastomas, fibroblasts, promyelocytic leukemia cells, and teratocarcinoma stem cells. In a number of studies

---

Fig. 1. **(A)** A model for the regulation of "SOS" DNA repair protein, from Gudas and Pardee (1975). **(B)** A model for the regulation of F9 teratocarcinoma stem cell differentiation induced by retinoic acid. F9 stem cells are analogous to the early embryonic stem cells of the mouse blastocyst; after retinoic acid addition (and cyclic AMP analogs), the cells resemble extraembryonic parietal endoderm.

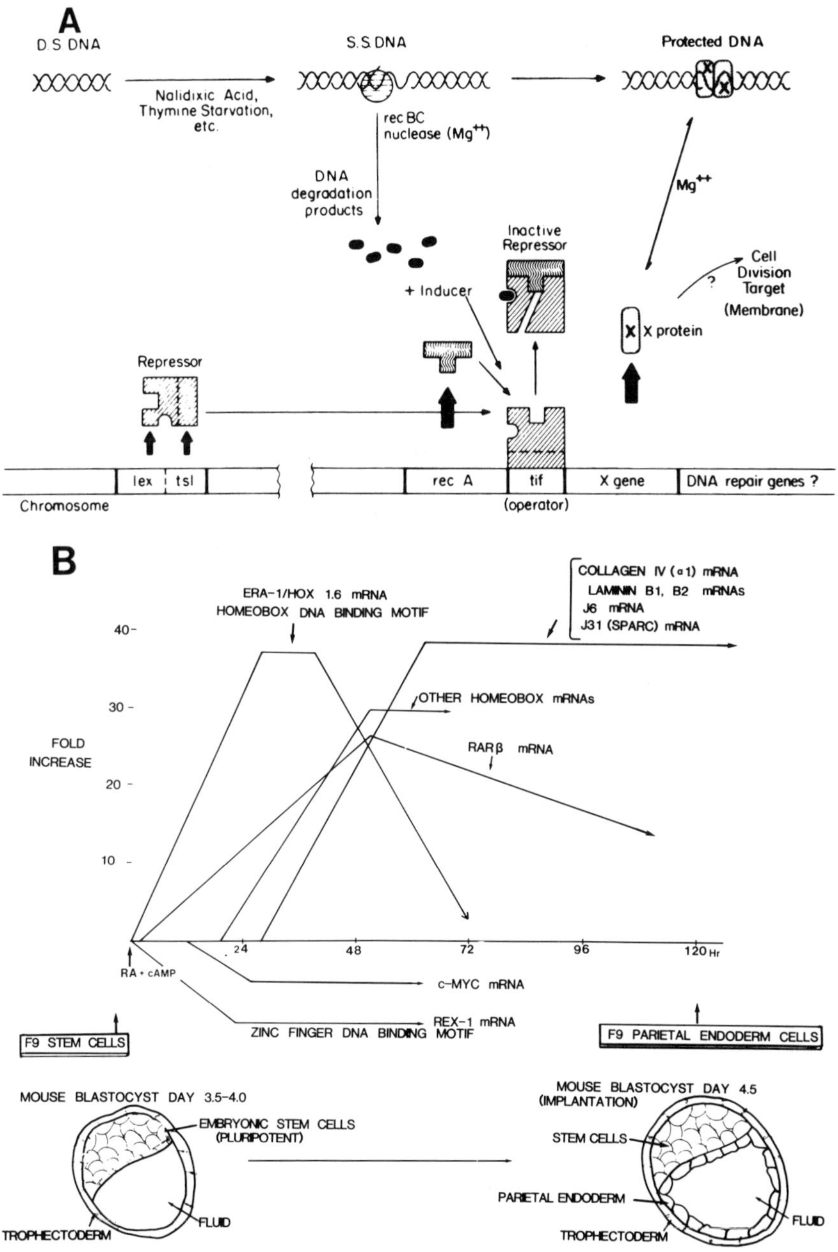
A
D.S. DNA
Nalidixic Acid,
Thymine Starvation,
etc.
S.S. DNA
rec BC
nuclease (Mg++)
DNA
degradation
products
Protected DNA
Mg++
Inactive
Repressor
+ Inducer
Cell
Division
Target
(Membrane)
X protein
Repressor
lex
tsl
rec A
tif
X gene
DNA repair genes ?
Chromosome
(operator)
B
ERA-1/HOX 1.6 mRNA
HOMEOBOX DNA BINDING MOTIF
COLLAGEN IV (α1) mRNA
LAMININ B1, B2 mRNAs
J6 mRNA
J31 (SPARC) mRNA
OTHER HOMEOBOX mRNAs
RARβ mRNA
FOLD
INCREASE
40
30
20
10
24
48
72
96
120 Hr
RA + cAMP
c-MYC mRNA
REX-1 mRNA
ZINC FINGER DNA BINDING MOTIF
F9 STEM CELLS
F9 PARIETAL ENDODERM CELLS
MOUSE BLASTOCYST DAY 3.5–4.0
EMBRYONIC STEM CELLS
(PLURIPOTENT)
TROPHECTODERM
FLUID
MOUSE BLASTOCYST DAY 4.5
(IMPLANTATION)
STEM CELLS
PARIETAL ENDODERM
TROPHECTODERM
FLUID

using experimental animals, partial retinoid deficiency has been associated with an increased incidence in several types of tumors, especially carcinomas (Sporn and Roberts, 1983). Retinoid deficiency can cause bronchial squamous metaplasia, a premalignant lesion, and increases the incidence of lung cancer in experimental animals. The cancer chemopreventive actions of retinoids are presumably related to the ability of retinoids to influence cell differentiation.

Another area in which great strides have been made concerning the role of retinoic acid is that of pattern formation in development. Pattern formation is defined as the three-dimensional spatial organization of the different types of cells in the developing embryo. There is strong evidence that retinoic acid is a morphogen in the developing vertebrate limb bud; a morphogen is defined as a substance whose concentration is measured by cells in a developing embryo so that the cells can determine their position relative to a specific landmark. Accurate positional information is essential for development to proceed, and morphogen gradients can provide cells with positional information. Retinoic acid is the first compound to be identified that has the characteristics of a vertebrate morphogen. Moreover, retinoic acid may act as a morphogen in developing structures in addition to the limb bud.

Many different theories concerning the mechanism of action of retinoids were developed over the past 30 years. For example, since retinoids are lipophilic, some researchers suggested that retinoids might damage the plasma membranes of cells, leading to cytotoxicity. This is almost certainly not true, at least for the low retinoic acid concentrations (10^{-9}–10^{-8} M) that lead to biological effects. The effects of retinoic acid were also presumed to be mediated by intracellular receptors and/or binding proteins.

The first high-affinity retinoic acid binding protein to be identified was the cellular retinoic acid-binding protein (CRABP), a small cytoplasmic protein (approximately 15 kDa) with no obvious DNA-binding motif (Chytil and Ong, 1979; Stoner and Gudas, 1989). A second group of high-affinity retinoic acid receptors (RARs) has also been identified; these proteins exhibit some sequence homologies with the nuclear receptors for steroid hormones, thyroid hormone, and vitamin D_3, suggesting that the newly identified receptors are RA-dependent transacting enhancer factors that regulate the transcription of RA-responsive genes. Stated slightly differently, by analogy with steroid hormone receptors, RARs must act in the nucleus by binding to specific DNA regulatory sequences to activate specific sets of genes in the presence of retinoic acid. The first retinoic acid receptor to be cloned and sequenced was RAR-α (Petkovich et al., 1987; Giguere et al., 1987). Two other RARs have since been identified, RAR-β and RAR-γ. Whether additional RARs exist and whether proteins within this receptor family that use retinol as a ligand exist are two important questions at present. Why there are three receptor proteins that utilize the same ligand, retinoic acid, is another question that it is crucial to answer.

How might RARs be involved in the control of processes as complex as pattern formation and cell differentiation? It is likely that the RARs, themselves transcription factors, regulate the genes for other, different types of transcription factors. The work from our laboratory within the past few years has shown that the genes for several transcription factors are regulated by retinoic acid (LaRosa and Gudas, 1988a,b; Hosler et al., 1989; Hu and Gudas, 1990). For example, the ERA-1 gene, a gene that encodes a protein that contains a homeobox domain, is rapidly and transiently induced in response to retinoic acid (LaRosa and Gudas, 1988a,b). Proteins that contain homeobox domains are thought to be transcription factors involved in regulating pattern formation in the developing embryo. Another gene that we have characterized, REX-1, exhibits a rapid decrease in transcription rate upon retinoic acid addition to F9 stem cells (Hosler et al., 1989). The putative protein encoded by the REX-1 gene has been found to contain an acidic domain and four repeats of a "zinc finger" motif. These structural features, each found in several eukaryotic transcription factors, imply a possible regulatory function for REX-1 as well. In addition, one of the retinoic acid receptor genes, RAR-β, is rapidly activated by retinoic acid (Hu and Gudas, 1990). Thus, genes encoding three types of DNA binding proteins exhibit either increased or decreased expression quite rapidly after retinoic acid addition to F9 cells.

Although retinoic acid responsive genes have been described in a variety of different retinoic acid responsive cell types, retinoic acid response elements (RAREs) (i.e., portions of promoter DNA that are involved in the retinoic acid response of the gene) have been identified in the 5′ flanking regions of two genes, the murine laminin B1 gene, a gene encoding a subunit of the extracellular matrix protein laminin (Vasios et al., 1989) and the human RAR-β gene (de The et al., 1990). These RAREs are presumably the sequences that interact with the retinoic acid receptors. Certainly, RAREs will be identified in many other retinoic acid responsive genes over the next few years. The specificity of the retinoic acid response will presumably be determined by both the levels at which the three RARs are expressed in different cell types, and the affinities of the RARs for the RAREs present in different RA-responsive genes. A wide variety of such interactions might be expected if one considers the striking effects of retinoic acid on the differentiation of many cell types.

We propose the following model for the retinoic acid-induced differentiation of F9 teratocarcinoma stem cells. Retinoic acid would diffuse through the plasma membrane and enter the cell. Once it is intracellular, it would either bind to the cytoplasmic retinoic acid-binding protein (CRABP) or move to the nucleus as free retinoic acid. Higher levels of the CRABP protein in the cell would result in larger amounts of retinoic acid bound to this cytoplasmic protein, and lower levels of free retinoic acid in the nucleus (Boylan and Gudas, 1991). Once retinoic acid enters the nucleus, retinoic acid could bind to either the RAR-α or RAR-γ, retinoic acid receptors that are constitutively expressed in F9 stem cells (Hu and Gudas, 1990). When retinoic acid is bound to the RAR-α and/or RAR-γ protein(s), these transcription factors may activate the ERA-1 gene and, conversely, inhibit the expression of the REX-1 gene. The ERA-1 gene would then activate other cellular genes expressed at later times after retinoic acid addition. The retinoic acid-bound RAR-β or the RAR-γ would also activate the laminin B1, B2, and A genes, genes involved in encoding basement membrane proteins (Vasios et al., 1989). This differentiation model is depicted in Figure 1B.

The questions that we are currently focusing on in the laboratory are the following. Why are there three retinoic acid receptors, and what is the role of each receptor in the regulation of teratocarcinoma cell differentiation and in the regulation of retinoic acid action during mouse embryogenesis? What is the role of the CRABP in mediating retinoic acid's effects? What is the role of the ERA-1 homeobox-containing gene in F9 stem cell differentiation? Are any other retinoic acid metabolites involved in interacting with receptors and activating or inhibiting gene transcription? Does the action of retinoic acid in activating gene transcription differ from the mechanism by which retinoic acid inhibits gene transcription? What are the enzymes involved in the conversion of vitamin A (retinol) to retinoic acid? All of these questions should keep the laboratory occupied for a number of years to come.

These differentiation-related regulatory circuits and DNA–protein interactions in mammalian cells in many respects are not that different from the regulatory pathways governing DNA repair that Dr. Pardee and I studied in *E. coli* in the 1970s, prior to the development of molecular cloning techniques. This interest of mine in regulatory circuitry within the cell that initially developed under Dr. Pardee's guidance has grown into a very active research program into the mechanisms of mammalian cell differentiation in 1990.

References

Boylan JF, Gudas LJ (1991): Overexpression of the cellular retinoic acid binding protein (CRABP) results in a reduction in the differentiation specific gene expression in F9 teratocarcinoma cells. J Cell Biol, in press.

Chytil F, Ong DE (1979): Cellular retinol and retinoic acid binding proteins in vitamin A action. Fed Proc 38:2510–2514.

de The HM, del Mar Vivanco-Ruiz M, Tiollais P, Stunnenberg H, Dejean A (1990): Identification of a retinoic acid responsive element in the retinoic acid receptor β gene. Nature (London) 343:177–180.

Giguere V, Ong ES, Sequi P, Evans RM (1987): Identification of a receptor for the morphogen retinoic acid. Nature (London) 330:624–629.

Gudas LJ, Mount DW (1977): Identification of the recA (tif) gene product of *Escherichia coli*. Proc Natl Acad Sci USA 74:5280–5284.

Gudas LJ, Pardee AB (1975): Model for regulation of *Escherichia coli* DNA repair functions. Proc Natl Acad Sci USA 72:2330–2334.

Hosler BA, LaRosa GJ, Grippo JF, Gudas LJ (1989): The expression of REX-1, a gene containing zinc finger motifs, is rapidly reduced by retinoic acid in F9 teratocarcinoma cells. Mol Cell Biol 9:5623–5629.

Hu L, Gudas LJ (1990): Retinoic acid and cyclic AMP analogs influence the expression of retinoic acid receptor α and β mRNAs in F9 murine teratocarcinoma cells. Mol Cell Biol 10:391–396.

LaRosa GJ, Gudas LJ (1988a): An early effect of retinoic acid: Cloning of an mRNA (ERA-1) exhibiting rapid and protein synthesis independent induction during teratocarcinoma stem cell differentiation. Proc Natl Acad Sci USA 85:329–333.

LaRosa GJ, Gudas LJ (1988b): The early retinoic acid-induced F9 teratocarcinoma stem cell gene ERA-1: Alternate splicing creates both a transcript for a homeobox-containing protein and one lacking the homeobox. Mol Cell Biol 8:3906–3917.

Petkovich M, Brand NJ, Krust A, Chambon P (1987): A human retinoic acid receptor which belongs to the family of nuclear receptors. Nature (London) 330:444–450.

Sporn MB, Roberts AB (1983): Role of retinoids in differentiation and carcinogenesis. Cancer Res 43:3034–3040.

Stoner CM, Gudas LJ (1989): Mouse cellular retinoic acid binding protein: Cloning, cDNA sequence and mRNA expression during the retinoic acid-induced differentiation of F9 wild type and RA-3-10 mutant teratocarcinoma cells. Cancer Res 49:1497–1504.

Vasios GW, Gold JD, Petkovich M, Chambon P, Gudas LJ (1989): A retinoic acid responsive element is present in the 5′ flanking region of the laminin B1 gene. Proc Natl Acad Sci USA 86:9099–9103.

Perspectives on Cellular Regulation:
From Bacteria to Cancer, pages 129–141

From the Cell Surface to the Nucleus: A Journey Through Multiple Signaling Pathways

Enrique Rozengurt

Imperial Cancer Research Fund, Lincoln's Inn Fields, London WC2A 3PX, United Kingdom

Introduction

I had the good fortune of arriving in Art Pardee's laboratory at a specially interesting time, when his major scientific interest was moving from bacterial to animal cells. Little was known about growth regulation of animal cells, whereas some fundamental concepts regarding the bacterial cell cycle were already established (Pardee and Rozengurt, 1975). One of the key concepts was that the cell surface plays a central role in the production of bacterial cells. Thus, the logical extension of this notion to animal cells was that the cell surface could play a critical role in the regulation of cell proliferation.

In a multicellular organism, the fundamental feature of growth control is that the proliferation of individual cells must be adjusted to the functional requirements of the whole organism. The breakdown of such rule would lead to neoplasia. Thus, a plausible hypothesis was that the cell membrane of higher animals governs the decision of the cell either to make DNA and divide or to not divide, depending on signals that reach it from the outside environment and from other cells. This was discussed in detail in a review article that I had the privilege to write with Art in early 1973: ''Role of the surface in the production of new cells'' (Pardee and Rozengurt, 1975). The preparation of this article stimulated many discussions that had a profound influence on my subsequent research activities. A central question raised was how the environmental information received by the plasma membrane is transferred to the intracellular milieu to produce a growth response. It was evident that a key step in elucidating the bases of growth control was to identify the signal-transduction pathway(s) involved in the generation of the mitogenic response.

Many studies on the mechanism of action of growth factors have used cultured fibroblasts, such as murine 3T3 cells, as a model system. These cells cease to proliferate when they deplete the medium of its growth-promoting activity and can be stimulated to reinitiate DNA synthesis and cell division either by replenishing the medium with fresh serum or by the addition of purified growth factors or pharmacological agents in serum-free medium. For years many investigators have proposed and searched for a single key signal governing the initiation of cell proliferation. Extensive analysis of early signaling events, however, has led to the formulation of an alternative model: the existence of multiple growth factor-activated signaling pathways that synergistically led to a mitogenic response (Rozengurt, 1986; Rozengurt et al., 1988). In this article I shall focus on the findings of my laboratory concerning the cyclic AMP signal-transduction pathway in the mitogenic response of Swiss 3T3 cells. The choice of this subject is easy to justify in the context of this book: my interest in cyclic AMP-mediated cell regulation started when I was a postdoctoral fellow in the laboratory of Art Pardee and it continues up to date. In fact, the first paper that I published with Art was on the role of cyclic AMP in the growth of transformed hamster cells (Rozengurt and Pardee, 1972).

Cyclic Nucleotides and Initiation of DNA Synthesis

The possibility that cyclic nucleotides, cyclic AMP and cyclic GMP, may regulate the proliferative response of quiescent fibroblastic cells has been the subject of a large and controversial literature. In 3T3 cells and other fibroblastic cells, increased levels of cyclic AMP were widely thought to reduce the rate of growth and inhibit the simulation of DNA synthesis promoted by adding serum to quiescent cells (Rozengurt, 1981). Using a sensitive radioimmunoassay and a rapid and convenient method to extract cyclic AMP from cultured cells (Rozengurt et al., 1981), it was possible to examine the effect of multiple pharmacological and physiological agents on both cyclic AMP levels and reinitiation of DNA synthesis by either serum or defined growth-promoting factors in serum-free medium. The results of these experiments provided compelling evidence in favor of a radically different role of cyclic AMP in the regulation of cell growth, namely, that increased cellular concentrations of cyclic AMP act synergistically with growth-promoting agents to stimulate DNA synthesis in quiescent cultures of 3T3 cells. In what follows, I summarize the experimental evidence leading to this conclusion.

Cholera Toxin Stimulates Initiation of DNA Synthesis by 3T3 Cells

It is well established that cholera toxin catalyzes the transfer of the ADP-ribosyl moiety of NAD to the G protein that stimulates adenylyl cyclase (G_s) and thereby stimulates cyclic AMP accumulation in many cell types. We found that cholera toxin, at concentrations that increase cyclic AMP levels in intact 3T3 cells, promoted (rather than inhibited) initiation of DNA synthesis in serum-stimulated Swiss 3T3 cells. Furthermore, cholera toxin added with insulin, or with phorbol esters, synergistically stimulated DNA synthesis in cultures of

Swiss 3T3 cells (Rozengurt et al., 1981). Cholera toxin elevated cyclic AMP and stimulated DNA synthesis in a concentration-dependent fashion. The shape of the dose–response curves for inducing DNA synthesis (in the presence of insulin) and for increasing cyclic AMP was similar (Rozengurt et al., 1981).

If the primary effect of cholera toxin on the initiation of DNA synthesis by 3T3 cells was due to its activation of the adenylate cyclase and cellular accumulation of cyclic AMP, inhibitors of cyclic nucleotide phosphodiesterase activity should be expected to potentiate the stimulation of DNA synthesis and the cellular accumulation of cyclic AMP produced by cholera toxin. Addition of either isobutylmethylxanthine (IBMX) or Ro 20-1724, potent inhibitors of phosphodiesterase activity, dramatically potentiated the ability of cholera toxin to stimulate DNA synthesis and to increase cellular cyclic AMP levels (Rozengurt et al., 1981). These findings strongly suggested that an increase in the intracellular levels of cyclic AMP acts synergistically with other mitogenic agents to stimulate DNA synthesis in Swiss 3T3 cells (Rozengurt, 1986). This conclusion has been substantiated by further experiments showing that other agents that activate the adenylyl cyclase and induce cyclic AMP accumulation are also mitogenic for Swiss 3T3 cells maintained in serum-free medium.

Enhancement of Cyclic AMP Levels and Stimulation of DNA Synthesis by Adenosine Agonists, Prostaglandin E_1, and Forskolin

Adenosine and prostaglandins bind to separate cell surface receptors of a variety of cell types and either stimulate or inhibit the activity of the adenylyl cyclase. A potent adenosine analog that binds selectively to stimulatory receptors is 5′*N*-ethylcarboxamide-adenosine (NECA). Addition of either this adenosine analog or prostaglandin E_1 (PGE_1) stimulated adenylyl cyclase activity, increased the cellular levels of cyclic AMP, and triggered reinitiation of DNA synthesis in Swiss 3T3 cells maintained in the presence of insulin (Rozengurt, 1985). The mitogenic effects of NECA and PGE_1 were closely related to their ability to increase the intracellular levels of cyclic AMP. Indeed, PGE_1 induced DNA synthesis and cyclic AMP accumulation at concentrations that were orders of magnitude lower than those used in previous studies to elicit an inhibitory effect.

Recently, we found that the diterpene forskolin, which directly activates adenylate cyclase in membrane preparations and intact cells from a variety of tissues, caused a marked increase in cyclic AMP levels and stimulates DNA synthesis acting synergistically with insulin in quiescent cultures of 3T3 cells (Rozengurt et al., 1988). In addition, 8Br cyclic GMP failed to initiate DNA synthesis in 3T3 cells under conditions in which 8Br cyclic AMP was effective (Rozengurt, 1985). These findings with adenosine agonists, prostaglandin E_1, forskolin, cholera toxin, and exogenously added cyclic nucleotide derivatives clearly demonstrate that an increase in the cellular levels of cyclic AMP (but not cyclic GMP) acts as a mitogenic signal for Swiss 3T3 cells (Fig. 1).

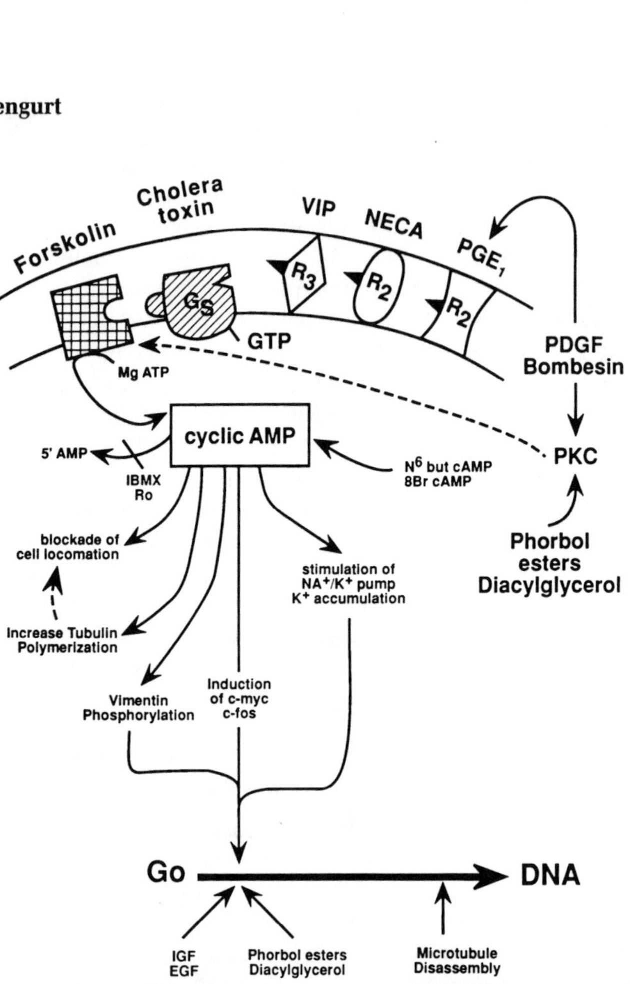

Fig. 1. The cyclic AMP signal transduction pathway in 3T3 mitogenesis. A variety of extracellular agents including PGE_1, VIP, and NECA bind to specific receptors on the plasma membrane and promote G_s-mediated increase in the synthesis of cyclic AMP. Cholera toxin and forskolin act at the level of G_s and catalytic subunit of the adenylyl cyclase, respectively. The accumulation of cyclic AMP by these agents is potentiated by inhibitors of cyclic AMP degradation (IBMX; Ro 20-1724). Cyclic AMP synthesis is also modulated in an indistinct manner involving PKC activation and prostaglandin synthesis (cross-talk). An increase in cyclic AMP, acting via cyclic AMP-dependent protein kinase, causes cytoskeletal changes, ionic fluxes, and selective gene expression. In the presence of other growth-promoting factors (e.g., IGF) or agents that stimulate PKC phorbol esters stimulate cellular exit from G_0 and entry into DNA synthesis.

The Role of Cyclic AMP in the Action of Growth Factors

Since a sustained increase in the cellular level of cyclic AMP constitutes a mitogenic signal for Swiss 3T3 cells, it was important to evaluate whether physiological growth factors added to serum-free medium could alter cyclic AMP metabolism in 3T3 cells. In the course of these studies it was found that addition of platelet-derived growth factor (PDGF), one of the most potent mitogens for fibroblastic cells, induced a striking accumulation of cyclic AMP in confluent and quiescent cultures of 3T3 cells incubated in the presence of inhibitors of cyclic nucleotide degradation (Rozengurt et al., 1983b). The accumulation of cyclic AMP elicited by PDGF was mediated by increased synthesis of E-type prostaglandins, which in turn stimulated cyclic AMP production by 3T3 cells in a paracrine and autocrine fashion through their own receptor (Rozengurt et al., 1983b). These findings suggest that cyclic AMP may be one of the signals utilized by PDGF to stimulate initiation of cell proliferation in Swiss 3T3 cells (Fig. 1).

Recent evidence has demonstrated that PDGF occurs as three dimeric isoforms (AA, AB, and BB) and that the PDGF receptor is composed of two separate subunits, designated α and β, which dimerize to form three distinct receptor species with different binding specificities for the PDGF isoforms. The heterogeneous nature of PDGF receptors themselves raises the important question of whether each receptor subtype can mediate the full repertoire of PDGF-induced intracellular signals. Since Swiss 3T3 cells possess equivalent numbers of α- and β-subunits, this question was investigated in detail by comparing the early biochemical events induced by PDGF in parallel cultures treated with highly purified preparations of PDGF-AA and PDGF-BB. Mehmet et al. (1990b) found that either PDGF-AA or PDGF-BB stimulated arachidonic acid release and cyclic AMP accumulation in a dose-dependent manner. Thus, α- and β-receptor subunits do not differ in their ability to transduce PDGF-mediated signals including cyclic AMP accumulation.

The Role of Cyclic AMP in the Mitogenic Action of Neuropeptides

It is increasingly recognized that small regulatory peptides act as molecular messengers in a complex network of information processing by cells throughout the body. They may act on postganglionic receptors (neurotransmitters), nearby cells (paracine hormones), or distant target organs (endocrine hormones). The classical role of these peptides as fast-acting neurohumoral signalers has recently been challenged by the discovery that they also stimulate slow-acting mitogenesis (Rozengurt, 1986; Zachary et al., 1987; Woll and Rozengurt, 1989). In particular, bombesin, vasopressin, and bradykinin can act as growth factors for cultured 3T3 cells. The early cellular and molecular responses elicited by bombesin and structurally related peptides (listed in Table 1) have been elucidated in detail. The cause–effect relationships and temporal organization of these early signals and molecular events have been reviewed (Rozengurt, 1988; Rozengurt and Sinnett-Smith, 1988; Rozengurt et al., 1988). They provide a paradigm for the study of other growth factors and mitogenic neuropeptides and illustrate the activation and interaction of a variety of signaling pathways. Like bombesin, vasopressin and bradykinin also induce rapid polyphosphoinositide breakdown and Ca^{2+} mobilization in Swiss 3T3 cells. The receptors for these neuropeptides are not directly coupled to cyclic AMP synthesis,

TABLE 1. Events in the Action of Bombesin in Swiss 3T3 Cells[a]

Binding to specific receptors
Cross-linking to M_r 75,000–85,000 glycoprotein
Ligand internalization and degradation
Activation of PKC (intact cells)
Activation of PKC (permeabilized cells)
Elevation of DAG levels
Ins (1,4,5)P_3 production
Ca^{2+} mobilization
Na^+ influx and Na^+/K^+ pump
Transmodulation of EGF receptor
Arachidonic acid release and prostaglandin synthesis
Enhancement of cAMP accumulation
Increase in c-*fos* and c-*myc* mRNA levels
Elevation of c-*fos* protein
Stimulation of DNA synthesis

[a]Detailed references can be obtained from the following review articles: Rozengurt (1986), Rozengurt and Sinnett-Smith (1988), and Rozengurt et al., (1988).

though they can influence cyclic AMP accumulation through an indirect mechanism discussed later.

Vasoactive intestinal peptide (VIP), a 28 amino acid polypeptide closely related to secretin and glucagon, binds to specific receptors and stimulates adenylate cyclase activity in target cells. VIP stimulates DNA synthesis in Swiss 3T3 cells in the presence of insulin and modulators of cyclic AMP metabolism such as forskolin or inhibitors of cyclic AMP phosphodiesterase (Zurier et al., 1988). In contrast to bombesin, vasopressin, and bradykinin, the mitogenic effects of VIP are mediated by elevation of cyclic AMP. Collectively, these findings demonstrate that the mitogenic actions of regulatory peptides are mediated by multiple signaling pathways and imply that the participation of regulatory peptides in the control of cell proliferation may be broader and more fundamental than previously thought.

Microtubules, Cell Locomotion, and Cyclic AMP

For many years, cell biologists have been interested in the relationship between cell locomotion and reentry into the cell cycle of quiescent animal cells (Pardee and Rozengurt, 1975). It was well known that addition of cyclic AMP-increasing agents to cultured cells changes the arrangement and length of microtubules, thereby causing morphological changes and reducing cell movement (see Rozengurt, 1981 for early literature). It was thought that there was a causal connection between contact inhibition of cell locomotion and density-dependent inhibition of cell growth. This hypothesis was attractive because it provided an elegant mechanism by which the cell surface could transduce external signals into cellular responses. The discovery that cyclic AMP acted as a mitogenic signal for 3T3 cells, as described in the preceding section, had a considerable impact in defining the interplay between microtubule organization, cell locomotion, cyclic AMP, and cell growth.

Cyclic AMP-elevating agents caused a small increase in the state of polymerization of tubulin in 3T3 cells (Wang and Rozengurt, 1983). Antimicrotubule drugs, which enhanced the mitogenic response of 3T3 to various growth factors (Rozengurt, 1986), also potentiated the mitogenic effects of cyclic AMP. These results indicated that cyclic AMP does not act via microtubules but that cyclic AMP and microtubule disassembly constitutes separate mitogenic signals.

Using time-lapse cinematography, O'Neill et al. (1985) found that an increase in the cellular level of cyclic AMP caused a profound inhibition of cell locomotion in the presence of insulin. In the same experiments, it was verified that the combination of cyclic AMP and insulin stimulated cell division. This experiment provided a conclusive dissociation of cell locomotion from initiation of cell cycle in quiescent cells: cyclic AMP blocked cell locomotion but stimulated cell division. The next step was to define the molecular events initiated by an increase in the cellular level of cyclic AMP and to determine the relationship between cyclic AMP and other signal-transduction pathways leading to mitogenesis.

Cyclic AMP-Mediated Phosphorylation in Quiescent 3T3 Cells

It is well established that protein phosphorylation is a fundamental mechanism by which extracellular stimuli regulate intracellular events and that cyclic AMP-dependent protein kinase, a tetramer containing two regulatory and two catalytic subunits, is the intracellular mediator of the biological effects of cyclic AMP. In spite of its potential importance as one of the transmembrane signaling pathways leading to mitogenesis, little is known of the effects of cyclic AMP on protein phosphorylation in quiescent Swiss 3T3 cells. Recent results demonstrate that an elevation in the cellular levels of cyclic AMP rapidly increased the phosphorylation of an M_r 58,000 cellular protein in these cells (Escribano and Rozengurt, 1988). Detergent extraction, immunoblotting, and immunoprecipitation identified the M_r 58,000 phosphoprotein as vimentin, the main protein subunit of the intermediate filaments of mesenchymal cells, including Swiss 3T3 cells (Escribano and Rozengurt, 1988).

A possible physiological role for the cyclic AMP-mediated phosphorylation of vimentin is indicated by two pieces of evidence. Phosphorylation of purified vimentin by cyclic AMP-dependent protein kinase in vitro blocks the assembly of this protein into filaments. Moreover, in intact 3T3 cells an increase in the intracellular level of cyclic AMP induced a marked redistribution and collapse of the intermediate filaments (Escribano and Rozengurt, 1988). These striking changes in intermediate filament distribution in vivo are indeed caused by cyclic AMP-mediated phosphorylation of vimentin at specific sites, since the phosphorylation profile of vimentin obtained from intact cells incubated with ^{32}P and cyclic AMP-increasing agents is comparable to that of purified vimentin phosphorylated in vitro by the catalytic subunit of cyclic AMP-dependent protein kinase.

The question remains as to how the cyclic AMP accumulation and subsequent vimentin phosphorylation induced by cyclic AMP-increasing agents might function to regulate the mitogenic response. One attractive possibility is that intact intermediate filaments transmit a negative signal that restricts the initiation of the cell cycle in 3T3 cells. The dissociation and redistribution of these filaments induced via specific vimentin phosphorylation by cyclic AMP-dependent protein kinase might remove a constraint for the initiation of cell proliferation.

Independence of Cyclic AMP and Other Signal-Transduction Pathways

The phosphorylation of the M_r 58,000 vimentin protein is useful as an indicator of the activation of the cyclic AMP-dependent signaling pathway in Swiss 3T3 cells. It was important therefore to determine the effect of cyclic AMP on other phosphorylation substrates that serve as indicators of the activation of other growth factor-induced signaling pathways.

An acidic protein, which on SDS polyacrylamide gels migrates with an apparent size of 80 kDa, has been identified as a major and specific substrate for protein kinase C (PKC) in quiescent mouse 3T3 cells (Rozengurt et al., 1983a). The phosphorylation of this protein is stimulated by a variety of growth-promoting agents including phorbol esters, diacylglycerols, serum, PDGF, bombesin, and vasopressin (Erusalimsky et al., 1988; Rodriguez-Pena and Rozengurt, 1986). Removal of phorbol esters or mitogenic peptides results in rapid dephosphorylation of this protein. These findings raise the possibility that the 80-kDa protein could be involved in PKC-mediated mitogenic signal transduction. Although the physiological role of the M_r 80,000 phosphoprotein remains to be elucidated, its phosphorylation provides a specific marker for assessing which mitogenic agents activate PKC in intact cells.

The phosphorylation of the M_r 80,000 protein was not increased either by pharmacological agents used to increase cyclic AMP (Escribano and Rozengurt, 1988) or by the neuropeptide VIP (Zurier et al., 1988). Furthermore, down-regulation of PKC (Rozengurt, 1986), which blocks the cellular response to mitogens that act via the PKC signal-transduction pathway (Rozengurt and Sinnett-Smith, 1988), does not prevent mitogenesis in response to agents that increase cyclic AMP levels. These results indicate that the PKC and cyclic AMP-mediated signal-transduction pathways represent distinct and independent mechanisms for mitogenic stimulation.

Cyclic AMP and Ionic Responses

The hypothesis that cyclic AMP, presumably acting through cyclic AMP-dependent protein kinase, activates a pathway leading to mitogenesis that is clearly separate from that utilized by other growth factors or mitogenic neuropeptides has been further substantiated by studies of the effect of cyclic AMP-increasing agents on early ionic responses. In contrast to the neuropeptides bombesin, vasopressin, or bradykinin, addition of either PGE_1 or VIP did not induce a rapid increase in the cytoplasmic concentration of Ca^{2+}, as measured in 3T3 cells loaded with the fluorescent Ca^{2+} indicator Fura-2 (Rozengurt et al., 1988).

It is well established that the stimulation of the monovalent K^+, H^+, and Na^+ ion fluxes is a general early response seen in most types of quiescent cells stimulated to proliferate by multiple combinations of growth promoting factors (Rozengurt and Mendoza, 1986). This ubiquity suggests a possible role for enhanced ion fluxes in the mitogenic response. Addition of cyclic AMP-increasing agents to quiescent 3T3 cells also increases the activity of the ouabain-sensitive Na^+/K^+ pump. In contrast to the stimulation of the Na^+/K^+ pump within 1–2 min after the addition of serum, PDGF, bombesin, or activators of protein kinase C, the increase in the Na^+/K^+ pump activity by cyclic AMP reached a maximum after 60 min of incubation (Rozengurt and Mendoza, 1986). Thus, the effects of cyclic AMP on divalent and monovalent cation fluxes further support the proposition that cyclic AMP activates a distinct pathway leading to mitogenesis.

Induction of the Protooncogenes c-*fos* and c-*myc* by Cyclic AMP

Serum and other mitogens rapidly and transiently induce the cellular oncogenes c-*fos* and c-*myc* in quiescent fibroblasts (reviewed in Rozengurt and Sinnett-Smith, 1988). Since these cellular oncogenes encode nuclear proteins, it is plausible that their transient expression may play a role in the transduction of growth factor-induced mitogenic signals.

The transcriptional regulation of the c-*fos* gene itself is complex. The c-*fos* promoter contains several upstream enhancer elements that bind sequence-specific protein factors and thereby control the transcription of the gene. Among these is the cyclic AMP-responsive element (CRE), which is required for c-*fos* induction by cyclic AMP *in vitro*. Despite the fact that the c-*fos* promoter contains a CRE, the effect of cyclic AMP on c-*fos* mRNA levels in intact cells remains controversial. Mehmet et al. (1988) demonstrated that an increase in cellular cyclic AMP causes only a slight increase in c-*fos* mRNA levels. For example, 3T3 cells treated with forskolin exhibit a 20-fold lower level of c-*fos* mRNA than those stimulated with PKC-activating agents. However, the role of synergistic interactions in the regulation of c-*fos* induction has not been investigated in detail. Recently, we have demonstrated that cyclic AMP can potentiate the induction of c-*fos* in intact Swiss 3T3 cells by other intracellular signals that act through PKC-dependent or independent signal-transduction pathways (Mehmet et al., 1990a). It is likely that these synergistic effects are mediated through the phosphorylation of specific transactivating proteins. The effect of cyclic AMP on gene expression is mediated by cyclic AMP-dependent protein kinase, which presumably phosphorylates CRE-binding proteins and thereby increases their affinity for the specific enhancer sequences.

While cyclic AMP potentiates other signals to promote c-*fos* mRNA expression, it does not act synergistically with insulin. Indeed, the combination of forskolin and insulin, which is mitogenic for 3T3 cells, induces low levels of c-*fos* mRNA. These results show that a large induction of c-*fos* mRNA is not necessary for cyclic AMP-mediated mitogenesis (Mehmet et al., 1988).

In contrast to the effects on c-*fos* expression, an increase in cellular cyclic AMP causes a striking increase in c-*myc* mRNA levels in the absence of any other synergistic signal (Rozengurt et al., 1988). In fact, the time course and magnitude of these effects are similar to those induced by bombesin or PDGF (Rozengurt and Sinnett-Smith, 1988). Thus, cyclic AMP causes differential induction of the cellular oncogenes of c-*fos* and c-*myc*.

Cross-Talk Between PKC and Cyclic AMP Signal Pathways

While, as shown above, cyclic AMP and PKC represent separate signal-transduction pathways, recent results indicate the existence of interactions between these major transmembrane signaling systems. Specifically, activation of PKC by either phorbol esters or diacylglycerols markedly enhances the accumulation of cyclic AMP in response to forskolin or cholera toxin, while down-regulation of PKC blocks this enhancing effect (Rozengurt et al., 1987).

A further example of cyclic AMP/PKC signal pathway interactions has come from studies with the neuropeptides of the bombesin family. These peptides bind to specific receptors in Swiss 3T3 cells and stimulate inositol polyphosphate formation, mobilization of Ca^{2+} from intracellular stores, activation of PKC, inhibition of epidermal growth factor (EGF) binding, secretion of the E-type prostaglandins, and induction of the cellular oncogenes c-*fos* and c-*myc* (Table 1). Bombesin caused a marked enhancement of cyclic AMP accumulation in the presence of forskolin; this increase is partially diminished both by down-regulation of PKC and by the cyclooxygenase inhibitor indomethacin (Millar and Rozengurt, 1988). The inhibitory effects are additive in nature, suggesting the existence of two mechanisms by which bombesin can enhance cyclic AMP accumulation. These findings suggest that cyclic AMP could contribute to the signaling of growth factors that primarily act through the PKC pathway (Fig. 1).

One of the most intriguing areas of PKC/cyclic AMP "cross-talk" involves the molecular basis for these pathway interactions. Further studies using phorbol esters and cyclic AMP-increasing agents have shown that this ‘‘cross-talk’’ is abolished by treatment with pertussis toxin in a time- and dose-dependent fashion. Since pertussis toxin does not itself promote cyclic AMP accumulation in Swiss 3T3 cells (Rozengurt et al., 1987), it is unlikely to act by removing a tonic inhibitory influence on the adenylate cyclase via G_i. An attractive possibility is that a novel pertusssis toxin substrate mediates the ‘‘cross-talk’’ between the PKC and the cyclic AMP pathways.

Conclusions

The central question posed at the outset of this chapter was how environmental signals received by the plasma membrane of animal cells are transmitted to the intracellular milieu to produce an appropriate growth response. The results discussed here provide evidence for the existence of multiple signal-transduction pathways in mitogenesis (Fig. 1). Specifically, cyclic AMP and PKC represent two separate signal-transduction pathways for mitogenic stimulation. In the presence of insulin, an increase in the cellular level of cyclic AMP (by multiple pharmacological agents, prostaglandins, or the neuropeptide VIP) leads to DNA synthesis without an early activation of PKC, Ca^{2+} mobilization, or a large induction of c-*fos*. Reciprocally, direct activation of a PKC pathway by phorbol esters or diacylglycerols causes reinitiation of DNA synthesis without an early increase in cyclic AMP; simultaneous activation of the PKC and cyclic AMP pathways stimulates DNA synthesis in the absence of insulin or any other ligand that occupies a tyrosine kinase receptor. The synergistic effect between PKC and cyclic AMP provides a clear example of the elicitation of mitogenesis by the integration of defined, independent intracellular signal-transduction pathways.

In addition to the cyclic AMP and PKC-mediated pathways discussed here, there is ample evidence for the existence of other signal-transduction mechanisms. For example, insulin, which acts synergistically with agents that utilize either PKC or cyclic AMP pathways, but does not itself activate PKC or elevate cyclic AMP, is envisaged to initiate yet another mitogenic pathway. Indeed, stimulation of DNA synthesis in 3T3 cells by insulin and EGF occurs without activation of either the PKC or cyclic AMP pathways. Since the receptors for both insulin and EGF possess tyrosine kinase activity, it is possible that specific tyrosine phosphorylations of key intracellular proteins are involved in the mitogenic signal pathway utilized by these ligands. Further variations in the signaling theme are revealed by recent studies with *Pasteurella multocida* toxin, a novel and extremely potent mitogen for cultured cells (Rozengurt et al., 1990). In conclusion, mitogenesis can be stimulated through multiple signal-transduction pathways that act in a synergistic fashion for transducing environmental signals from the whole organism into the mitogenic response of a specific cell.

References

Erusalimsky JD, Friedberg I, Rozengurt E (1988): Bombesin, diacylglycerols, and phorbol esters rapidly stimulate the phosphorylation of an Mr = 80,000 protein kinase C substrate in permeabilized 3T3 cells. J Biol Chem 263:19188–19194.

Escribano J, Rozengurt E (1988): Cyclic AMP increasing agents rapidly stimulated vimentin phosphorylation in quiescent cultures of Swiss 3T3 cells. J Cell Physiol 137:223–234.

Mehmet H, Sinnett-Smith J, Moore JP, Evan GI, Rozengurt E (1988): Differential induction of c-*fos* and c-*myc* by cyclic AMP in Swiss 3T3 cells: Significance for the mitogenic response. Oncogene Res 3:281–286.

Mehmet H, Morris C, Rozengurt E (1990a): Multiple synergistic signal transduction pathways regulate c-*fos* expression in Swiss 3T3 cells: The role of cyclic AMP. Cell Growth Diff 1: 293–298.

Mehmet H, Nånberg E, Lehmann W, Murray MJ, Rozengurt E (1990b): Early signals in the mitogenic response of Swiss 3T3 cells: A comparative study of purified PDGF homodimers. Growth Factors 3:83–95.

Millar JB, Rozengurt E (1988): Bombesin enhancement of cAMP accumulation in Swiss 3T3 cells: Evidence of a dual mechanism of action. J Cell Physiol 137:214–222.

O'Neill C, Riddle P, Rozengurt E (1985): Stimulating the proliferation of quiescent 3T3 fibroblasts by peptide growth factors or by agents which elevate cellular cyclic AMP level has opposite effects on motility. Exp Cell Res 156:65–78.

Pardee AB, Rozengurt E (1975): The role of the cell surface in the production of new cells. In C.F. Fox (ed): Biochemistry of Cell Walls and Membranes. London: Butterworth.

Rodriguez-Pena A, Rozengurt E (1986): Phosphorylation of an acidic mol. wt. 80,000 cellular protein in a cell-free system and intact Swiss 3T3 cells: A specific marker of protein kinase C activity. EMBO J 5:77–83.

Rozengurt E (1981): Cyclic AMP: A growth-promoting signal for mouse 3T3 cells. In J. Dumont, P. Greengard, and G. Robinson (eds): Advances in Cyclic Nucleotide Research, Vol. 14. New York: Raven Press, pp. 429–442.

Rozengurt E (1985): The mitogenic response of cultured 3T3 cells: Integration of early signals and synergistic effects in a unified framework. In P. Cohen and M. Houslay (eds): Molecular Mechanisms of Transmembrane Signalling, Vol. 4. Amsterdam: Elsevier Science Publishers BV, pp. 429–452.

Rozengurt E (1986): Early signals in the mitogenic response. Science 234:161–166.

Rozengurt E (1988): Bombesin-induction of cell proliferation in 3T3 cells. Ann NY Acad Sci 547:277–292.

Rozengurt E, Pardee AB (1972): Opposite effects of dibutyryl adenosine 3′:5′ cyclic monophosphate and serum on growth of Chinese hamster cells. J Cell Physiol 80:273–280.

Rozengurt E, Mendoza SA (1986): Early stimulation of Na^+/H^+ antiport, Na^+/K^+ pump activity and Ca^{2+} fluxes in fibroblast mitogenesis. In L. Mandel and D. Benos (eds): Current Topics in Membrane and Transport, Vol. 27. New York: Academic Press, pp. 163–191.

Rozengurt E, Sinnett-Smith J (1988): Early signals underlying the induction of the c-*fos* and c-*myc* genes in quiescent fibroblasts: Studies with bombesin and other growth factors. Progr Nucleic Acid Res Mol Biol 35:261–295.

Rozengurt E, Legg A, Strang G, Courtenay-Luck N (1981): Cyclic AMP: A mitogenic signal for Swiss 3T3 cells. Proc Natl Acad Sci USA 78:4392–4396.

Rozengurt E, Rodriguez-Pena M, Smith KA (1983a): Phorbol esters, phospholipase C, and growth factors rapidly stimulate the phosphorylation of a M_r 80,000 protein in intact quiescent 3T3 cells. Proc Natl Acad Sci USA 80:7244–7248.

Rozengurt E, Stroobant P, Waterfield MD, Deuel TF, Keehan M (1983b): Platelet-derived growth factor elicits cyclic AMP accumultin in Swiss 3T3 cells: Role of prostaglandin production. Cell 34:265–272.

Rozengurt E, Murray M, Zachary I, Collins M (1987): Protein kinase C activation enhances cAMP accumulation in Swiss 3T3 cells: Inhibition by pertussis toxin. Proc Natl Acad Sci USA 84:2282–2286.

Rozengurt E, Erusalimsky J, Mehmet H, Morris C, Nånberg E, Sinnett-Smith J (1988): Signal transduction in mitogenesis: Further evidence for multiple pathways. Cold Spring harbor Symp Quant Biol 53:945–954.

Rozengurt E, Higgins T, Chanter N, Lax AJ, Staddon JM (1990): Pasteurella multocida toxin: Potent mitogen for cultured fibroblasts. Proc Natl Acad Sci USA 87:123–127.

Wang Z-W, Rozengurt E (1983): Interplay of cAMP and microtubules in modulating the initiation of DNA synthesis in 3T3 cells. J Cell Biol 96:1743–1750.

Woll PJ, Rozengurt E (1989): Neuropeptides as growth regulators. Br Med Bull 45:492–505.

Zachary I, Woll PJ, Rozengurt E (1987): A role for neuropeptides in the control of cell proliferation. Dev Biol 124:295–308.

Zurier RB, Kozma M, Sinnett-Smith J, Rozengurt E (1988): Vasoactive intestinal peptide synergistically stimulates DNA synthesis in mouse 3T3 cells: Role of cAMP, Ca^{2+}, and protein kinase C. Exp Cell Res 176:155–161.

**Perspectives on Cellular Regulation:
From Bacteria to Cancer, pages 143–151**

The Effects of Cell Anchorage on Signal Transduction Contributing to Proliferation

Robert W. Tucker

The Johns Hopkins Oncology Center, Baltimore, Maryland 21205

Introduction

When I came to Art Pardee's laboratory in 1976 as a Medical Oncology fellow at the Sidney Farber Cancer Institute, I had little interest in the cell cycle, and I had barely learned how to grow cultured cells during my studies at NIH. However, I did have experience with and interest in comparisons between normal and neoplastic cells. Art quickly persuaded me that the cell cycle was important because the "restriction point" control of the cell cycle might explain the differences in growth behavior between normal and neoplastic cells in culture (Pardee, 1974). Indeed, the problems I first encountered during my work with Art still remain the focus of my research studies.

Both Art and I were curious about the different growth characteristics of normal and neoplastic cells in culture. Why do normal cells stop growing when confluent, and why do neoplastic cells continue to grow despite confluence and extensive cell–cell contact (Stoker and Rubin, 1967)? Why do normal cells fail to grow in a suspension culture that easily supports the continued growth of malignant cells? Initially, Art and I focused on the cytoskeleton, which was an obvious link between cell membrane events and nuclear/cytoplasmic events. Art suggested that the plasma membrane could determine how the extracellular environment controlled the intracellular biosynthetic machinery. This membrane, with its transport proteins and growth factor receptors, was most likely the initial transducer for the extracellular signals that ultimately reach the nucleus.

Art's viewpoint led to extremely productive studies of growth regulation and actually presaged later developments. Indeed, molecular biological techniques have recently shown that many, if not all, of the growth signals produced by polypeptide hormones are initiated at the plasma membrane, including receptor occupancy (e.g., tyrosine phosphorylation, autophosphorylation)

(Escobedo et al., 1988) and membrane turnover [phosphotidylinositol bisphosphate (PIP_2) hydrolysis, providing inositol trisphosphate and diacylglycerol] (Berridge, 1988; Nishizuka, 1989). However, when Art and I started our studies, membrane lipid turnover and receptor dynamics were relatively unexplored. Our studies on the possible role of the cytoskeleton in cell-cycle control revealed some interesting information about the relatedness of the cytoskeletal (centriole) and DNA synthesis cycles. Initially, the structural changes seemed only an epiphenomenon in the main processes controlling proliferation, but recently cytoskeletal influences on cellular regulation have become the focus of exciting research once again.

The Centriole and the Cell Cycle

Art and I first turned our attention to the possible relationship between the centriole and the cell cycle. Using antitubulin antibody to visualize microtubules during the G_0 to G_1 transition in fibroblasts (BALB/c 3T3 cells, clone A31), we found that centrioles form a primary cilium (9&0 microtubular structure) in all quiescent cells (Tucker et al., 1979). Most interestingly, the primary cilium transiently shortens in response to certain growth factors—i.e., competence growth factors (such as platelet-derived growth factor, PDGF)—which are required for growth primarily by nonneoplastic cells (e.g., 3T3). In contrast, neoplastic cells will often grow in plasma that does not contain any competence (or wound healing) factors (platelet-poor plasma, PPP) (Scher et al., 1978). Art's view of these morphological changes as indicators of important biochemical events controlling the cell cycle was an important concept. Moreover, these results indicated that DNA synthesis must be tightly correlated with centriole events; certainly both the centriole and DNA have to be precisely duplicated before mitosis can be successfully completed and the production of two daughter cells can occur. The centriole is a ''mitotic organelle,'' yet it clearly has some important relationship to DNA synthesis. The current biochemical expression of this relationship may be the *cdc2* gene product, which contributes to both G_1/S and G_2/M transitions (Norbury and Nurse, 1989).

Despite the ubiquitous presence of the primary cilium in nearly all nonhematopoietic cells, we have no definitive information about the role of the primary cilium in normal cellular physiology, and, in particular, about how centriole changes in G_1 influence the rest of the cell cycle. However, recently we have found that the primary cilium forms from the parent centriole in early G_1 at precisely 5 hr before the initiation of DNA synthesis; this is true in both serum-stimulated quiescent cells and the postmitotic cells of exponentially growing cultures (Ho, Tucker, 1989). In contrast to Zetterberg's hypothesis (Zetterberg and Larsson, 1985), this finding clearly indicates that cell variability (cell-to-cell differences in the timing of initiation of DNA synthesis) occurs in early G_1. This evidence for early G_1 variability is also consistent with Art Pardee's finding (Rossow et al., 1979) that inhibition of protein synthesis affects cells in early G_1, before the restriction point, after which cells become serum insensitive. Perhaps some of the processes controlled by serum are structural changes (i.e., changes in primary cilium) that either are important in the timing of DNA synthesis itself or reflect important changes in the metabolic machinery crucial in the initiation of DNA synthesis (i.e., kinase activity). Our results also show that serum stimulates quiescent cells to recapitulate early G_1 events, thereby repeating postmitotic events, at least with regard to the centriole cycle. Perhaps most interestingly, the effects of the two general classes of growth factors (competence and progression) occur in parallel paths (Ho and Tucker, 1989), not in series, as has been most commonly proposed. Thus, PDGF and plasma produce events that occur simultaneously and that ultimately synergize to produce DNA synthesis (Fig. 1). The main difference between the effects of competence and progression factors is the long "memory" (18–24 hr) for competence events, which allows an early treatment with a competence factor (PDGF) to synergize with a later addition of a progression factor. Intracellular events produced by PDGF and PPP might therefore differ qualitatively, perhaps by involving different intracellular messengers.

The ciliary shortening induced by growth factors suggests the involvement of a particular intracellular messenger, free cytosolic calcium (Ca_i), in the induction of DNA synthesis, because calcium is a prime regulator of microtubule polymerization in vitro (Weisenberg, 1972) and could therefore contribute to ciliary shortening in situ. A transient Ca_i increase can also induce prolonged phosphorylation and activation of calmodulin-dependent protein kinases (Thiel et al., 1988), thereby providing a biochemical substrate for the "memory" induced by competence factors. In fact, in BALB/c 3T3 cells, the calcium ionophore A23187 can induce ciliary shortening and Ca_i increases identical to those produced by PDGF. However, ciliary shortening induced by an increase in Ca_i does not participate directly with the induction of DNA synthesis. First, the stabilization of microtubules by taxol prevented ciliary shortening, but not the DNA synthesis induced by growth factors (Tucker, 1980). Second, calcium ionophore A23187 was not mitogenic by itself. Thus, ciliary/centriole changes represent central and important events leading to DNA syn-

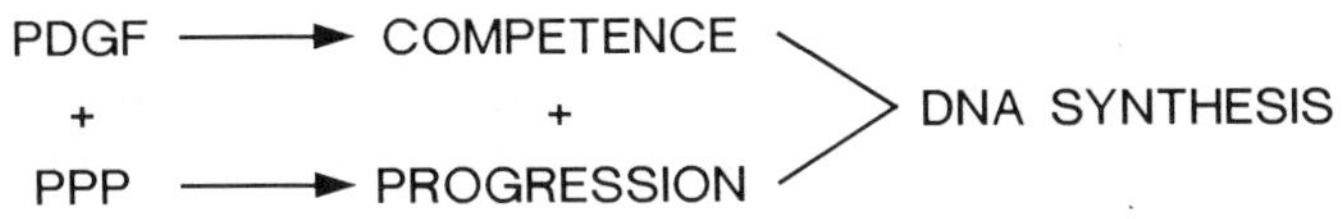

Fig. 1. PDGF and PPP (plasma) stimulate intracellular events in parallel, rather than in a series. These events then synergize to cause the initiation of DNA synthesis.

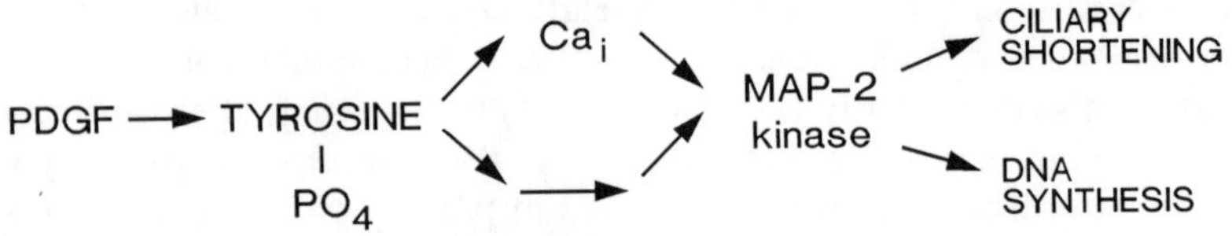

Fig. 2. Tyrosine phosphorylation of PDGF receptors stimulate multiple pathways, including Ca_i increase, which produce serine/threonine and tyrosine phosphorylation of MAP-2 kinase. Activated MAP-2 kinase then phosphorylates MAP-2 to destabilize ciliary microtubules, and activates other kinases that play a direct role in initiating DNA synthesis (e.g., S6 kinase).

thesis, but the changes themselves do not actually participate in the process. Centriole changes are therefore closely correlated with the timing of DNA synthesis, and may reflect early activation steps important in the macromolecular synthesis and organellar duplication necessary for cellular proliferation (Fig. 2). Because of the close correlation among Ca^{2+}, ciliary shortening, and DNA synthesis, the first steps crucial in mitogenesis could involve early ionic signals, such as Ca_i increases.

Calcium and Cell Growth

Calcium has been implicated in signal transduction for many years. In fact, both calcium and pH increases have been correlated with the stimulation of intracellular biosynthetic pathways by external signals that interact with the plasma membrane (Hesketh et al., 1988). Many external stimulators, such as sperm during fertilization, do not enter the cell, and therefore signal transduction has to occur through the external cell membrane. Even polypeptide growth factors, although eventually internalized, probably produce their mitogenic signal at the external cell membrane. The resulting membrane perturbations stimulate the hydrolysis of minor membrane components (PIP_2), causing the release of lipid-soluble (diacylglycerol) and water-soluble (inositol trisphosphate) messengers that activate protein kinase C and release internal Ca^{2+} stores, respectively. These principles formed the basis for research into the role of calcium in cell growth.

At first, the photoprotein aequorin was used to measure intracellular calcium transients, but aequorin had two disadvantages: 1) it was not taken up by cells and had to be microinjected or otherwise forced into cells, and 2) it was insensitive to local Ca_i changes. Although in large cells, such as amphibian eggs, the aequorin luminescence from a single cell was measurable, for small cells, such as fibroblasts, a bulk loading method using hypoosmolar shock had to be developed to load a large number of cells simultaneously. Using this technique, even with its inherent limitations, PDGF was indeed found to induce transient increases in Ca_i; however, other competence factors, such as fibroblast growth factor (FGF), did not produce Ca_i increases (Tucker et al.,

1986). Difficulties in interpreting these population measurements clouded the issue of whether all growth factors (including FGF) induced Ca_i increases, but in only a few cells or in only a local area of a given cell. Only when the fluorescent, calcium-sensitive indicator Fura-2 was used could Ca_i transients induced by FGF be measured in single cells. Fura-2 could be conveniently introduced into cells by incubating the cells in media containing an esterified derivative Fura-2/AM. Deesterification of Fura-2/AM by cellular enzymes trapped free Fura-2 intracellularly. We showed that PDGF induced a biphasic increase in Ca_i in Fura-2-loaded 3T3 fibroblasts: a transient release from Ca^{2+} stores, followed by Ca^{2+} entry through the plasma membrane. Surprisingly, we found that FGF also caused dose-related Ca_i increases in Fura-2-loaded cells, but these ionic signals were unrelated to mitogenic stimulation; in fact, large concentrations of FGF produced no Ca_I increase, yet still induced perfectly good ciliary shortening and mitogenesis (Tucker et al., 1989).

The most important question was whether the Ca_i increases induced by PDGF were required for PDGF-stimulated mitogenesis (initiation of DNA synthesis). Mitogenesis and Ca_i increases were both closely related to the concentration of PDGF used for stimulation: the percentage of cells with a 2-fold Ca_i increase correlated with the percentage of cells competent for DNA synthesis. In addition, as the concentration of PDGF was increased, the Ca_i per cell increased without any Ca_i oscillations. This latter point is pertinent because of the current enthusiasm for the idea that signaling is related to the period and amplitude of Ca_i oscillations. At least in BALB/c 3T3 cells, our experiments show that peak Ca_i increases do not oscillate in cells stimulated by PDGF. This correlation between Ca_i increases and DNA synthesis suggests that elevated Ca_i is important in the stimulation of DNA synthesis by growth factors. Since PDGF + EGTA, which effectively eliminates the Ca^{2+} entry through the plasma membrane, stimulates as many cells as does PDGF alone, only the brief (2 min) Ca_i increases from internal Ca^{2+} stores must contribute to growth stimulation by PDGF. In fact, an intracellular chelator like Quin-2 can inhibit Ca_i increases and dramatically reduce the number of competent cells produced by PDGF, but not by FGF (Tucker et al., 1989). Thus Ca_i increases are part of initial intracellular signals required for mitogenic action by some growth factors, such as PDGF. Much more work has to be done to define the long-lasting (hours) effects of transient Ca_i increases. Presumably, calmodulin-dependent protein kinase activity is important, since it undergoes sustained activation due to autophosphorylation following a transient calcium increase (Thiel et al., 1988). More recently, MAP-2 kinase has been shown to be the first protein kinase that is tyrosine phosphorylated by a growth factor receptor (Ely et al., 1990). MAP-2 kinase phosphorylates S6 kinase II, which presumably stimulates protein synthesis, which in turn is required for DNA synthesis. Furthermore, since MAP-2 kinase phosphorylates MAP-2, a microtubule-associated protein that can destabilize microtubules, MAP-2 kinase activation may be responsible for the ciliary shortening previously described. In fact, mitogenic stimulation could involve a cascade of protein kinase activations, including activation of MAP-2 (ciliary shortening), S6 (protein synthesis), and cdc2 kinases. The spectrum of activated kinases may depend on the particular mitogen. For example, progression factors never cause ciliary shortening, and probably do not activate MAP-2 kinase.

While Ca_i increases are certainly important for mitogenesis, the Ca_i increases induced by calcium ionophore are not sufficient to induce competence for DNA synthesis. The roles of other ionic factors in inducing competence have also been studied. pH_i, for instance, probably does not contribute to PDGF-stimulated mitogenesis, because, if the pH_i changes induced by growth factors are prevented by stimulating the cells in bicarbonate-containing medium (pH 7.4), mitogenic stimulation is unaffected (Bierman et al., 1988). Cyclic AMP, however, may affect Ca_i increases and/or synergize with a Ca_i signal in inducing DNA synthesis. Another obvious candidate for synergy with calcium is diacylglycerol (DAG), the second member of the bifurcating signaling pathway induced by growth factors like PDGF. Indeed, prolonged (1–2 hr) incubation with PDGF induces a sustained increase in diacylglycerol in fibroblasts stimulated with thrombin, and the increase correlates very closely with mitogenesis (Wright et al., 1990). This DAG increase, possibly produced via hydrolysis of phosphatidylcholine, may be an important contributor to the mitogenesis produced by many growth factors, but calcium increases resulting from release of intracellular Ca^{2+} stores may be important only for certain growth factors (e.g., PDGF but not FGF).

Thus, our studies showed that Ca_i increases are important for PDGF-stimulated mitogenesis but must synergize with other intracellular signals to induce maximal cell proliferation (Fig. 3). Once these aspects of the intracellular signals required for mitogenesis had been defined, the next step was to examine whether a change in these pathways could explain why nonneoplastic cells fail to grow in suspension culture.

Calcium and Cell Anchorage

In examining the relationship between cell anchorage and cell growth, it is important to distinguish between the initiation of one or a few rounds of DNA synthesis, and the continued growth required for the development of a colony or a group of cells. Ca_i increases are required for the initiation of DNA synthesis in attached cells, so we next examined the relationship between Ca_i increases and DNA synthesis in 3T3 cells in suspension culture. Do 3T3 cells fail to grow in suspension culture because early ionic signals cannot be stimulated or maintained in suspended, unattached cells?

We first discovered that serum can induce a Ca_i transient in every suspended 3T3 cell; the Ca_i increases have the same magnitude and duration as those

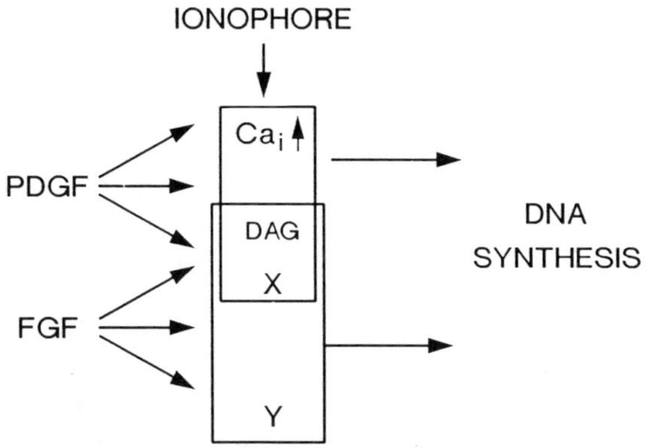

Fig. 3. PDGF and FGF induce different sets of intracellular messengers leading to the initiation of DNA synthesis. Only PDGF requires Ca_i increases, whereas both growth factors may need DAG increases for maximal mitogenesis.

produced by serum in attached cells (Tucker et al., 1990). Interestingly, serum can also induce most suspended 3T3 cells to initiate DNA synthesis, although few cells continue to grow progressively into a colony. In contrast, PDGF, even at high (100 ng/ml) concentrations, did not induce a Ca_i increase in most (>70%) suspended cells (Tucker et al., 1990). In preliminary studies, the same proportion of PDGF-stimulated, suspended cells (15–30%) that showed an increase in Ca_i also initiated DNA synthesis. Thus the Ca_i increase and DNA synthesis are tightly correlated in suspended, as well as in attached, cells. More importantly, however, 70% of suspended cells failed to initiate DNA synthesis after PDGF stimulation. Although there may be other blocks in the progression events that contribute to the failure of nonneoplastic cells to grow in suspension, these results clearly indicate that the early membrane perturbations induced by growth factors are defective in cells lacking attachment to a solid substratum (Fig. 4).

The mechanism by which cell attachment potentiates the signal transduction pathways stimulated by PDGF remains unexplained. Preliminary evidence suggests that the unattached cells retain PDGF receptors, and can still bind exogenous PDGF. Current work is examining whether PDGF receptors are phosphorylated and IP_3 is generated in suspension-grown cells. Future work will focus on the role of the integrin receptors in the stimulation or potentiation of the signaling pathways. However, it is already clear that cell attachment can stimulate IP_3 generation (Breuer and Wagener, 1989) and gene expression (Dike and Farmer, 1988). The fascinating possibility remains that occupancy of the cell attachment receptor (e.g., integrins) and/or the associated cytoskeletal adhesion plaque complex (actin, actinin, integrin, profilin) can modulate or even control the very start of the signal transduction process

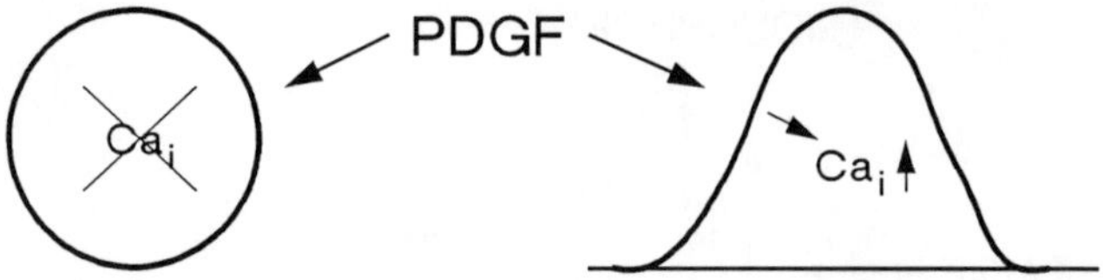

Fig. 4. PDGF induces Ca_i increase in attached BALB/c 3T3 cells, but fails to change Ca_i in the same cells when they are detached from the substratum. PDGF binds to its receptor in the rounded cells, but is unable to initiate the signaling cascade.

at the membrane receptor. In some cells the contact site for the cell has increased amounts of protein kinase C (Jaken et al., 1989) and profilin, an actin-binding protein that binds PIP_2 and inhibits PIP_2 hydrolysis (Goldschmidt-Clermont et al., 1990). Truly, then, the role of the cytoskeleton in controlling signal transduction will be an exciting and fruitful area of study in the near future.

Conclusion

Art Pardee has had a major influence on the approach we have taken in the study of growth regulation in mammalian cells. Our studies have focused on the role of the cytoskeleton in the control of cell proliferation by polypeptide growth factors. Early results demonstrated strict correlation between cytoskeletal perturbations (ciliary shortening) and the initiation of DNA synthesis. More recent biochemical evidence has suggested that activation of specific kinases (e.g., MAP-2 kinase) by growth factors may explain these close correlations. In addition, components of the adhesion plaque (i.e., integrin receptors) may control the initiation of the growth signal by the plasma membrane receptor. Thus, there is increasing evidence that the structure of the cell plays an essential role in the signaling processes responsible for the growth and eventual duplication of that structure and the cell. Disruption of the crucial relationships between cell shape and growth control may underlie many of the growth-regulatory differences between normal and neoplastic cells.

References

Berridge Mj (1988): The Croonian lecture. Inositol lipids and calcium signalling. Proc Roy Soc London 234:359–378.

Bierman AJ, Cragoe EJ, de Laat SW, Moolenaar WH (1988): Bicarbonate determines cytoplasmic pH and suppressed mitogenic-induced alkalinization in fibroblastic cells. J Biol Chem 263:15253–15256.

Breuer D, Wagener C (1989): Activation of the phosphatidylinositol cycle in spreading cells. Exp Cell Res 182:659–663.

Dike LE, Farmer SR (1988): Cell adhesion induces expression of growth-associated genes in suspension-arrested fibroblasts. Proc Natl Acad Sci USA 85:6792–6796.

Ely CM, Oddie KM, Litz JS, Rossomando AJ, Kanner SB, Sturgill TW, Parsons SJ (1990): A 42-kD tyrosine kinase substrate linked to chromaffin cell secretion exhibits an associated MAP kinase activity and is highly related to a 42-kD mitogen-stimulated protein in fibroblasts. J Cell Biol 110:731–742.

Escobedo JA, Keating MT, Ives HE, Williams LT (1988): Platelet-derived growth factor receptors expressed by cDNA transfection couple to a diverse group of cellular responses. J Biol Chem 263:1482–1587.

Goldschmidt-Clermont PJ, Machesky LM, Baldassare JJ, Pollard TD (1990): The actin-binding protein profilin binds to PIP_2 and inhibits its hydrolysis by phospholipase C. Science 247:1575–1578.

Hesketh TR, Morris JDH, Moore JP, Metcalfe JC (1988): Ca^{2+} and pH responses to sequential additions of mitogens in single 3T3 fibroblasts: Correlations with DNA synthesis. J Biol Chem 263:11879–11886.

Ho PTC, Tucker RW (1989): Centriole ciliation and cell cycle variability during G_1 phase of BALB/c 3T3 cells. J Cell Physiol 139:398–406.

Jaken S, Leach K, Klauck T (1989): Association of type 3 protein kinase C with focal contacts in rat embryo fibroblasts. J Cell Biol 109:697–704.

Nishizuka Y (1989): Studies and prospectives of the protein kinase C family for cellular regulation. Cancer 63:1892–1903.

Norbury CJ, Nurse P (1989): Control of the higher eukaryote cell cycle by p34*cdc2* homologues. Biochim Biophys Acta 989:85–95.

Pardee AB (1974): A restriction point for control of normal animal cell proliferation. Proc Natl Acad Sci USA 71:1286–1290.

Rossow PW, Riddle VGH, Pardee AB (1979): Synthesis of labile, serum-dependent protein in early G_1 controls animal cell growth. Proc Natl Acad Sci USA 76:4446–4450.

Scher CD, Pledger WJ, Martin P, Antoniades H, Stiles CD (1978): Transforming viruses directly reduce the cellular growth requirement for a platelet-derived growth factor. J Cell Physiol 97:371–380.

Stoker MG, Rubin H (1967): Density dependent inhibition of cell growth in culture. Nature (London) 215:171–172.

Thiel G, Czernik AJ, Gorelick F, Nairn AC, Greengard P (1988): Ca^{2+}/calmodulin-dependent protein kinase II: Identification of threonine-286 as the autophosphorylation site in the alpha subunit associated with the generation of Ca^{2+}-independent activity. Proc Natl Acad Sci USA 85:6337–6341.

Tucker RW (1980): cytoplasmic microtubules, centriole ciliation and mitogenesis in BALB/C 3T3 cells. In M. DeBrabander and J. DeMey (eds): Microtubules and Microtubule Inhibitors. Amsterdam: Elsevier/North Holland Biomedical Press, pp. 497–508.

Tucker RW, Pardee AB, Fujiwara K (1979): Centriole ciliation is related to quiescence and DNA synthesis in 3T3 cells. Cell 17:527–535.

Tucker RW, Snowdowne K, Borle AB (1986): Cytosolic free calcium and DNA synthesis in BALB/c 3T3 cells: Aequorin luminescence studies. Eur J Cell Biol 41:347–351.

Tucker RW, Chang DT, Meade-Cobun K (1989): Effects of platelet-derived growth factor and fibroblast growth factor on free intracellular calcium and mitogenesis. J Cell Biochem 39:139–151.

Tucker RW, Meade-Cobun K, Ferris D (1990): Cell shape and free cytosolic calcium (Ca_i) increases induced by growth factors. Cell Calcium 11:201–209.

Weisenberg RC (1972): Microtubule formation *in vitro* in solutions containing low calcium concentrations. Science 177:1104–1105.

Wright TM, Shin HS, Raben DM (1990): Sustained increases in 1,2-diacylglycerol precedes DNA synthesis in epidermal-growth-factor-stimulated fibroblasts. Biochem J 267:501–507.

Zetterberg A, Larsson O (1985): Kinetic analysis of regulatory events in G_1 leading to proliferation or quiescence of Swiss 3T3 cells. Proc Natl Acad Sci USA 82:5365–5369.

Perspectives on Cellular Regulation:
From Bacteria to Cancer, pages 153–163

Growth Control Mechanisms and the State of Differentiation

Judith Campisi

Division of Cell and Molecular Biology, Lawrence Berkeley Laboratory, University of California, Berkeley, California 74720

This paper, like the others in this book, is dedicated to Arthur Pardee in celebration of his 70th birthday. I spent four years as a postdoctoral fellow in Art's laboratory. When I arrived, I had been trained as a biochemist and had a background in membrane biochemistry and biophysics. I joined Art's lab because I wanted to learn cell biology from someone who was a biochemist at heart. When I left, I felt privileged to have experienced the fertile intellectual climate he created and nurtured. And I had learned much more than cell biology.

Anyone scanning Art's curriculum vitae cannot help but be struck by his continual ability to make important contributions in diverse areas of biochemistry and cell biology. His contributions span more than four decades, and continue to mark the present. What is not apparent from a casual reading of his curriculum vitae is the quality of the guidance that he gives to his students and fellows: there is strength, but an overriding gentleness; enthusiasm, yet an unassuming tranquility; and a naivety that does not lose its critical view. Under his guidance, it is possible to learn how to harness intuition and the scientific method in order to approach problems with a fresh view. And for many of us—certainly for myself—Art was and remains a friend, someone who has always been a source of support in matters both personal and professional.

When I joined Art's laboratory, there was much excitement about the events that appeared to control the cell cycle. Typical of Art's style, some of his students and fellows focused on the transition from quiescence into an active growth state, some on the G_1 to S phase transition, some on G_2 and M, and some on the related problems of repairing, replicating, and segregating damaged DNA. There were studies on growth factors, the synthesis and degradation of proteins, the synthesis of DNA and the enzymes responsible for its metabolism, and the failure and consequences of DNA repair mechanisms. Although by this time the lab was entirely "eukaryotic," the projects were not limited to higher eukaryotes. I remember with warmth and pleasure the days (and nights!)

peppered with lively discussions that covered all aspects of the cell cycle and, indeed, much of cell physiology.

Growth Control in Fibroblasts

As a postdoctoral fellow in Art Pardee's laboratory, I began studying the control of cell proliferation using murine fibroblasts (3T3 cells) in culture. Much of our understanding about the control of the cell cycle in higher eukaryotes derives from studies of fibroblasts. This cell type is perhaps the easiest to propagate in culture. In developed tissues, fibroblasts exist primarily in a reversible, proliferatively quiescent growth state (G_0); tissue injury or wounding stimulates them to leave G_0, progress through the cell cycle, and divide (Baserga, 1985). In culture, fibroblasts readily enter G_0 when grown to confluence or deprived of growth factors for a few days. Several laboratories, including Art's, have established that fibroblasts in culture require the sequential action of multiple growth factors in order to leave G_0 and resume proliferation (Pardee et al., 1978; Scher et al., 1979; Campisi and Pardee, 1984).

Growth-Regulated Genes

The tools of molecular biology have provided a powerful means with which to identify and study the genes that may function to regulate cell proliferation. Initially, a number of laboratories searched for genes whose expression changed when cells were stimulated to leave G_0 or traverse a particular phase of the cell cycle. In Art's lab, Harry Gray and I, in collaboration with Michael Dean and Gail Soneneshein, were among the first to show a marked increase in expression of the c-*myc* and c-*ras* protooncogenes after quiescent cells were stimulated by growth factors, and a disruption of this control in tumorigenically transformed cells (Campisi et al., 1984). A number of laboratories screened cDNA libraries for mRNAs that are specifically induced or suppressed by mitogenic growth factors (reviewed in Baserga, 1985; Denhardt et al., 1986).

By now, several dozen genes have been identified as growth- or cell cycle-regulated, principally in fibroblasts (Baserga, 1985; Denhardt et al., 1986; Pardee, 1989). Some of these (e.g., the protooncogenes) almost certainly play a regulatory role in cell proliferation. Others may be necessary for proliferation, but are unlikely to have regulatory functions (e.g., the replication-dependent histone genes). In addition, a number of mitogen-inducible genes probably have nothing to do with the proliferative response; rather, they are differentiated cell products.

The identification of growth-regulated genes has enabled molecular and cell biologists to approach an understanding of growth control from three vantage points. First, by studying the cytoplasmic events by which these genes are regulated, much has been learned about the signal transduction mechanisms used by different growth factors. Second, many of these genes are controlled by complex mechanisms. As a general rule, in fibroblasts, mitogenic growth factors induce the expression of growth-regulated genes by stimulating nuclear events such as transcription initiation or posttranscriptional processing. Thus, by studying the nuclear events responsible for regulating these genes it has been possible to identify key nuclear proteins that are candidates for coordinating gene expression and regulating progression through the cell cycle. As described elsewhere in this volume, this approach has and continues to be used with great success in Art Pardee's laboratory. Finally, transfection and microinjection experiments have shed some light on the functions of, and requirements for, certain growth-regulated genes in cell proliferation.

Figure 1 summarizes several years' work from a number of laboratories (reviewed by Pardee, 1989). It shows the organization of the cell cycle in fibroblasts stimulated to proliferate from G_0, the points at which three important mitogenic growth factors act, and some of the genes whose expression is dependent on the cell's growth state or position in the cell cycle.

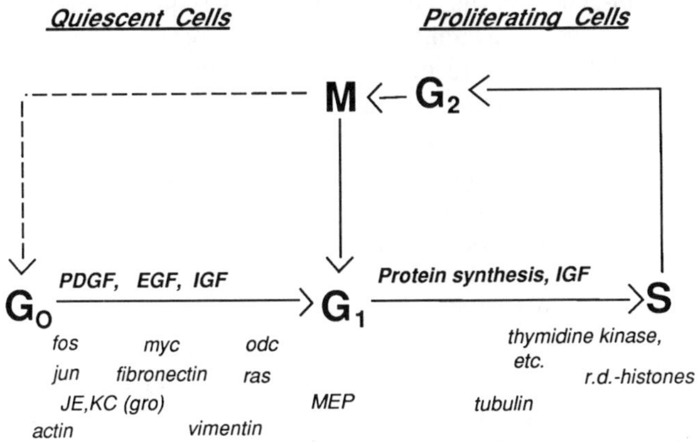

Fig. 1. Growth control in murine 3T3 fibroblasts. The broad aspects of much of what has been learned about growth control in untransformed 3T3 cells also pertain to fibroblasts from a variety of tissues and organisms. Fibroblasts enter a reversible, proliferatively quiescent state, G_0, when they grow to confluence or are deprived of serum growth factors for a few days. Fresh medium containing growth factors will reactivate proliferation. Some growth factors (e.g., PDGF, EGF) stimulate the cells to leave G_0, and these growth factors induce the expression of several genes, some of which are indicated in the figure. Other growth factors (e.g., IGF-I) act primarily (although not exclusively) to stimulate cells in G_1 to enter the S phase, and the genes induced during this period generally require the prior induction of gene expression during the G_0 to G_1 transition. The G_1 to S transition is also highly dependent on a rapid rate of protein synthesis. In general, once a cell has initiated DNA synthesis, it is committed to complete the remainder of the cycle through mitosis. After mitosis, cells may initiate another cell cycle, or, if the environmental conditions are not permissive for proliferation, may return to G_0. Details of this scheme can be found in Baserga (1985), Campisi (1989), Denhardt et al. (1986), Pardee (1989), and Scher et al. (1979).

Growth-Regulated Protooncogenes

Among the genes that are induced by mitogenic growth factors are several protooncogenes. By virtue of their role in transformation, protooncogenes are believed to act as positive (stimulatory) regulators of cell proliferation. A large number of gene transfection experiments have generally supported this idea. Fibroblasts transfected with mutated or unregulated forms of many protooncogenes often show a loss of growth control and acquire at least some of the properties of transformed cells (reviewed in Bishop, 1983). There is substantial experimental support for the idea that at least some protooncogenes are the critical, intracellular mediators of the action of the mitogenic growth factors.

The regulation and function of three protooncogenes have been particularly well studied in growth factor-stimulated fibroblasts: c-*fos*, c-*myc*, and c-*ras* (reviewed in Campisi, 1989). c-*fos* and c-*myc* encode nuclear proteins that may act to control the transcription or processing of other genes. In the case of c-*fos*, it is now clear that the fos protein forms a heterodimer with the product of another protooncogene, c-*jun*, and that the *fos/jun* dimer stimulates the transcription of genes bearing specific regulatory sequences in their 5′ flanking regions. c-*ras* genes, a family of three protooncogenes, encode membrane-associated GTPases that are thought to participate in transducing growth factor-generated signals to the nucleus.

The c-*fos*, c-*myc*, and c-*ras* genes are expressed at low or undetectable levels in quiescent cells. Within a few minutes (c-*fos*) to a few hours (c-*ras*) after quiescent cells are stimulated by growth factors, c-*fos*, c-*myc*, and c-*ras* expression is induced manyfold. These genes are not necessarily induced by the same growth factors or mechanisms. For example, we have found that platelet-derived growth factor (PDGF) and epidermal growth factor (EGF) each induce c-*fos* and c-*myc* expression, albeit by different intracellular signalling pathways (Ran et al., 1986; McCaffrey et al., 1987). By contrast, insulin-like growth factor-I (IGF-I) had no effect on the expression of c-*fos* or c-*myc*, but it induced the expression of c-*ras* (Lu et al., 1989).

The induction of the c-*fos*, c-*myc*, and c-*ras* protooncogenes by mitogenic growth factors appears to be a critical step in the cell's transition from a quiescent to a proliferating growth state. Antibody microinjection experiments and antisense RNA experiments strongly suggest that the expression of all three of these protooncogenes is essential in order for fibroblasts to leave G_0 and progress throug the cell cycle (Mulcahy et al., 1985; Holt et al., 1986; Heikkila et al., 1987). Thus, of the many growth factor-inducible genes, the c-*fos*, c-*myc*, and c-*ras* protooncogenes are most likely critical, regulatory genes.

Growth-Regulated Genes Also Encode Differentiated Cell Products

It is a common observation that there is an intimate association between cell proliferation and differentiated function, and that this relationship is often abnormal in transformed cells. Fibroblasts are, of course, differentiated cells.

This fact is often overlooked in studies that focus on understanding the basic mechanisms of growth control. However, a number of genes whose expression changes with the growth state of fibroblasts code for differentiated cell products. For example, differential screening of fibroblast cDNA libraries have identified a number of growth-regulated genes that encode small, secreted polypeptides; most of these are not mitogenic, but appear to mediate the inflammatory response (Wolpe and Cerami, 1989). In addition, a number of growth-regulated genes encode cytoskeletal and extracellular matrix proteins (Denhardt et al., 1986). In vivo, fibroblasts are often induced to proliferate in response to local trauma. Thus, one might expect that increased synthesis of differentiated products, such as cytokines or extracellular matrix components, should be coupled to the proliferative response.

Fibroblasts are most commonly studied in the context of growth control, and less commonly in the context of differentiation. There are certainly some features of growth control in fibroblasts that appear to be nearly universal. For example, many of the genes and biochemical reactions that control the onset of mitosis are common from yeast to human fibroblasts (Nurse, 1990). On the other hand, as discussed below, the differentiated state of fibroblasts can change, and this may alter the regulation of some but not all growth-regulated genes. In addition, at least some epithelial cells may show a pattern of growth-regulated gene expression that differs from that shown by fibroblasts. These findings suggest that some of the molecular events that control proliferation may very much depend on the differentiated state of the cell. The variations and constraints on growth control that are imposed by a program of cellular differentiation will undoubtedly occupy us for several years to come.

Cellular Senescence and Terminal Differentiation in Fibroblasts

All normal cells, when placed in culture, have only a limited ability to proliferate. Even highly proliferative cells such as fetal fibroblasts, which grow vigorously when first placed in culture, eventually lose the ability to undergo cell division in culture. This progressive decline in capacity to progress through the cell cycle has been termed the finite life span phenotype or cellular senescence.

In cell cultures established from many rodent species, variant cells having an infinite life span (immortal cells) arise spontaneously from cultures of senescing cells at a low but measurable frequency ($\sim 10^{-5}$–10^{-6}). 3T3 cells are in fact immortal, fibroblastic cell lines that arise spontaneously from cultures of 14- to 16-day-old mouse embryos. Immortal cells are not necessarily tumorigenic, but they can be readily transformed to tumorigenicity by radiation, carcinogenic chemicals, or oncogenic viruses.

In striking contrast to rodent cell cultures, cultures established from humans, and certain other species, virtually never spontaneously give rise to immortal variants (frequency $<10^{-12}$). Thus, for human cells, senescence is essentially complete and irreversible. Relative to rodent cells, human cells are also exceedingly resistant to tumorigenic transformation, and many, but not all, tumor cells are immortal (Sager, 1989). Thus, one view of cellular senescence holds that it may be an important mechanism for curtailing transformation or the establishment of metastases. In addition, cell cultures established from older individuals senesce after fewer population doublings (PD) than do cultures from younger individuals. Thus, another view of senescence holds that it may reflect, at a cellular level, some of the processes of aging in vivo (Hayflick, 1965). In either case, a particularly interesting feature of senescence is that the finite life span phenotype is dominant (i.e., immortality is recessive). This feature, and the extreme rarity of spontaneous immortalization in human cell cultures, has led to the suggestion that the senescence of human cells may be controlled by tumor suppressor genes (also referred to as antioncogenes), which presumably act as dominant inhibitors of cell proliferation and transformation (Sager, 1989).

Cellular senescence has been studied most extensively in human fibroblasts. In general, human fibroblasts senesce after 20 (adult) to 60 (fetal) PD in culture. Senescent cultures remain viable for long periods of time, during which they synthesize and turn over RNA and protein and, in fact, express many of the mRNAs and proteins expressed by earlier passage cells. However, although senescent cells possess apparently normal receptors for the major mitogenic growth factors, they cannot be stimulated to synthesize DNA (i.e., enter the S phase of the cell cycle) by any combination of growth factors or physiological mitogens (Phillips and Cristofalo, 1985). Nonetheless, senescent cells fully induce the mRNAs for several growth factor-inducible genes, including c-*myc, odc, JE,* and *vimentin,* after stimulation by serum (Rittling et al., 1986). Therefore, it is unlikely that the failure to proliferate is due to a general breakdown in growth factor signal transduction.

Despite these similarities between early and late passage cells, we have found a number of changes gene expression that occur when human fetal lung fibroblasts senesce (Seshadri and Campisi, 1990). Taken together, our data suggest that senescence entails a partial reprogramming of the pattern of gene expression in fibroblasts, and that arrested growth comprises just one part of the new program. In their broadest interpretation, our data are consistent with the idea that senescence is the expression of the terminally differentiated phenotype of fibroblasts.

First, we found that a ''control'' mRNA detected by a plasmid designated pHE-7 was markedly underexpressed in senescent cells. pHE-7 detects a ~1.5-kb mRNA whose abundance, in many cell types, does not change with growth state or position in the cell cycle. Indeed, in early passage cultures (PD <30; >80% proliferative cells), the level of pHE-7 mRNA remained constant whether the cells were proliferating, made quiescent by serum deprivation, or stimulated to leave G_0 by providing quiescent cells with 15% serum. However, once the culture reached senescence (PD >48; <10% proliferative cells), pHE-7 mRNA declined 5- to 10-fold. Because the pHE-7 gene was regulated by senescence—and not by growth state per se—we set out to learn more about it. Recently, we cloned and sequenced a nearly full-length cDNA. The sequence analysis showed that pHE-7 encodes the human homologue of the rat ribosomal protein L7 (Seshadri and Campisi, unpublished). At present, we can only speculate on the significance of this result. Partial repression of a major ribosomal protein may explain why the overall rate of protein synthesis falls (by 3- to 5-fold) in senescent cultures (a decline in the overall rate of protein degradation ensures that senescent cells do not become smaller). In addition, it is intriguing that tissues from older animals synthesize protein at a lower rate than tissues from younger animals. it remains to be seen whether there is a down-regulation of L7 in aged tissues; if so, it would support the idea that senescence in culture and aging in vivo entail some common mechanisms. Finally, several lines of evidence, largely from Art Pardee's laboratory, indicate that a reduction in protein synthetic rate can inhibit the proliferation of normal cells (Rossow et al., 1979; Pardee, 1989). Thus, the down-regulation of L7 expression in senescent cells may contribute to their inability to proliferate. Many questions remain. Are other ribosomal protein genes down-regulated in senescent cells? What is the mechanism for the partial repression of L7? Do senescent cells synthesize a different complement of proteins as a result of the reduced expression of L7?

A second intriguing set of changes shown by senescent human fibroblasts is a repression of the replication-dependent histones and the induction of a novel histone mRNA. Once human fibroblasts have undergone senescence, none of the replication-dependent histones (H2A, H2B, H3, H4) can be induced. This is too not surprising, since the expression of these four genes is generally tightly coupled to DNA synthesis (Stein et al., 1984), and senescent cells are unable to synthesize DNA. However, in senescent fibroblasts only (never in quiescent fibroblasts), we detected an unusual histone-related mRNA that cross hybridized with our murine histone 3 probe. This unusual mRNA was 400–500 bases larger than the replication-dependent histone 3 mRNA, and, in sharp contrast to the replication-dependent mRNA, it was polyadenylated. We have not yet fully characterized this polyadenylated histone mRNA, nor do we know anything about the mechanism responsible for its senescence-dependent expression. However, we speculate that the mRNA may originate from a gene encoding a replacement or variant histone. The significance of this idea is that expression of the replacement or variant histone genes is generally confined to terminally differentiated cells.

The most interesting change shown by senescent cells was a suppression of c-*fos* expression. As noted earlier, c-*fos* expression is induced to high levels within minutes after quiescent cells are stimulated by mitogenic growth factors. This is true for all (proliferative) fibroblasts examined. However, once fibroblasts had undergone senescence, c-*fos* expression could not be induced by any mitogen that we tested. This suppression was virtually complete, was due to a transcriptional mechanism, and was selective because c-*myc,* c-*ras*, and c-*jun* remained fully inducible in senescent cells. Thus, our results suggest that the c-*fos* gene is under specific, dominant transcriptional repression in senescent cells.

As discussed earlier, c-*fos* expression is essential for the proliferation of fibroblasts (Holt et al., 1986). Therefore, the c-*fos* repression we have described may at least partially explain the inability of senescent cells to proliferate. Recently, we found that we can efficiently introduce c-*fos* expression vectors into senescent cells by manual microinjection; we will now determine whether and to what extent c-*fos* expression can stimulate DNA synthesis in senescent cells. It would not be surprising, however, if c-*fos* expression were insufficient to drive senescent cells into S phase, as our preliminary results indicate (Surmacz and Campisi, unpublished). If it were, one might expect that mutations in the c-*fos* promoter should allow human cells to escape from senescence at a measurable frequency; however, as discussed above, spontaneous immortalization of human cells virtually never occurs. We favor the hypothesis that a *trans*-acting repressor is induced in senescent cells, which is responsible for suppressing c-*fos* expression and at least one other gene essential for proliferation. This idea would explain the dominance of the senescent phenotype (a *trans*-acting repressor could repress c-*fos* and other genes in a proliferative cell) and the low frequency of escape from senescence (since both alleles of the repressor must be inactivated to escape the repression). An intriguing corollary of this hypothesis is that a repressor of this sort should have tumor suppressor or antioncogene activity.

In summary, senescent human fibroblasts express and regulate some genes in a manner typical of proliferative fibrobasts (e.g., c-*myc*, c-*ras, JE*), but they also show striking, irreversible changes in the expression of other genes. Moreover, they do not behave like quiescent fibroblasts that are arrested at any point in the G_0/G_1 interval. Two temporally separated, growth-regulated genes were repressed: one normally induced early in the G_0/G_1 interval (c-*fos*) and one normally induced at the G_1/S boundary (histone); a non-growth-regulated gene (L7) was constitutively underexpressed; and a unique histone mRNA was expressed. These data suggest that senescent fibroblasts arrest growth in a state that is distinct from G_0. In this state, they show irreversible changes in gene expression, only some of which may play a role in the block to proliferation. Therefore, fibroblast senescence has several features of terminal differentiation. This idea was first proposed nearly two decades ago, based on morphology, and our data provide molecular support for the hypothesis. Our data further suggest that an important step in immortalization, and perhaps also in the loss of growth control that is the hallmark of transformation, is the ability to overcome the action of transcriptional repressors that, at least in part, act on protooncogenes.

Growth Control in Differentiated Epithelial Cells

We have recently begun to study growth control and immortalization in a very different type of cell—an epithelial cell that has only a very limited proliferative potential, both in vivo and in culture. This is the type 2 cell of the lung alveolus. Lung alveoli are lined by two types of simple epithelial cells: type 1 cells, which are terminally differentiated and carry out the gas exchange; and type 2 cells, which synthesize and secrete surfactant, a proteolipid complex that prevents alveolar collapse during exhalation.

Type 2 cells normally do not proliferate in vivo. However, they can undergo limited proliferation and differentiation into type 1 cells after partial pneumonectomy, lung injury, or during neonatal development. When isolated from the lungs of adult animals, type 2 cells do not proliferate in culture. However, 30–40% of type 2 cells isolated from neonatal lungs undergo 2–3 rounds of cell division in culture (Clement et al., 1990). Thus, type 2 cells are not typical epithelial stem cells, which generally are capable of extensive proliferation and differentiation both in vivo and, under appropriate conditions, in culture. Rather, they are in an intermediate stage between a proliferative stem cell and a terminally differentiated cell.

To begin to understand the mechanisms that control the proliferation of type 2 cells, we studied the expression of several growth-regulated genes in type 2 cells isolated from the lungs of adult or neonatal rats. Two interesting results emerged from these studies. First, four genes that have been described as growth-regulated in fibroblasts showed constitutive, non-growth-dependent expression in the cultured cells (Clement et al., 1990). These genes were c-*myc*, c-*ras, odc,* and *histone 3*. The corresponding mRNAs were present at identical levels in adult (nonproliferating) and neonatal (proliferating) cells, and the mRNA levels did not change when the proliferation of the neonatal cells was arrested by serum deprivation. However, in the case of *histone 3* and *odc,* translation of the mRNAs did depend on growth state. Thus, adult cells and serum-deprived neonatal cells expressed no *odc* activity and did not synthesize histone proteins, despite high levels of mRNA; proliferating neonatal cells, by contrast, expressed both mRNA and protein. These results suggest that, in contrast to fibroblasts wherein growth-regulated genes are controlled largely by transcriptional mechanisms, growth-regulated gene expression may be controlled at the level of translation in differentiated type 2 epithelial cells. We do not yet know anything about the mechanism of this translational control.

The finding that c-*myc*, c-*ras, odc,* and *histone 3* mRNAs were not growth-regulated in type 2 cells, as they are in fibroblasts, raises the possibility that some of the growth control mechanisms described for fibroblasts may not be applicable to these epithelial cells. We do not know whether type 2 cells really arrest growth in G_0, at least as this growth state has been described for fibroblasts. By flow microfluorimetry, it appears that most of the adult type 2 cells arrest growth with a G_1 DNA content, and that a substantial fraction of proliferating neonatal cells have an S phase or G_2 DNA content. However, when neonatal cells were deprived of serum for 48–72 hr, the cells ceased proliferation, but they arrested growth with a random cell cycle distribution and not with a G_1 DNA content.

A second interesting finding in cultured type 2 cells was that under no conditions could we induce the expression of c-*fos* (Oshima and Campisi, unpublished). However, we have recently derived immortal cell lines from neonatal rat type 2 cells by introducing a wild-type or transformation-defective simian virus 40 (SV40) T antigen. T antigen is the potent transforming protein of the oncogenic DNA virus SV40. It is a multifunctional nuclear protein that, recently, has been shown to bind and presumably inactivate two putative tumor suppressor proteins (the retinoblastoma susceptibility gene product and the p53 protein). In all the T antigen-immortalized type 2 cell derivatives, c-*fos* was highly inducible by mitogenic growth factors. One intriguing possibility is that c-*fos* repression may be responsible for the very limited proliferative potential shown by type 2 cells in vivo and in culture. Whether or how the mechanism of this repression relates to the c-*fos* repression shown by senescent human fibroblasts remains to be seen. In addition, this result suggests that T antigen may have the ability to alleviate the c-*fos* repression that occurs at the end of the lifespan of human fibroblasts and type 2 epithelial cells.

Conclusion

When I began my studies on growth control in Art Pardee's laboratory, the discovery of protooncogenes and antioncogenes was still in its infancy and there were few molecular bases for understanding the complexities of differentiation. It has been exhilarating to witness, and in a small way participate in, the progress that has been made in recent years in establishing a framework for understanding growth control and differentiation at a molecular level. In looking back, it is clear to me that Art Pardee is responsible for many of the ideas that form the foundation for our current thinking.

References

Baserga R (1985): The Biology of Cell Reproduction. Cambridge: Harvard University Press.

Bishop JM (1983): Cellular oncogenes and retroviruses. Annu Rev Biochem 52:301–354.

Campisi J (1989): Growth factors, protooncogenes and the control of cell proliferation: Lessons from the fibroblast. In A.N. Hirshfield (ed): Growth Factors and the Ovary. New York: Plenum, pp. 61–73.

Campisi J, Gray HE, Pardee AB, Dean M, Sonenshein GE (1984): Cell cycle control of c-myc but not c-ras expression is lost following chemical transformation. Cell 36:241–247.

Campisi J, Pardee AB (1984): Posttranscriptional control of the onset of DNA synthesis by an insulin-like growth factor. Mol Cell Biol 4:1807–1814.

Clement A, Campisi J, Farmer S, Brody J (1990): Constitutive expression of growth-related mRNAs in proliferating and nonproliferating lung epithelial cells in primary culture: Evidence for growth-dependent translational control. Proc Natl Acad Sci USA 87:318–322.

Denhardt DT, Edwards DR, Parfett CJ (1986): Gene expression during the mammalian cell cycle. Biochim Biophys Acta 865:83–125.

Hayflick L (1965): The limited in vitro lifetime of human diploid cell strains. Exp Cell Res 37:614–636.

Heikkila R, Schwab G, Wickstrom E, Loke SL, Pluznik DH, Watt R, Neckers LM (1987): A c-myc antisense oligonucleotide inhibits entry into S phase but not progression from G_0 to G_1. Nature (London) 328:445–449.

Holt JT, Venkat-Gopal T, Moulton AD, Nienhuis AE (1986): Inducible production of c-fos antisense RNA inhibits 3T3 cell proliferation. Proc Natl Acad Sci USA 83:4794–4798.

Lu K, Levine RA, Campisi J (1989): c-ras-Ha gene expression is regulated by insulin or insulin-like growth factor and by epidermal growth factor in murine fibroblasts. Mol Cell Biol 9:3411–3417.

McCaffrey P, Ran W, Campisi J, Rosner MR (1987): Two independent growth factor-generated signals regulate c-fos and c-myc mRNA levels in Swiss 3T3 cells. J Biol Chem 262:1442–1445.

Mulcahy LS, Smith MR, Stacey D (1985): Requirement for ras protooncogene function during serum-stimulated growth of NIH 3T3 cells. Nature (London) 313:241–243.

Nurse P (1990): Universal control mechanisms regulating onset of M-phase. Nature (London) 344:503–508.

Pardee AB (1989): G1 events and regulation of cell proliferation. Science 246:603–606.

Pardee AB, Dubrow R, Hamlin JL, Kletzien RF (1978): Animal cell cycle. Annu Rev Biochem 47:715–750.

Phillips PD, Cristofalo VJ (1985): A review of recent cellular aging research: Regulation of cell proliferation. Rev Biol Res Aging 2:339–357.

Ran W, Dean M, Levine RA, Henkle C, Campisi J (1986): Induction of c-fos and c-myc mRNA by epidermal growth factor or calcium ionophore is cAMP-dependent. Proc Natl Acad Sci USA 83:8216–8220.

Rittling SR, Brooks KM, Cristofalo VJ, Baserga R (1986): Expression of cell cycle-dependent genes in young and senescent WI-38 fibroblasts. Proc Natl Acad Sci USA 83:3316–3320.

Rossow P, Riddle VGH, Pardee AB (1979): Synthesis of labile, serum-dependent protein in early G_1 controls animal cell growth. Proc Natl Acad Sci USA 76:4446–4450.

Sager R (1989): Tumor suppressor genes: The puzzle and the promise. Science 246:1406–1411.

Scher CD, Shepard RC, Antoniades HN, Stiles CD (1979): Platelet-derived growth factor and the regulation of the fibroblast cell cycle. Biochim Biophys Acta 560:217–241.

Seshadri T, Campisi J (1990): Repression of c-fos transcription and an altered genetic program in senescent human fibroblasts. Science 247:205–209.

Stein GS, Stein JL, Marzluff WF (1984): Histone Genes—Structure, Organization and Regulation. New York: John Wiley.

Wolpe SD, Cerami A (1989): Macrophage inflammatory proteins 1 and 2: Members of a novel superfamily of cytokines. FASEB J 3:2565–2573.

**Perspectives on Cellular Regulation:
From Bacteria to Cancer, pages 165–174**

Defined Medium Studies of Regulated Growth and Differentiation

Van Cherington, Ron Krieser, and Rinku Chatterjee

Departments of Pathology (V.C., R.K.), Anatomy and Cellular Biology (V.C.), and Physiology (V.C., R.C.), Tufts University School of Medicine, New England Medical Center, Boston, Massachusetts 02111

Introduction

Early studies on the control of cell proliferation in tissue culture recognized that serum contained growth regulatory factors (Holley and Kiernan, 1968), and that neoplastic transformation altered the requirements for these factors (Holley and Kiernan, 1968; Clarke et al., 1970; Dulbecco, 1970). To determine the mechanisms underlying growth control and the transformed phenotype, numerous studies focused on quantitating the reduced serum requirement shown by transformed cells, the role of serum in regulating growth, and the identification of the growth regulatory factor(s) in serum that were no longer required by transformed cells. To identify individual growth factors it was necessary to eliminate or reduce the serum supplement in the medium and replace it with alternatives. Alternative supplements included tissue extracts or plasma, each of which contains only a subset of the serum components, or purified growth factors and hormones. For example, Scher et al. (1974) demonstrated that serum could be divided into plasma-derived and platelet-derived components, which functioned differently with respect to initiation of DNA synthesis and proliferation in cells arrested in G_0. The requirements for the plasma-derived and the platelet-derived components of serum were not equally sensitive to transformation; cells transformed by viral oncogenes grew in media supplemented with either plasma or serum (Scher et al., 1978), whereas untransformed cells grew well in serum but arrested growth in plasma. In general, studies of growth control focused on growth factors because they regulated the initiation of DNA synthesis by cells in G_0 and because the requirements for one or more growth factors were altered in malignantly transformed cells.

When I was asked to contribute to this volume honoring Arthur Pardee, it occurred to me that defined media in tissue culture studies, a recurring theme in my work ever since I was a graduate student in Art's laboratory, would be an appropriate framework for discussion. As part of my thesis project I developed a serum-free medium to examine differences in the growth requirements of normal and transformed cells. At the time, this approach represented a new and, it was hoped, more informative way to study the regulation of growth and differentiation in vitro. As has been so often the case in Art's career, work since that time in a number of laboratories has shown this approach to be valuable and insightful. In those experimental systems where a suitable serum-free medium has been developed, a range of supplements, in addition to growth factors, has been utilized. These findings have accentuated the complex interaction of nutrients, extracellular matrix components, hormones, and growth factors in regulating cells in vitro and also, presumably, in vivo. I will focus on a few of these studies in this paper. My own data that I present, and the work that I refer to in the text are intended to be illustrative; they are by no means a comprehensive representation of all experimental systems that have benefitted from the application of serum-free, hormonally defined media. Such studies have been reviewed more extensively elsewhere (Barnes, 1987; Barnes, 1984; Mather, 1984).

Transformed Cells Selectively Lose Their Requirement for Individual Factors

Part of the incentive for developing defined growth media for normal and transformed cells was to determine (1) which growth factors were required by a given cell type and (2) whether the reduced serum requirement shown by transformed cells was due to a reduction in the requirement for one or multiple growth factors. Identification of the factor(s) that could regulate the growth of normal but not transformed cells would, it seemed, provide clues into mechanisms of neoplastic transformation.

Studies using defined media show that changes in the requirement for an individual growth factor is responsible for individual attributes of the transformed phenotype, such as loss of density-dependent growth inhibition or decreased serum-dependency. Using a defined medium, Powers et al. (1984) demonstrated an association between changes in the platelet-derived growth factor (PDGF) requirement and density-dependent growth inhibiton in Swiss murine 3T3 cells, SV40-transformed 3T3 cells, and revertants of the transformants. It was found that reacquisition of density-dependent growth regulation by the revertants cosegregated with reacquisition of a PDGF requirement. In addition, reacquisition of serum dependency cosegregated with reacquisition of insulin dependency. Therefore, individual growth factors influenced different parameters of growth regulation.

Genetic changes leading to transformation in vitro can specifically eliminate requirements for individual growth factors. In our studies, we compared the growth requirements of normal Chinese hamster embryo fibroblasts (CHEF;

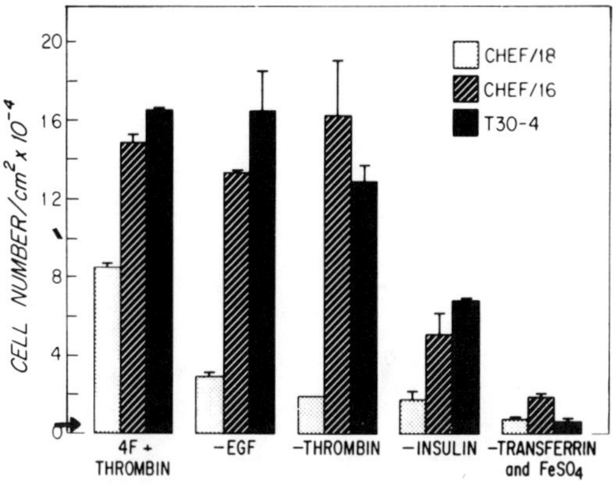

Fig. 1. Spontaneously and chemically transformed Chinese hamster fibroblasts have lost a requirement for thrombin and EGF in order to grow in a serum-free medium. Cells were plated into 35-mm dishes at 3×10^3 cells/cm^2 in α-MEM/F12 (1:1) plus 10% fetal calf serum. The following day the cells were rinsed three times with serum-free α/F12 and fed with α/F12 supplemented with thrombin (10 ng/ml), EGF (10 ng/ml), FGF (10 ng/ml), insulin (10 μg/ml), and transferrin (5 μg/ml) with individual factors omitted as indicated (4F + thrombin = complete serum-free medium). Duplicate cell counts (bar = SEM) were made at the time of the shift (arrow) and 3 days later. From Cherington and Pardee (1980).

Sager and Kovac, 1978) with the requirements of spontaneously and chemically transformed derivative cell lines. We found that transformed CHEF cells grew equally as well in a defined medium in the presence or absence of thrombin and epidermal growth factor (EGF). The growth controlled CHEF/18 parent cells required thrombin and EGF, as well as insulin and transferrin for optimal growth (Fig. 1). The transformed cells grew more rapidly than CHEF/18 cells in both fully supplemented and insulin-deficient media, suggesting a relaxed requirement for all factors, albeit partial retention of the requirements for insulin and/or insulin-like growth factor type 1 (IGF-1) was observed. Omission of transferrin from the medium resulted in limited growth of both normal and transformed cells (Fig. 1). Cells stopped growing randomly throughout the cell cycle when deprived of transferrin (not shown), consistent with transferrin's role as a nutritional factor and not a regulatory factor.

By comparing the growth factor requirements after spontaneous or chemically induced transformation, as outlined above, it appears that transformed cells can become completely independent of certain growth factors. A logical extension of these studies was to establish a genetic basis for transformation and related changes in growth factor requirements by examining the growth factor requirements of cells expressing an exogenous, activated oncogene. In an example of such a study, the cDNA for middle T antigen, the principal transforming protein of the polyoma DNA tumor virus, was introduced into NIH3T3

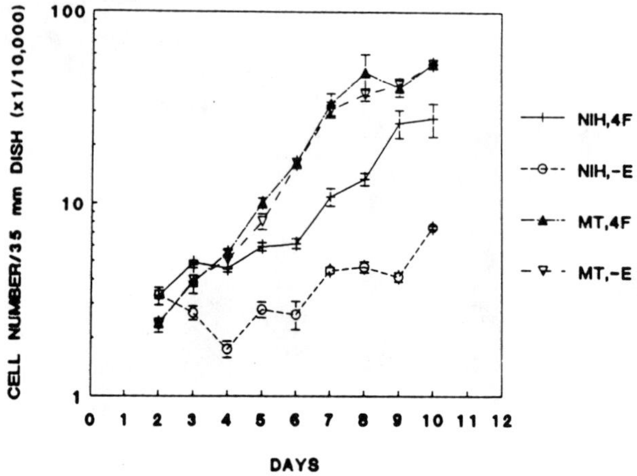

Fig. 2. Polyoma middle tumor (T) antigen expression in NIH3T3 cells eliminates the growth requirement for EGF in a serum-free medium. NIH3T3 (NIH) cells or NIH3T3 cells expressing polyoma middle T antigen (MT) were trypsinized using cold (4°C) trypsin/EDTA and were suspended in cold α/F12. Cells were washed once with α/F12 by centrifugation and seeded at 2×10^4 cells/35-mm dish in α/F12 supplemented with bovine fibronectin (5 μg/ml), v-*sis* supernatant (5 μl/ml; a serum-free medium conditioned by v-*sis* transformed NRK cells; courtesy of C. Stiles, DFCI), insulin (10 μg/ml), and transferrin (5 μg/ml). In addition, EGF (10 ng/ml) was either included (4F) or omitted (−E) from the medium. Cells from duplicate dishes were counted daily starting at day 2 (bar = range between duplicates). Fresh medium was added at day 5. From Cherington et al. (1986).

mouse fibroblasts using a recombinant retroviral vector (Cherington et al., 1986). The transformants grew well in defined media whether or not EGF was present, whereas control NIH3T3 cells showed a stringent requirement for EGF (Fig. 2). The transformants also grew more rapidly than control cells in defined media lacking a PDGF-containing supplement or insulin, although growth in these deficient media was less rapid than in fully supplemented, defined media (data not shown). These results are similar to the results obtained with the spontaneous and chemically induced CHEF cell transformants, where certain requirements were completely eliminated, while growth of transformed cells, even in suboptimal media, was more rapid than controls. The middle T antigen transformants grew in plasma as well as serum-supplemented media, indicating a diminished requirement for PDGF, in addition to the ablated EGF requirement observed in the defined medium (Cherington et al., 1986).

The advantage of studying transformation by well-characterized gene products like polyoma middle T antigen is that a molecular basis for its effects may be analyzed. For example, the basis for the middle T antigen effect on the PDGF requirement lies in part in its ability to stimulate a specific phosphatidylinositol kinase activity through pp60$^{c\text{-}src}$ tyrosine kinase activation, a process also regulated by the PDGF receptor tyrosine kinase (Kaplan et al., 1987). The mechanism for the middle T antigen effect on the EGF requirement has not been determined, but it is probably different from its effect on the PDGF requirement, since no EGF mediated effect on phosphatidylinositol kinase has been demonstrated in NIH3T3 cells.

Defined media have been used as a screening method for oncogenes with growth factor activity. Zhan et al. (1988) identified a human oncogene that encodes a member of the fibroblast growth factor (FGF) family by selecting NIH3T3 cells, transfected with human genomic DNA, in a defined medium lacking FGF and PDGF. It is surprising that this approach has not been utilized more as a selection method for genes for other growth factors, or for regulatory components of growth factor signal transduction pathways.

Cooperating Factors Are Required for Differentiation as Well as for Proliferation: Current and Future Studies

Although it is clear that multiple factors participate in the control of cell proliferation, it also appears that multiple factors cooperate to control the conversion of stem cells into differentiated populations. Recently my laboratory has been using a defined medium to study the regulated differentiation of mesenchymal cells into adipocytes. One murine preadipocyte cell line we use is 3T3-F442A (Green and Kehinde, 1976) which is a well-characterized model system for the study of terminal differentiation. These cells, when injected subcutaneously into athymic nude (nu/nu) mice, form white fat pads indistinguishable from native adipose tissue (Green and Kehinde, 1979). In vitro differentiation is induced (Fig. 3) by incubating confluent monolayers of the cells for 7–10 days in 10% fetal calf serum plus 5 μg/ml insulin, or in a defined medium (DM) supplemented with EGF, insulin, and human growth hormone (hGH). When induced to differentiate, these cells convert from fibroblast-like cells to functional adipocytes, as determined morphologically (Fig. 4) or us-

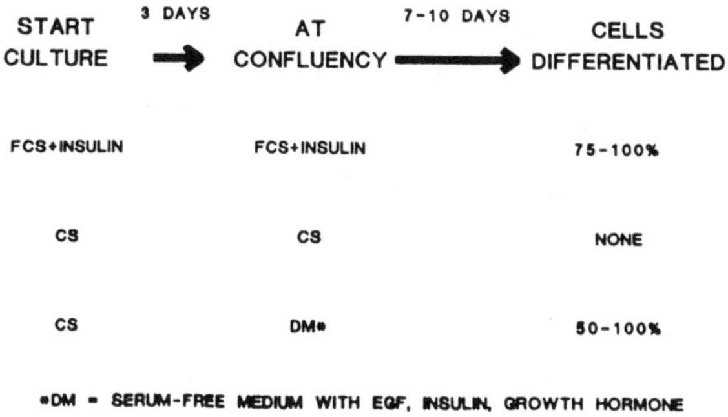

Fig. 3. Induction of differentiation of 3T3-preadipocytes.

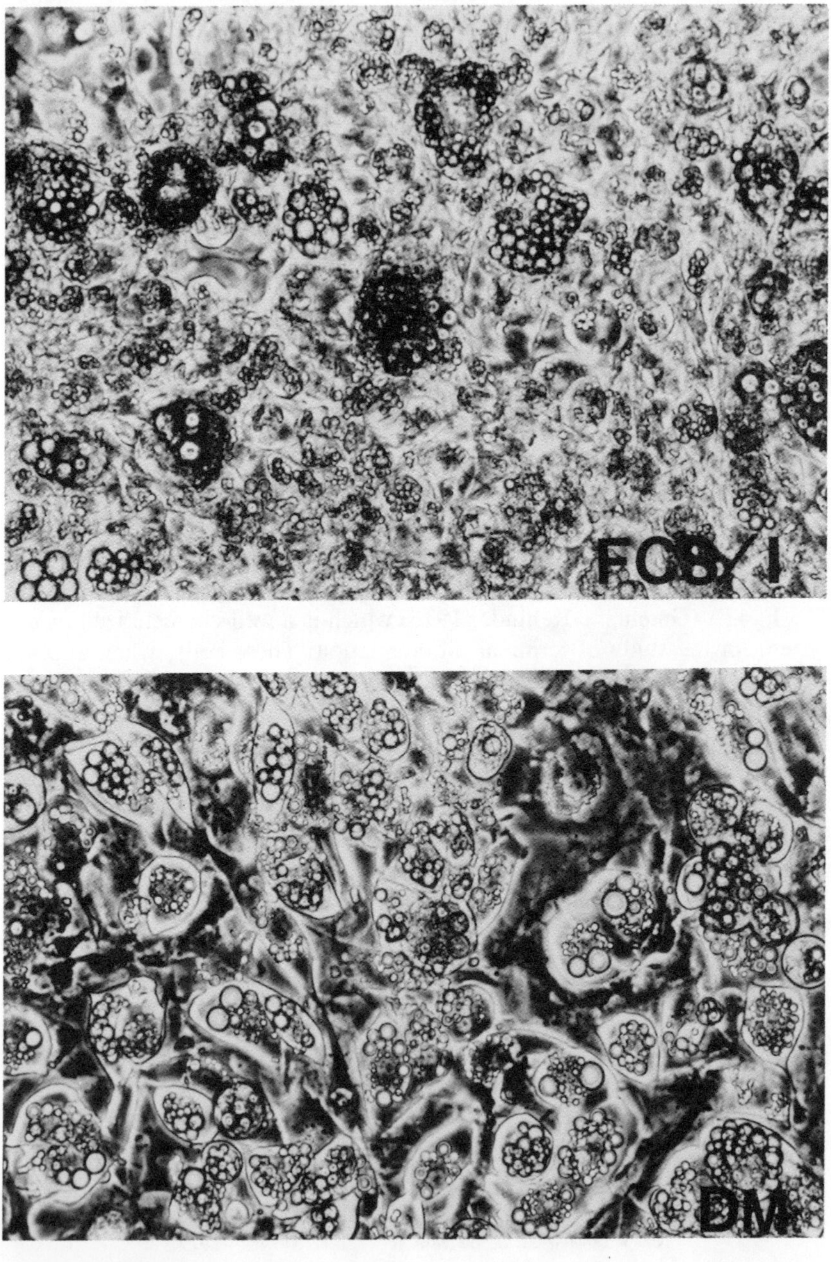
FCS/I
DM

ing a number of molecular and functional assays for adipocyte properties (Green, 1978; Spiegelman et al., 1988). If calf serum is used in place of fetal bovine serum, or if insulin is omitted from serum-supplemented media, differentiation is greatly reduced (not shown). Likewise, if any one the supplements present in DM is omitted, little or no differentiation occurs, as indicated morphologically (not shown) or by using glycerophosphate dehydrogenase (GPD) activity as an indicator of differentiated function (Fig. 5).

Although the molecular regulation of adipose gene expression is being investigated in a number of laboratories (see Spiegelman et al., 1988), the genetic determinants of preadipocytes and the mechanisms regulating the initiation of adipocyte differentiation have not been identified. Green et al. (1985) described a model for the regulation of adipocyte differentiation that suggests that hGH is at least partly involved in the early regulation of the differentiation program. Preadipocytes exposed to hGH become more sensitive to the mitogenic effects of IGF-1, resulting in clonal expansion of differentiating cells; this has also been observed by Guller et al. (1988), using a defined medium. These results may explain why hyperphysiological levels of insulin are usually required for the in vitro differentiation of 3T3-preadipocytes, since insulin presumably functions through IGF-1 receptors, as well as through its cognate receptor. The molecular events induced by hGH early in the differentiation of these cells are not known, nor is it known whether hGH functions alone or in concert with other hormones. Using defined media, we plan to precisely recreate hormone/growth factor interactions to resolve their roles in adipogenesis.

Using studies of growth control as an analogy, we have also introduced expression vectors for exogenous genes into preadipocytes to determine the effect of their products on differentiation. Certain genes, such as the large T antigens of polyoma and simian virus 40 (SV40) viruses, suppress differentiation of 3T3-F442A preadipocytes (Cherington et al., 1986, 1988). Identification of the critical cellular target(s) for these gene products will help to identify some of the normal cellular regulators of differentiation. In addition, these studies will establish the relationship between oncogenesis and differentiation.

Fig. 4. Differentiation of 3T3-F442A mouse preadipocytes in serum-supplemented and serum-free media. Cells were seeded into 35-mm wells at 7×10^4 cells/wells in DME + 10% calf serum. Upon reaching confluency (3–4 days) the monolayers were rinsed extensively (3–4 times) with serum-free DME and the medium was replaced with either DME + 10% fetal calf serum + 5 μg/ml insulin (FCS/I: the medium conventionally used to induce differentiation of these cells), or with DME/F12 (1:2) supplemented with EGF (50 ng/ml), insulin (5 μg/ml), and human growth hormone (22 ng/ml) (DM). This medium is adapted from one described by Guller et al. (1988). Mouse submaxillary EGF was purchased from Collaborative Research, Inc., porcine insulin was purchased from Sigma, and recombinant human met growth hormone was a gift from Genentech, Inc. The media were replaced on day 4 and day 7 following shift to experimental conditions and cells were photographed on day 10 using phase contrast microscopy (200 ×).

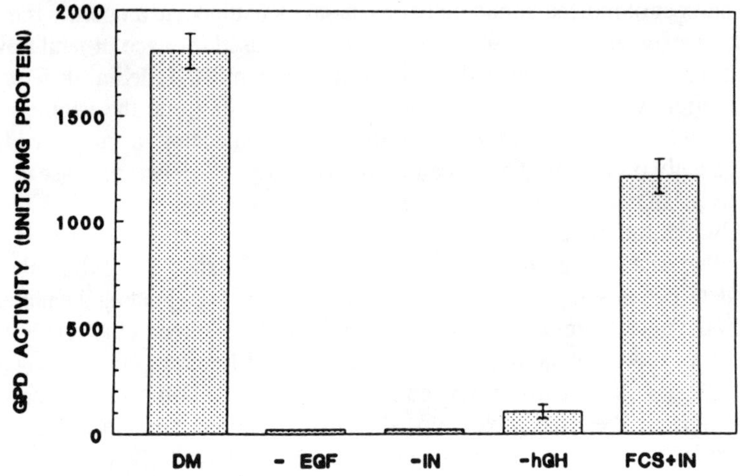

Fig. 5. Requirements for differentiation of 3T3-F442A mouse preadipocytes in a serum-free medium. Conditions were essentially as described in Fig. 4, except DM supplements were omitted as indicated. Extracts for glycerophosphate dehydrogenase assays (Wise and Green, 1979) were prepared in duplicate (bar = range between duplicates) on day 10.

We also can identify exogenous genes that, when expressed in preadipocytes, enhance differentiation. For example, preadipocytes expressing the polyoma middle T antigen differentiate more rapidly than controls (data not shown). One reason for this may be that the T antigen transformed cells have a greatly reduced requirement for fetal bovine serum in order to differentiate, as assayed using GPD activity as an enzymatic marker for adipose development (Table 1). Future studies in defined media will determine whether individual factors are no longer required for the induction of differentiation in cells expressing polyoma middle T antigen. As in our studies of polyoma and SV40 large T

TABLE 1. Diminished Serum Requirement for the Differentiation of Polyoma Middle T Antigen Transformed 3T3-Preadipocytes

	Units/mg protein Glycerophosphate dehydrogenase	
Fetal calf serum (%)	3T3-F442A	PyMT transf.
0	0	3954
0.1	1214	5542
1.0	2686	4565
10.0	3093	2127

Cells were grown to confluency in 35-mm wells in DME + 10% fetal calf serum + 5 μg/ml insulin. On reaching confluency the medium was replaced with DME + 5 μg/ml insulin plus the indicated levels of serum. The media were replaced after 4 days with identical media, and duplicate wells were extracted 8 days after confluency for enzyme assays.

antigen suppression of adipogenesis, we will be able to use a genetic and biochemical approach to identify cellular targets for middle T that are critical for this function, thereby identifying important endogenous regulators of adipocyte differentiation.

Acknowledgments

The portions of this work that have not previously been published have been supported in part by grants from the American Cancer Society (MV-367, JFRA-197), the U.S. Public Health Service (CA44761), the Weight Watchers Foundation, and the Life and Health Insurance Medical Research Fund.

References

Barnes D (1987): Serum-free animal cell culture. Biotechniques 5:534–542.

Barnes D, Sirbasku D, Sato G (eds) (1984): Cell culture Methods for Molecular and Cell Biology, Vols. 1–4. New York: Alan R. Liss.

Cherington V, Brown M, Paucha E, St Louis J, Spiegelman BM, Roberts TM (1988): Separation of simian virus 40 large-T-antigen-transforming and origin-binding functions from the ability to block differentiation. Mol Cell Biol 8:1380–1384.

Cherington V, Morgan B, Spiegelman BM, Roberts TM (1986): Recombinant retroviruses that transduce individual polyoma tumor antigens: Effects on growth and differentiation. Proc Natl Acad Sci USA 83:4307–4311.

Cherington PV, Pardee AB (1980): Synergistic effects of epidermal growth factor and thrombin on the growth stimulation of diploid Chinese hamster fibroblasts. J Cell Physiol 105:25–32.

Cherington PV, Smith BL, Pardee AB (1979): Loss of epidermal growth factor requirement and malignant transformation. Proc Natl Acad Sci USA 76:3937–3941.

Clarke GD, Stoker MGP, Ludlow A, Thornton, M (1970): Requirement of serum for DNA synthesis in BHK/21 cells: Effects of density, suspension and virus transformation. Nature (London) 227:798–801.

Dulbecco R (1970): Topoinhibition and serum requirement of transformed and untransformed cells. Nature (London) 277:802–806.

Green H (1978): The adipose conversion of 3T3 cells. In F. Ahmad et al. (eds): Tenth Miami Winter Symposium on Differentiation and Development. New York: Academic Press, pp. 13–29.

Green H, Kehinde O (1976): Spontaneous and heritable changes leading to increased adipose conversion in 3T3 cells. Cell 7:105–113.

Green H, Kehinde O (1979): Formation of normally differentiated fat pads by an established preadipose cell line. J Cell Physiol 101:169–171.

Green H, Morikawa M, Nixon T (1985): A dual effector theory of growth hormone action. Differentiation 29:195–198.

Guller S, Corin RE, Mynarcik DC, London BM, Sonenberg M (1988): Role of insulin in growth hormone-stimulated 3T3 cell adipogenesis. Endocrinology 122:2084–2089.

Holley RW, Kiernan JA (1968): "Contact inhibition" of cell division in 3T3 cells. Proc Natl Acad Sci USA 60:300–304.

Kaplan DR, Whitman M, Schaffhausen B, Pallas DC, White M, Cantley L, Roberts TM (1987): Common elements in growth factor stimulation and oncogenic transformation: 85 kd phosphoprotein and phosphatidylinositol kinase activity. Cell 50:1021–1029.

Mather JP, ed (1984): Mammalian Cell Culture. The Use of Serum-Free Hormone-Supplemented Media. New York: Plenum Press.

Powers S, Fisher P, Pollack R (1984): Analysis of the reduced growth factor dependency of simian virus 40-transformed 3T3 cells. Mol Cell Biol 4:1572–1576.

Sager R, Kovac P (1978): Genetic analysis of tumorigenesis: I. Expression of tumor forming ability in hamster hybrid cell lines. Som Cell Genet 4:375–392.

Scher CD, Pledger WJ, Martin D, Antoniades HN, Stiles CD (1978): Transforming viruses directly reduce the cellular growth requirement for a platelet-derived growth factor. J Cell Physiol 97:371–380.

Scher CD, Stathakos D, Antoniades HN (1974): Dissociation of cell division stimulating capacity for Balb/c-3T3 from the insulin-like activity in human serum. Nature (London) 278: 279–281.

Spiegelman BM, Distel RJ, Ro H-S, Rosen BS, Satterberg B (1988): *fos* protooncogene and the regulation of gene expression in adipocyte differentiation. J Cell Biol 107:829–832.

Wise LS, Green H (1979): Participation of one isozyme of cytosolic glycerophosphate dehydrogenase in the adipose conversion of 3T3 cells. J Biol Chem 254:273–275.

Zhan X, Bates B, Hu X, Goldfarb M (1988): The human FGF-5 oncogene encodes a novel protein related to fibroblast growth factors. Mol Cell Biol 8:3487–3495.

Perspectives on Cellular Regulation:
From Bacteria to Cancer, pages 175–184

Protease Nexins: Regulation of Proteases in the Extracellular Environment

Dennis D. Cunningham

Department of Microbiology and Molecular Genetics, University of California College of Medicine, Irvine, California 92717

Introduction

When the suggestion was first made that we honor Art Pardee with a book on his 70th birthday, I helped phone some of his former graduate students and postdoctoral fellows to circulate the idea. Each conversation was similar. Everyone agreed that it was an excellent idea. We then reflected on the many ways that Art influenced our careers. We discussed his keen insight into biological problems and his long record of major achievements. We also emphasized the many stimulating discussions with Art and how much we enjoyed working with him. There were also a few chuckles about the thrift that Art emphasized in his lab!

My studies with Art from 1968 to 1970 at Princeton University focused on early cellular events associated with initiation of proliferation of cultured fibroblasts. We found cellular changes that represented early markers—ones that occurred within several minutes after adding the growth stimulus. These included rapid increases in the uptake of phosphate and uridine (Cunningham and Pardee, 1969) as well as changes in the turnover of certain phospholipids (Cunningham, 1972). When I moved to the University of California at Irvine in 1970, I continued to study the changes in nutrient uptake. By manipulating the concentrations of phosphate, glucose, and uridine in the culture medium it was possible to regulate their uptake and examine rates of uptake in relation to initiation of cell proliferation. We came to the conclusion that the rapid increases in phosphate, uridine, and glucose uptake that occurred after adding the growth stimulus were neither necessary nor sufficient for the subsequent growth response (Thrash and Cunningham, 1974; Naiditch and Cunningham, 1977; Barsh and Cunningham, 1977).

Prompted by the finding of Chen and Buchanan (1975) that thrombin is a potent mitogen for fibroblasts in serum-free medium, I began studies on the mechanism by which this protease initiates cell proliferation. There were two features about thrombin as a mitogen that suggested that its study would be interesting. First, it was probably important physiologically since thrombin is released at sites of tissue injury. Second, the well-characterized proteolytic activity of thrombin indicated that it might be possible to identify cellular substrates whose cleavage was necessary for mitogenic stimulation. In the initial phases of these studies, we showed that internalization of thrombin was not necessary for the proliferative response (Carney and Cunningham, 1978a). We also studied cell surface binding sites for thrombin and their relationship to mitogenic stimulation (Carney and Cunningham, 1978b). Later we examined cell surface substrates of thrombin, but did not find proteolytic cleavages that always correlated with the mitogenic response (Thompson et al., 1987; Thompson and Cunningham, 1987).

As we were studying the mechanisms by which thrombin binds to the cell surface, we identified a cellular component that is secreted into the culture medium that forms SDS-stable complexes with thrombin (Fig. 1) (Baker et al., 1980). We named this protein protease nexin (PN). (Nexin is from the Latin *nexus,* a tying or binding together.) The complexes then bind back to the cells via the PN moiety of the complex. After binding, the complexes are

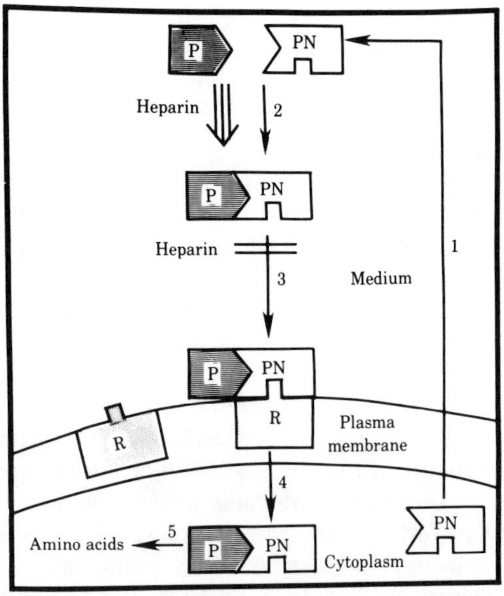

Fig. 1. Summary of interactions of protease nexin (PN) with serine protease (P) and with receptor (R) on cells. Step 1, release of PN into the culture medium; step 2, complex formation between P and PN; step 3, cellular binding of P–PN complexes to cell surface receptors; step 4, internalization of complexes; step 5, degradation of complexes.

rapidly internalized and degraded (Low et al., 1981). We subsequently identified another PN with similar properties but that inhibits and complexes different proteases. This chapter will summarize studies on PN-1 and PN-2 that have shown that they are protease inhibitors that regulate and clear proteases in the extracellular environment. It will also summarize studies that showed that PN-1 and PN-2 modulate cell proliferation and differentiation. Our recent studies have shown that PN-1 and PN-2 are relatively abundant in human brain and that they are altered in Alzheimer's disease.

Protease Nexin-1 (PN-1)

General Properties of PN-1

PN-1 is a 43-kDa protein that is released by a variety of cultured cells including fibroblasts, smooth muscle cells, and brain astrocytes. Studies on the purified protein in solution showed that it is an effective inhibitor of several serine proteases including thrombin, urokinase, and plasmin (Baker et al., 1980; Scott et al., 1985). However, PN-1 binds with a high affinity to the extracellular matrix (ECM) (Farrell et al., 1988) and this interaction regulates both its activity and its target protease specificity. First, binding of PN-1 to the ECM greatly accelerates its inactivation of thrombin (Farrell and Cunningham, 1986). This is due to ECM heparan sulfate since selective removal of heparan sulfate from the ECM abolishes its acceleration of the reaction between PN-1 and thrombin (Farrell and Cunningham, 1987). Second, binding of PN-1 to the ECM blocks its ability to inhibit urokinase or plasmin (Wagner et al., 1989a). Thus, in the extracellular environment, PN-1 is a specific and potent thrombin inhibitor. Indeed, it is the fastest mammalian thrombin inhibitor described. These results indicate that the physiological function of PN-1 is to inhibit thrombin. This conclusion is in accord with findings, described below, that the biological effects of PN-1 on cultured cells depend on its ability to inhibit thrombin.

Modulation of Cell Proliferation by PN-1

When we first learned that PN-1 is an effective thrombin inhibitor, we guessed that one of its functions might be to modulate thrombin-stimulated cell proliferation. This was based on findings that the mitogenic activity of thrombin requires its proteolytic activity (Glenn et al., 1980). Moreover, the cells that are most responsive to the mitogenic action of thrombin are ones that secrete little PN-1. To determine if PN-1 could play a role in the regulation of cell proliferation, purified PN-1 was added to the medium of cells under various conditions. These experiments showed that PN-1 regulates the mitogenic response to thrombin but not to EGF (Low et al., 1982). It is noteworthy that thrombin stimulates the secretion of PN-1 from cells (Eaton and Baker, 1983). This occurs after initiation of cell proliferation. Thus, the secreted PN-1 may attenuate further rounds cell division.

Regulation of Neural Cell Differentiation by PN-1

New insights into the functions of PN-1 were obtained from the amino acid sequence that was deduced from the sequence of its cloned cDNA (McGroggan et al., 1988). The deduced sequence of PN-1 is identical to the sequence of a glial-derived neurite-promoting factor that had been identified and studied by Monard and his colleagues (Gloor et al., 1986). After confirming that purified PN-1 indeed possessed neurite outgrowth activity on cultured neuroblastoma cells, we decided to pursue the mechanism by which this occurred. Starting with the finding that PN-1 is a specific thrombin inhibitor when bound to the cell surface and ECM, we determined if thrombin itself played a role in the regulation of neurite outgrowth.

These studies turned up the unexpected finding that thrombin brings about retraction of neurites on cultured neuroblastoma cells (Gurwitz and Cunningham, 1988). The effect is highly specific for thrombin since several other serine proteases including trypsin, urokinase, and plasmin do not produce neurite retraction even at much higher concentrations. Thus, the effect is not a result of general proteolysis that might, for example, simply detach neurites from the culture dish. Moreover, neurite retraction occurs at very low thrombin concentrations: a half maximal effect requires only 50 pM thrombin. Neuroblastoma cells possess specific cell surface binding sites for thrombin (Snider et al., 1986) that may mediate neurite retraction.

The ability of PN-1 to stimulate neurite outgrowth depends on its ability to inhibit thrombin (Gurwitz and Cunningham, 1988, 1990). This was shown in studies with hirudin, a specific and potent thrombin inhibitor obtained from leeches, that is structurally different from PN-1. Addition of hirudin to cultured neuroblastoma cells leads to neurite outgrowth. The kinetics and extent of neurite outgrowth produced by PN-1 and hirudin are indistinguishable. Also, removing thrombin from the culture medium of neuroblastoma cells leads to neurite outgrowth. Finally, added PN-1 does not stimulate neurite outgrowth in the absence of thrombin. Thus, thrombin brings about the primary response and the neurite outgrowth activity of PN-1 results from inhibiting thrombin.

We conducted similar experiments on astrocytes, because these cells also have processes that extend from their cell body. Frequently, a process on an astrocyte extends to a capillary and another process on the same astrocyte extends to a neuron. This has led to the suggestion that these cells may play some nutritional role for neurons. Whatever their specific role, it seems likely that these processes are critical for the normal functions of neurons. Accordingly, we examined the effects of PN-1 and thrombin on the processes of cultured early passage fetal rat astrocytes. These studies showed that thrombin retracts the processes and PN-1 leads to their extension (Cavanaugh et al., 1990). The effect of thrombin was highly specific since much higher concentrations of other proteases did not produce a similar effect. The retraction of astrocyte processes occurred at low thrombin concentrations; the half-maximal effect required only 0.5 pM thrombin. These studies showed that thrombin and PN-1 could reciprocally regulate the processes on cultured astrocytes and that this was generally similar to the regulation by these molecules of neurites on neuroblastoma cells.

In the above neuroblastoma and astrocyte cell cultures, the thrombin came from serum present in the culture medium. An important question is whether thrombin is present in brain under physiological conditions and might regulate the processes on neurons and astrocytes in vivo. This is an interesting possibility, since neurite retraction occurs during development (Schreyer and Jones, 1988). Moreover, binding sites for thrombin have been detected in primary cultures of rat brain (Means and Anderson, 1986) and in homogenates of human brain and spinal cord (McKinney et al., 1983). Experiments are in progress to determine if prothrombin mRNA can be detected in cultured neural cells or in brain. If it is, this would provide evidence for a possible regulatory role of thrombin in the brain. It is not feasible to evaluate the presence of thrombin biochemically because homogenization of the brain tissue leads to release of prothrombin from blood vessels and its subsequent conversion to thrombin.

Alterations of PN-1 in Alzheimer's Disease

In an effort to probe the possible physiological or pathological significance of the effects of PN-1 and thrombin on cultured neural cells, we measured PN-1 activity and PN-1 levels in autopsy brain samples of individuals who died of various neurological diseases. These studies showed that PN-1 activity and free PN-1 protein were reduced about 7-fold in Alzheimer's disease brain compared to brain from age-matched controls with similar postmortem times (Wagner et al., 1989b). Interestingly, Western blots using an anti-PN-1 polyclonal antibody showed that most of the Alzheimer's disease brain samples contained PN-1–thrombin complexes. These complexes were not detected in the control brain tissue. PN-1 mRNA levels in the Alzheimer's disease and control tissue appeared to be about the same. These results indicated that the decrease in PN-1 in Alzheimer's disease might be a result of increased thrombin that leads to the formation of PN-1-thrombin complexes and thus a decrease in free PN-1 and PN-1 activity. Studies are now underway to test this hypothesis. It is known that alterations in the blood–brain barrier occur in Alzheimer's disease. This could lead to extravasation of prothrombin from the circulation into the brain tissue where it would be converted to active thrombin. If this occurs in Alzheimer's disease, then the effects of thrombin on neural cells, observed in the above cell culture studies, could possibly play a role in the pathology of the disease.

Recent unpublished immunohistochemical studies on the distribution of PN-1 in normal human brain have shown that much of the PN-1 in brain is localized around capillaries and larger blood vessels. We suggest that a role of this PN-1 is to protect neurons and astrocytes from thrombin that might be produced under conditions where the blood–brain barrier is compromised. This can occur in a variety of conditions including trauma, infection, stroke, high blood pressure, and Alzheimer's disease. This suggestion is based on the findings, described above, that PN-1 is a specific thrombin inhibitor and that thrombin can retract processes on both neurons and astrocytes.

Protease Nexin-2 (PN-2)

General Properties

PN-2 is a 110-kDa protein that shares with PN-1 the property of inhibiting and clearing certain serine proteases in the extracellular environment by the mechanism shown in Figure 1 (Knauer and Cunningham, 1982; Van Nostrand and Cunningham, 1987). In other respects the protein is different. It does not share significant homology with PN-1. It retains activity after treatment with SDS or pH 1.5; these conditions inactivate PN-1. PN-2 also inactivates different serine proteases. The proteases it inhibits listed in order of decreasing effectiveness are: blood clotting Factor XIa, trypsin, chymotrypsin, EGF binding protein, and the γ-subunit of NGF (Van Nostrand et al., 1990a). PN-2 has been more difficult to study than PN-1 because it is secreted by cultured cells in lower quantities and less purified PN-2 is available for experiments.

Identity to Amyloid β-Protein Precursor

When we first purified PN-2 and reported some of its properties and the sequence of its 33 amino-terminal amino acids, a search of a protein sequence data base did not reveal similar sequences (Van Nostrand and Cunningham, 1987). During attempts to clone the cDNA for PN-2 we obtained amino acid sequence data from additional PN-2 peptides, and this was entered into a protein sequence data base. By this time, the deduced sequence of the amyloid β-protein precursor had been entered. There was complete identity between the PN-2 peptides analyzed and the amyloid β-protein precursor (Van Nostrand et al., 1989; Oltersdorf et al., 1989). The one exception was an amino acid that we reported in the amino-terminal sequence as a questionable phenylalanine (Van Nostrand and Cunningham, 1987).

The β-protein is a 4.2-kDa peptide that is the major constituent of amyloid neuritic plaques that characterize Alzheimer's disease (Glenner and Wong, 1984). The density of neuritic plaques in the brain correlates with the severity of the disease. Studies on the cloned cDNA for the β-protein showed that it is synthesized as part of a larger precursor that contains not only the β-protein domain, but also a domain that is homologous to the Kunitz family of protease inhibitors (Ponte et al., 1988; Tanzi et al., 1988; Kitaguchi et al., 1988). The amyloid β-protein precursor has a hydrophobic domain and appears to be a transmembrane protein. A secreted form of the precursor has also been identified. PN-2 appears identical to this secreted form (Van Nostrand et al.,

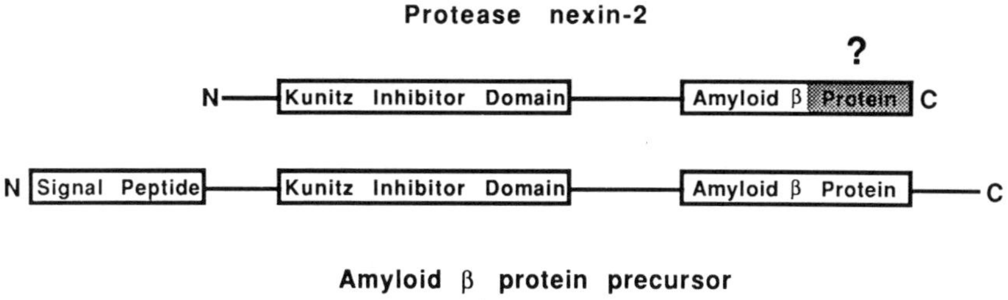

Fig. 2. Model showing identity between PN-2 and the secreted form of the amyloid β-protein precursor. The complete amyloid β-protein precursor is shown. The secreted form of it lacks the signal sequence. The precise cleavage site that gives rise to the secreted form (identical to PN-2) is unknown but appears to reside within the β-protein region as indicated.

1989; Oltersdorf et al., 1989). Figure 2 depicts the relationship between PN-2 and the amyloid β-protein precursor and makes the point that the precise cleavage site that gives rise to PN-2 is not yet known. However, PN-2 contains at least a part of the β-protein since it cross-reacts with antibodies made against synthetic β-protein. Further evidence for identity between PN-2 and the secreted form of the amyloid β-protein precursor came from the finding that an anti-PN-2 monoclonal antibody binds to neuritic plaques in Alzheimer's disease brain tissue (Van Nostrand et al., 1989).

The above findings brought together information obtained from studies on the PN-2 protein and on the cloned cDNA for the amyloid β-protein precursor. Studies are now in progress to understand the mechanisms by which the β-protein is cleaved from the precursor in Alzheimer's disease and the processes that lead to its deposition in neuritic plaques. Studies are also in progress, as described below, to understand the physiological functions of PN-2. We hope this will shed some light on altered proteolytic mechanisms that may be important in Alzheimer's disease.

Presence of PN-2 in Platelets—Possible Role in Wound Healing

At about the same time we realized that PN-2 and the amyloid β-protein precursor were identical, a report appeared that suggested a possible circulating form of the β-protein (Joachim et al., 1989). Using antibodies against the synthetic β-protein, it had not been possible to detect a source in plasma. We decided to check platelets, based on the finding that PN-1 is contained in platelets, and the general observation that PN-1 and PN-2 are frequently found in the same places. These studies showed that PN-2 is contained in the α-granules of platelets and is released upon platelet activation (Van Nostrand et al., 1990b). The platelet PN-2 was immunopurified and shown to be identical in its properties to fibroblast PN-2.

The presence of PN-2 in α-granules of platelets along with other recent findings suggest that it plays a role in wound healing. As noted earlier, the protease that PN-2 inhibits most effectively is blood clotting Factor XIa ($K_i = 3 \times 10^{-10}$ M) (Van Nostrand et al., 1990a). This suggests that PN-2 might modulate the blood clotting process at wound sites where platelets release the contents of their α-granules. PN-2 also possesses growth activity; to detect this, the target cells must be transfected with an antisense cDNA for the amyloid β-protein/PN-2 (Saitoh et al., 1989). Finally, PN-2 promotes cell adhesion (Schubert et al., 1989). Together, these results suggest that PN-2 might play several roles in the complex events that occur after tissue injury.

Conclusions

The PNs inhibit certain serine proteases and clear them from the extracellular environment by the mechanism depicted in Figure 1. In some ways PN-1 and PN-2 resemble certain plasma protease inhibitors, which also form complexes with their target proteases. The plasma protease inhibitors, however, are generally synthesized in the liver and the complexes between plasma protease inhibitors and their target proteases are cleared in the liver. In contrast, the PNs do not occur in plasma, and instead are found in certain tissues where they provide a localized mechanism to inhibit and clear regulatory serine proteases. This localized mechanism is particularly well suited for the brain where the blood–brain barrier normally prevents plasma protease inhibitors from entering the brain. Indeed, both PN-1 and PN-2 are present in brain. PN-1 stimulates the outgrowth of processes on neurons and astrocytes as a result of its ability to inhibit thrombin; thrombin brings about retraction of these processes. The physiological significance of this is not yet clear. In Alzheimer's disease brain PN-1 is reduced and thrombin-PN-1 complexes are increased. It seems likely that an alteration in the balance between PN-1 and thrombin could alter neurons and astrocytes and contribute to the pathology of the disease. Changes in PN-2/amyloid β-protein precursor also occur in Alzheimer's disease. The mechanism by which this occurs remains to be established, but alterations in the proteolysis of the precursor lead to the formation of the β-protein, which is then deposited in neuritic plaques and the cerebrovasculature. Insights into the normal physiological functions of PN-2 have come from studies that showed that it is a potent inhibitor of blood clotting Factor XIa and that it is contained in the α-granules of platelets and is released on platelet activation. This suggests that PN-2 might modulate the blood clotting process. Studies have shown that PN-2 has growth activity and that it promotes cell adhesion. Thus, it may play several roles in the complex wound healing process. Taken together, these studies underscore the fundamental roles that proteolytic mechanisms play in biological control. Altered proteolytic mechanisms are involved in several diseases (e.g., emphysema and arthritis) and might also play some role in the pathology of Alzheimer's disease.

References

Baker JB, Low DA, Simmer RL, Cunningham DD (1980): Protease-nexin: A cellular component that links thrombin and plasminogen activator and mediates their binding to cells. Cell 21:37–45.

Barsh GS, Cunningham DD (1977): Nutrient uptake and control of animal cell proliferation. In C.F. Fox (ed): Molecular Aspects of Membrane Transport. New York: Alan R. Liss, pp. 425–431.

Carney DH, Cunningham DD (1978a): Cell surface action of thrombin is sufficient to initiate division of chick cells. Cell 14:811–822.

Carney DH, Cunningham DD (1978b): Role of specific cell surface receptors in thrombin-stimulated cell division. Cell 15:1341–1349.

Cavanaugh KP, Gurwitz D, Cunningham DD, Bradshaw RA (1990): Reciprocal modulation of astrocyte stellation by thrombin and protease nexin-1. J Neurochem 54:1735–1743.

Chen LB, Buchanan JM (1975): Mitogenic activity of blood components. I. Thrombin and prothrombin. Proc Natl Acad Sci USA 72:131–135.

Cunningham DD (1972): Changes in phospholipid turnover following growth of 3T3 mouse cells to confluency. J Biol Chem 247:2464–2472.

Cunningham DD, Pardee AB (1969): Transport changes rapidly initiated by serum addition to "contact inhibited" 3T3 cells. Proc Natl Acad Sci USA 64:1049–1056.

Eaton DC, Baker JB (1983): Phorbol ester and mitogens stimulate human fibroblast secretions of plasmin-activatable plasminogen activator and protease nexin, an antiactivator/antiplasmin. J Cell Biol 97:323–328.

Farrell DH, Cunningham DD (1986): Human fibroblasts accelerate the inhibition of thrombin by protease nexin. Proc Natl Acad Sci USA 83:6858–6862.

Farrell DH, Cunningham DD (1987): Glycosaminoglycans on fibroblasts accelerate thrombin inhibition by protease nexin. Biochem J 245:543–550.

Farrell DH, Wagner SL, Yuan RH, Cunningham, DD (1988): Localization of protease nexin-I on the fibroblast extracellular matrix. J Cell Physiol 134:179–188.

Glenn KC, Carney DH, Fenton JW II, Cunningham DD (1980): Thrombin active-site regions required for fibroblast receptor binding and initiation of cell division. J Biol Chem 255: 6609–6616.

Glenner GG, Wong CW (1984): Alzheimer's disease and Down's syndrome: Sharing of a unique cerebrovascular amyloid fibril protein. Biochem Biophys Res Commun 122:1131–1135.

Gloor S, Odink K, Guenther J, Nick H, Monard D (1986): A glia-derived neurite promoting factor with protease inhibitory activity belongs to the protease nexins. Cell 47:687–693.

Gurwitz D, Cunningham DD (1988): Thrombin modulates and reverses neuroblastoma neurite outgrowth. Proc Natl Acad Sci USA 85:3440–3444.

Gurwitz D, Cunningham DD (1990): Neurite outgrowth activity of protease nexin-1 requires thrombin inhibition. J Cell Physiol 142:155–162.

Joachim CL, Mori H, Selkoe D (1989): Amyloid β-protein deposition in tissues other than brain in Alzheimer's disease. Nature (London) 341:226–230.

Kitaguchi N, Takahashi Y, Tokushima Y, Shiojiri S, Ito H (1988): Novel precursor of Alzheimer's disease amyloid protein shows protease inhibitory activity. Nature (London) 331: 530–532.

Knauer DJ, Cunningham DD (1982): Epidermal growth factor carrier protein binds to cells via a complex formed with released carrier protein nexin. Proc Natl Acad Sci USA 79:2310–2314.

Low DA, Baker JB, Koonce WC, Cunningham DD (1981): Released protease-nexin regulates cellular binding, internalization and degradation of serine proteases. Proc Natl Acad Sci USA 78:2340–2344.

Low DA, Scott RW, Baker JB, Cunningham DD (1982): Cell regulate their mitogenic response to thrombin through release of protease nexin. Nature (London) 298:476–478.

McGrogan M, Gohari J, Li M, Hsu C, Scott RW, Simonsen CC, Baker JB (1988): Molecular cloning and expression of two forms of human protease nexin 1. Bio/Technology 6:172–177.
McKinney M, Snider RM, Richelson E (1983): Thrombin binding to human brain and spinal cord. Mayo Clin Proc 58:829–831.
Means ED, Anderson DK (1986): Thrombin interactions with central nervous system tissue and implications of these interactions. Ann NY Acad Sci 485:314–322.
Naiditch WP, Cunningham DD (1977): Hexose uptake and control of fibroblast proliferation. J. Cell Physiol 92:319–332.
Oltersdorf T, Fritz LC, Schenk DB, Lieberburg I, Johnson-Wood KL, Beattie FC, Ward PJ, Blacher RW, Dovey HF, Sinha S (1989): The secreted form of the Alzheimer's amyloid precursor protein with the Kunitz domain is protease nexin-II. Nature (London) 341:144–147.
Ponte P, Gonzalez-DeWhitt P, Schilling J, Miller J, Hsu D, Greenberg B, Davis K, Wallace W, Liebergurg I, Fuller F, Cordell B (1988): A new A4 amyloid mRNA contains a domain homologous to serine proteinase inhibitors. Nature (London) 331:525–527.
Saitoh T, Sundsmo M, Roch JM, Kimura N, Cole G, Schubert D, Oltersdorf T, Schenk DB (1989): Secreted form of amyloid p-protein precursor is involved in the growth regulation of fibroblasts. Cell 58:615–622.
Schreyer D, Jones EG (1988): Axon elimination in the developing corticospinal tract of the rat. Dev Brain Res 38:103–119.
Schubert D, Jin LW, Saitoh T, Cole G (1989): The regulation of amyloid β-protein precursor secretion and its modulatory role in cell adhesion. Neuron 3:689–694.
Scott RW, Bergman BL, Bajpai A, Hersh RT, Rodriguez H, Jones BN, Barreda C, Watts S, Baker JB (1985): Protease nexin: Properties and a modified purification procedure. J Biol Chem 260:7029–7034.
Snider RM, McKinney M, Richelson E (1986): Thrombin binding and stimulation of cyclic guanosine monophosphate formation in neuroblastoma cells. Semin Thromb Hemostasis 12:253–262.
Tanzi RE, McClatchey AI, Lamperth ED, Villa-Komaroff L, Gusella JF, Neve RL (1988): Protease inhibitor domain encoded by an amyloid protein precursor mRNA associated with Alzheimer's disease. Nature (London) 331:528–530.
Thompson JA, Cunningham DD (1987): Identification of cell surface proteins sensitive to proteolysis by thrombin. Methods Enzymol 147:157–161.
Thompson JA, Lau AL, Cunningham DD (1987): Selective radiolabeling of membrane proteins to a high specific activity. Biochemistry 26:743–750.
Thrash CR, Cunningham, DD (1974): Dissociation of increased hexose transport from proliferation in density-inhibited 3T3 mouse fibroblasts. Nature (London) 252:45–47.
Van Nostrand W, Cunningham DD (1987): Purification of protease nexin-II from human fibroblasts. J Biol Chem 262:8508–8514.
Van Nostrand WE, Wagner SL, Farrow JS, Cunningham DD (1990a): Immunopurification and proteinase inhibitory properties of protease nexin-2/β-amyloid precursor protein. J Biol Chem 265:9591–9594.
Van Nostrand WE, Schmaier AH, Farrow JS, Cunningham DD (1990b): Protease nexin-2/amyloid β-protein precursor circulates in plasma as a platelet α-granule protein. Science 248:745–748.
Van Nostrand WE, Wagner SL, Suzuki M, Choi BH, Farrow JS, Geddes JW, Cotman CW, Cunningham DD (1989): Protease nexin-2, a potent anti-chymotrypsin, shows identity to β-amyloid precursor protein. Nature (London) 341:546–548.
Wagner SL, Lau AL, Cunningham DD (1989a): Binding of protease nexin-I to the fibroblast surface alters its target proteinase specificity. J Biol Chem 264:611–615.
Wagner SL, Geddes JW, Cotman CW, Lau AL, Isackson PJ, Cunningham DD (1989b): Protease nexin-I, a protease inhibitor with neurite outgrowth activity, is reduced in Alzheimer's disease. Proc Natl Acad Sci USA 86:8284–8288.

Perspectives on Cellular Regulation: From Bacteria to Cancer, pages 185–199

How Regulative Is Early Amphibian Development?

John Gerhart

Department of Molecular and Cell Biology, University of California, Berkeley, California 94720

Metabolic Regulation

While an undergraduate in 1958, I had the chance to attend Boris Megasanik's lectures on bacterial metabolism at Harvard Medical School. This was still the heyday of intermediary metabolism: many pathways of degradation and biosynthesis had been worked out, hundreds of reactions had been defined and their enzymes partially purified. Layered on top of the metabolic chart were the first inklings of regulatory mechanisms integrating and adapting the flow of carbon through this network of reactions: the Novick–Szilard effect, (the sparing of metabolic intermediates when the end product is externally provided to the cell), enzyme induction, catabolite repression, and enzyme derepression. "Feedback inhibition in vitro," that is, the inhibition in cell-free extracts of the initial biosynthetic enzyme of a pathway by its endproduct, had just been achieved by Richard Yates and Art Pardee for the pyrimidine pathway, and by Ed Umbarger for the isoleucine–valine pathway. Genetics was making its first inroads into proper biochemistry in the use of bacterial auxotrophic mutants to deduce pathways and the selection of mutants constitutive for the expression of inducible enzymes, β-galactosidase and penicillinase being the leading examples. Boris described a remarkable experiment and its result, not yet published, of transient expression of β-galactosidase during DNA transfer between mating *Escherichia coli*. This was the first direct evidence for a repressor, a gene product, maybe a protein, preventing the expression of the *lacZ* gene. The methods combined biochemistry and genetics in a new way: mutants, matings, and enzyme assays. This was of course Art's sabbatical work at the Pasteur Institute, soon known as the Pardee–Jacob–Monod or PAJAMO experiment. When I mentioned to Boris my plan for graduate study at Berkeley, he suggested I make every possible effort to work with Art.

I indeed asked for a place in Art's laboratory. He suggested that it would be timely to scrutinize the feedback inhibition of aspartate transcarbamylase

(ATCase), the initial enzyme of pyrimidine biosynthesis. Margaret Shephardson, a postdoctoral fellow with Art, had just purified several milligrams of the enzyme from derepressed bacteria. She was returning to England and it fell to me to study the kinetics of inhibition of this purified preparation. Nucleoside triphosphates had recently become available (only monophosphates such as CMP had been available to Yates) and it was easy to show in 1960 that CTP was a strong inhibitor of the enzyme, that ATP was an activator, and that GTP was a peculiar partial inhibitor. However, the substrate kinetics were odd: inescapably sigmoidal. In an effort to control conditions, I prepared a high dilution of the purified enzyme and kept it in the refrigerator from day to day to use for assays. Indeed after a few days the kinetics and activity stabilized but to my chagrin the CTP inhibition disappeared, while fresh dilutions of the enzyme showed high sensitivity to the inhibitor.

Just then Art returned from a seminar trip to the East Coast, where he had talked with Harris Moyed and Ed Umbarger about the possibility that enzymes capable of feedback inhibition might use separate sites for binding substrates and regulators. I told Art my experience with the loss of CTP inhibition but not enzymatic activity, and we realized that this is just what one might expect if ATCase had separate sites. I then tried intentionally to destroy the inhibitor sites by reacting freshly diluted enzyme with mercurials. On the first try, *p*-hydroxymercuribenzoate (pHMB) completely densensitized the enzyme to CTP inhibition without destroying catalytic activity; in fact, activity doubled and the kinetics changed from sigmoidal to the normal Michaelis–Menten type. ATCase might indeed have two kinds of sites! Specialized protein structure might underlie this specialized function. ATCase might really resemble other regulated proteins such as hemoglobin. And maybe metabolic regulation did arise without regard to the biochemical details of individual enzymatic reactions. What a wonderful research project Art had introduced me to!

I wrote a draft manuscript for publication; Art read it and marked on the front page: ''Is English your first language?'' I deconvoluted the prose, and we sent the manuscript to the *Journal of Biological Chemistry,* only to have it rejected on the grounds that we should continue our studies until we could make some of the irregularities go away. I think Art said it was the first manuscript of his that had been rejected. It did finally gain acceptance and was published in 1962 (Gerhart and Pardee, 1962) as ''The enzymology of control by feedback inhibition.''

Recognizing the competitive nature of research, Art advised me to talk at meetings and to write articles. He cautioned me to think of a good name for these two-sited regulatory enzymes. The best I could do was: ''regulatory and active sites'' and ''regulatory enzymes.'' It turned out that J.-P. Changeux, then a student in Jacques Monod's laboratory, was thinking along similar lines about the first enzyme of the isoleucine–valine pathway, and as we all know, the Changeux, Monod, and Jacob chapters of the 1961 Cold Spring Harbor Symposium introduced the term ''allosteric enzymes.''

When Art left Berkeley in 1961 to organize a new biochemical sciences program at Princeton, I went with him to finish my thesis work. Soon thereafter, an assistant professor position opened at Berkeley in a newly formed Department of Virology headed by Wendell Stanley, and Art very kindly recommended me for that position. I returned in 1962 and took up residence in one of Art's old laboratories. Art was very generous about ATCase research and allowed me to take the project to Berkeley. There I returned to the pHMB desensitization of the enzyme, to explore the changes in protein structure attending the changes in behavior. I knew from sucrose gradient centrifugation that the catalytically active unit decreased its sedimentation rate by a factor of two when the enzyme was desensitized. This seemed well worth further study. In collaboration with Howard Schachman at Berkeley, we found in 1964 that the mercurial-treated enzyme dissociated into two protein subunits of different size, one binding substrates (its catalytic subunit) and one binding the regulatory molecules, CTP and ATP (its regulatory subunit). Thus, the two kinds of sites were located on two kinds of proteins that interacted in a complex to generate feedback inhibition and sigmoidal substrate kinetics. I assumed Art would be a co-author of the early papers about the separation and characterization of these subunits, but he declined. By 1964 he was making major contributions to the understanding of bacterial transport proteins (''permeases'') and of the bacterial cell cycle.

Art's interest in the regulation and integration of the cell's complex chemistry was unusual at the time. Others considered it good science to define the substrates and products of a new metabolic reaction, to purify the enzyme just enough to verify the absence of side reactions, and to work gradually toward an understanding of the reaction mechanism. I preferred Art's emphasis, which seemed the opposite in direction; to go from the cell's chemical details outward to its global coordination, adaptation to new conditions, self-restoration after perturbation, and dynamic organization, namely, its life-like properties. Art was using chemistry to explore biology.

Soon I had to choose my own direction of research toward chemistry or biology, since the analysis of allosteric enzymes was heading toward rapid reaction kinetics, X-ray crystallography, amino acid sequencing, and the physical chemistry of conformational changes. Howard Schachman was making incisive progress on the conformational changes of ATCase, and indeed under his direction in the past 20 years this has become, of all the examples, the most completely understood allosteric enzyme (hemoglobin not being an enzyme). All this emerged from Art's early studies of feedback inhibition in vitro. For my part, I felt free by 1971 to begin to pursue new aspects of regulation that might be hidden in developmental biology.

Developmental Regulation in Vertebrates

How are the activities of individual cells integrated in a developing multicellular embryo? Might there be a parallel to the integration of individual enzyme activities in the adaptable and self-adjusting network of metabolism? I chose to study the early development of the frog, *Xenopus laevis,* which like other amphibia and other vertebrates is said to display "regulative development" or "developmental regulation," namely an ability of the embryo to generate a near-normal larva despite the removal or addition of parts of preceding developmental stages. In the earliest experiments of embryology a century ago, W. Roux obtained evidence against regulative development and H. Driesch obtained evidence for it. Roux blocked the cleavage and development of one cell of a two cell frog embryo by poking the cell with a hot needle. The other cell continued cleaving and developed into a full sized right or left half embryo, the very part it would have produced in the undamaged egg. Roux concluded that egg halves already differ at the two cell stage, each limited to the development of a lateral half of the organism. He suggested that in general an organism develops as a "mosaic" in which each part of the egg independently develops into one small specific part of the embryo, the mere sum of the development of these prearranged parts. Driesch, following a different experimental plan, separated the first two cells of a sea urchin egg and found that each developed a half-sized *whole* embryo. Thus, the separate egg halves produced a wider variety of embryonic parts than they would have in the intact egg. Frog eggs were soon thereafter separated at the two-cell stage with the same result. In related experiments, two two-cell frog eggs were pressed together to make one large four-cell egg, and in some cases a single double-sized embryo resulted; here, each of the four cells gave less of the total embryonic pattern than it would have in a normal egg. Under these conditions, parts of the egg seemed able to adjust their development to restore the entirety of embryonic pattern (see Gerhart, 1989, for a review).

I wanted to study "regulative development" as exemplified by this famous case, but alas, the phenomenon may amount to less than the word implies. To me, regulative development implies the restoration of some complex and well-organized system in the egg: that different materials are arranged precisely in the egg, that this arrangement is essential for later development, that it is disrupted by the operations of halving or doubling, and that it is reestablished thereafter. These assumed qualities would place developmental regulation alongside metabolic regulation, which is known to adjust itself to new conditions and to restore itself after imbalances are caused. However, the frog egg seems to have a very limited pattern in the first place, and what it does have, it restores poorly after a disturbance. In choosing the term "regulative development," early embryologists may have assumed that the egg starts development with a cryptic organization nearly as complex as that of the finished embryo.

The alternative view to this was ''epigenesis,'' according to which the egg starts with little or no organization and increases its complexity gradually during development. This was highly implausible at the time, though quite acceptable now. If there is hardly any organization at the early developmental stages, there is developmental regulation to accomplish after experimental perturbations.

What is the limited organization of the unfertilized frog egg and how is it restored in certain circumstances after the egg is halved or doubled? In the animal–vegetal direction, the egg's hemispheres differ in their constituents and developmental abilities: the animal hemisphere gives only ectodermal and mesodermal embryonic parts while the vegetal hemisphere gives only endodermal parts of the embryo. These differences cannot be altered when the hemispheres are separated. In an orthogonal direction, the constituents located along any of the meridians of the surface of the unfertilized egg are equally suited for dorsal, lateral, or ventral development, in keeping with the egg's cylindrical symmetry around its animal–vegetal axis. However, soon after fertilization the egg generates a unique bilateral symmetry and a dorsoventral axis by rotating its rigid cortical layer of cytoplasm (the egg surface) in an animal–vegetal direction over its rigid sphere of deeper cytoplasm, along tracks of parallel microtubules in the vegetal hemisphere (Gerhart et al., 1989). During rotation, the egg increases its two part hemispheric organization to three parts: on the meridian along which inner materials move farthest toward the vegetal pole, a narrow sector of the vegetal hemisphere becomes activated to form the ''dorsal vegetal region'' (DVR). This region's position, which is just below the externally visible ''grey crescent'', straddling the bilateral plane of the egg, predicts accurately the future dorsal midline of the embryo. The remainder of the vegetal hemisphere is unaffected by rotation and is known as the ''ventral vegetal region'' (VVR). The animal hemisphere is also unaffected by rotation.

Later when the DVR has cellularized, it secretes protein signals, probably members of the TGF-β growth factor family, which induce nearby animal hemisphere cells to form the dorsal marginal zone, or ''organizer,'' a source of inductive signals at a subsequent stage. Thus, dorsal development depends on cortical rotation in the fertilized egg to found the DVR, which in turn is needed in the blastula stage to found the organizer. In a parallel manner, the VVR later secretes other protein signals, perhaps members of the FGF growth factor family, which during the blastula stage induce nearby animal hemisphere cells to form the ventral marginal zone. The animal cap remains as the region of the animal hemisphere unaffected by these vegetal signals. Thus, by induction, the three part organization of the rotated egg becomes the five part organization of the late blastula (to be discussed later).

Returning to the Roux–Driesch experiments, we can note that two half-sized complete embryos are obtained from separated cells only if the egg has cleaved on the bilateral plane. Each half egg will then contain half the DVR, half the VVR, and half the animal hemisphere. But if the cleavage plane departs 30°

or more from the bilateral plane, one cell gets the entire DVR and the other gets none. The former develops a quite normal embryo, perhaps slightly enlarged at the anterior end, whereas the other develops a "ventralized embryo" containing only ventral cell types and organization: red blood cells, coelomic cavities, a simple gut, and ciliated epithelium, arranged with cylindrical symmetry around the original animal–vegetal axis. Thus, regulative development occurs only in the special circumstance when both egg halves receive a sampling of all three cytoplasmic regions present in the intact egg. This is a modest amount of pattern, and it has little or no capacity to regulate. To understand the apparent regulation in the special case of bilaterally split eggs, we need to discuss the "organizer," the essential signalling center of the gastrula.

The Organizer

Vertebrates are extreme in the extent to which egg organization is remodeled into embryonic organization by morphogenesis. The body axis of the vertebrate embryo arises practically de novo during gastrulation when the 20,000 cells that have cleaved from the egg begin to migrate, change shape, change neighbors, and repack into new arrays. Six hours later the anteroposterior and dorsoventral organization of the embryo is largely complete. Of the five part pattern of the late blastula, three parts provide the initial conditions for generating embryonic organization during gastrulation: (1) the dome-shaped animal cap ($N = 8000$ cells), and (2) and (3) the annulus shaped marginal zone containing two subregions, the dorsal marginal zone (the "organizer," $n = 2000$ cells arranged 20 cells high, 25 cells wide, and 4 cells deep) and the ventral marginal zone ($n = 6000$). The interactions of cells of these three zones will generate the body axis.

The organizer is a unique chordate strategem of development, and its cells usually differentiate as the notochord, a unique chordate anatomical structure. The notochord is surrounded by the neural tube and somites in the chordate body axis. Inductive signals from the organizer seem to guarantee this tripartite arrangement. Three special functions are ascribed to organizer cells: (1) They signal nearby ventral marginal zone cells to develop into somites (the source of skeletal muscle, bone, and dermis) and kidney, a process called "dorsalization of the ventral marginal zone." (2) They signal nearby animal cap cells to initiate neural development, a process called "neuralization" or "primary induction" of the ectoderm. (3) They signal neighboring cells of the ventral marginal zone and ectoderm to converge toward them on the dorsal midline and to repack into extended arrays, lengthening the embryonic axis; this is organizer-dependent morphogenesis. These functions probably involve numerous intercellular signals of positive and negative sorts, about which little is known.

In light of this information, the difference of the Roux–Driesch results probably reflects the two different geometric contexts in which the organizer was allowed to act. Roux on the one hand left the dead cell in place while the cleaving half went on to form a hemigastrula. Since the first cleavage plane usually coincides with the bilateral plane in frog eggs, the hemigastrula possessed a half organizer abutting the dead cell. In this case, responsive cells of the ventral marginal zone and animal cap are available to the organizer on only *one side*. In the Driesch case, the two cells were separated and each took on a spherical shape. In each, the half organizer was surrounded on *both sides* by responsive cells of the animal cap and marginal zone. The embryo thereafter developed bilaterally. (The organizer itself has no inherent bilaterality, as will be discussed later.) Pattern regulation in the Driesch case seems mostly to involve the restoration of bilateral symmetry to each egg half by the act of their rounding up after separation and by conversion of the newly exposed cleavage membrane to a tough egg surface. The blastocoel then forms internally as cleavage continues, and the organizer later finds responsive cells on both its sides. It thereafter induces a bilateral body axis. This reduces much of the mystery of the Roux–Driesch paradox.

Still, mystery abounds in the fact that the organizer manages to establish an anteroposterior and dorsoventral pattern within the bilateral context of responsive cells even in a normal embryo. How is this done, and can true regulative development be found at this stage? The understanding of vertebrate development still awaits insights into the formation and function of the organizer, 75 years after its discovery by Spemann and Mangold (Spemann, 1938). In attempting to tease out a few aspects of the problem here, I have chosen to ask the following: (1) How organized is the organizer at the start of gastrulation, and how organized at the end, in terms of the types and distributions of signals it emits? (2) How does it acquire and change this signal-emitting organization? (3) How does it impose its inductive pattern on neighboring cells? These questions are far from answered, but enough information about the organizer is available to allow the formulation of proposals. To this end I will outline a crude model for organizer function and dynamics, based on the assumption that the marginal zone consists initially of two distinct parts that will interact across their common boundary to generate additional parts. The assumed rules of interaction concern the propagation of signals by cells of these parts and the cells' responses to these signals.

Although by the end of gastrulation the marginal zone will contain 8000 cells and comprise all the mesoderm and much of the endoderm, it is far from complete at the start of gastrulation. There are probably only a few of these cells initially committed to such fates, namely, those already having received short range inductive signals from the vegetal hemisphere. As shown in the top panel of Figure 1, the lowest part of the marginal zone might contain two cell types, O cells and V cells designating *O*rganizer and *V*entral marginal zone cells induced, respectively, by the DVR and VVR. For the rest of the marginal zone

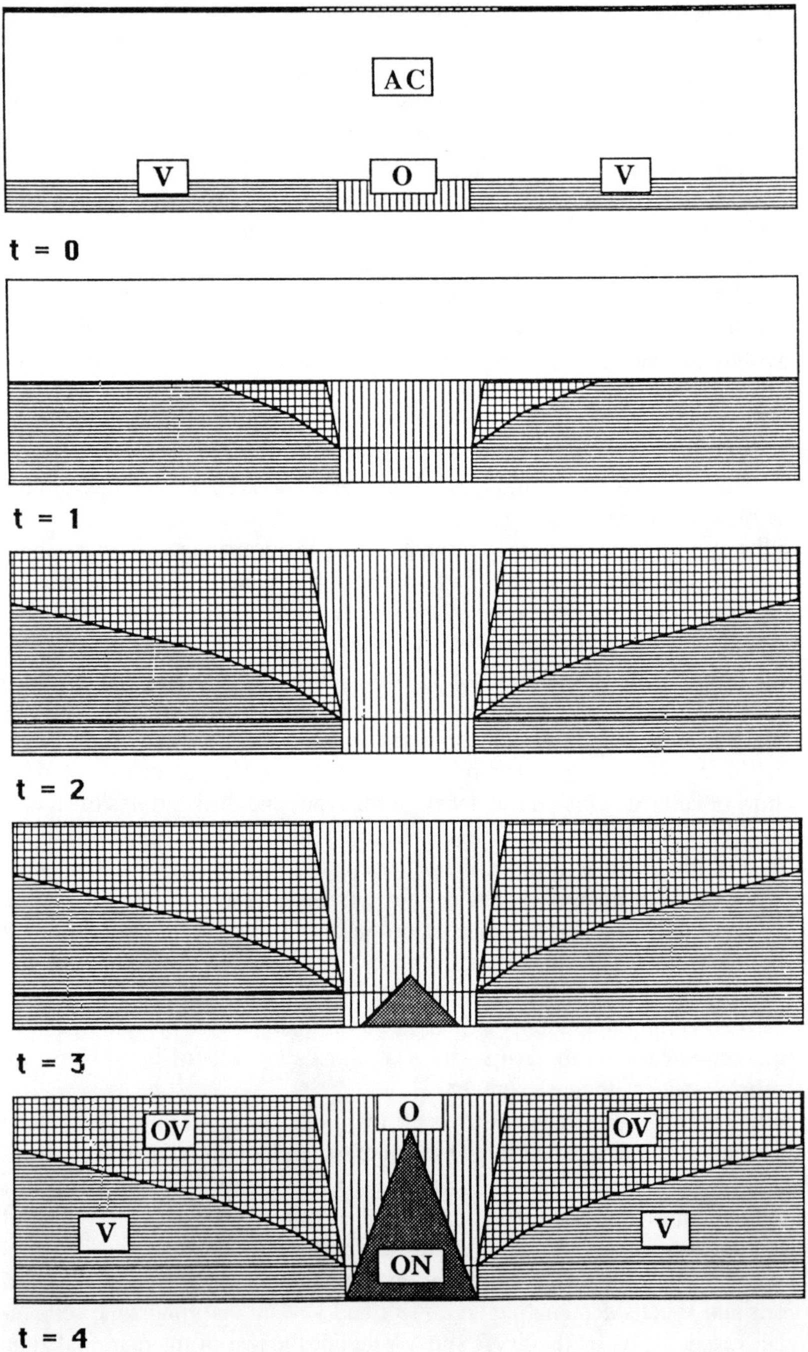
AC
V
O
V
t = 0
t = 1
t = 2
t = 3
O
OV
OV
V
V
ON
t = 4

(marked AC in Figure 1) to reach its full pattern, O and V cells are proposed to release O and V inductive signals, which reach the next row of yet-uninduced animal hemisphere cells, converting them also to marginal zone cells of the O, V, or yet other types. Converted cells would themselves become inductive, releasing O or V signals or both, and so the signals would propagate in the animal and lateral directions.

The patterning problem to be solved by the spreading of signals and conversion of cells is twofold: (1) to generate a new cell type, the somite cells, at the border between newly forming O and V cells. This is the "dorsalization of the marginal zone." And (2) to generate a subset of O cells, which release a neuralizing signal (ON cells), and have them be located mostly near the vegetal (lower) border. In this limited treatment of the problem,

Fig. 1. Dorsalization and neuralization by the amphibian organizer. The marginal zone is shown as a rectangle of tissue cut from the ring-shaped marginal (equatorial) zone of the early gastrula and opened at the prospective ventral midline, which now lies on the right and left edges of the rectangle. The prospective dorsal midline lies on the vertical centerline. Five times are shown at which the patterning of the marginal zone increases in complexity. At $t=0$, the lowest part of the zone has already been patterned by inductions from the vetegal hemisphere: the center part is a strong dorsalizing–mesodermalizing cell population (O cells, see text) established by dorsal vegetal cells (perhaps by an activin-type induction). O cells later become head mesoderm and notochord if there are no interferences with subsequent steps of their development. Lateral parts (V cells) are prospective ventral mesoderm, established by ventral vegetal cells (perhaps by FGF-type inductions). These two regions will now emit O and V signals, which pass vertically and diagonally upward, affecting nearby marginal zone cells that are not yet induced to form mesoderm. By $t=1$, signals have spread halfway up. The O signal converts cells to O cells unless a V signal is received at about the same time, in which case it becomes an OV cell (prospective somite mesoderm), which emits O and V signals. OV cells tend to promote the formation of more OV cells. If a cell receives only V signals, it later becomes lateral plate mesoderm. Cells become refractory to further signaling at some time after receiving signals. Since the O signal spreads faster than the V signal, the OV (somite) domains expand laterally, eventually reaching to the ventral midline. The notochord region expands slightly laterally also. By $t=2$ the patterning of the mesoderm is complete. In $t=3$, the initial O region has begun to convert to an ON region: O cells lose the dorsalizing signal function and begin to emit a neuralizing signal. This is a spontaneous coversion open to all O cells, but one that can be blocked by OV cells, which also prevent their own conversion. As shown in $t=4$, the upper region of the O cell domain is unable to gain neuralizing signaling and stays an O source. Thus, neuralizing signals are most strongly emitted from the leading (early) part of the O domain, and progressively more weakly in the trailing (later) parts. The circuitry of interaction can be represented as:

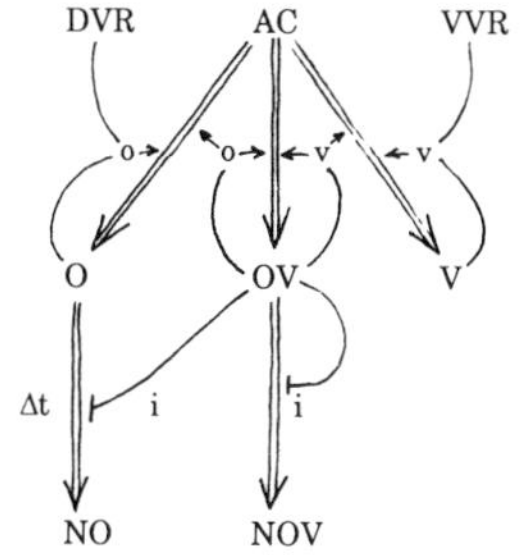

it will be necessary to ignore morphogenesis of the marginal zone, even though cell movement and the repacking of cells in populations are important responses to intercellular signals.

Several "rules of interaction" can be proposed for the response of cells to O and V signals. The first category of these concerns dorsalization of the marginal zone, as illustrated in the top three panels of Figure 1. First, a cell receiving only V signals becomes a V cell, releasing V signals. These cells later differentiate as lateral plate mesoderm. Second, a cell receiving only O signals will become an O cell, releasing O signals. These later differentiate as notochord and head mesoderm. And third, a cell receiving both O and V signals will become an OV cell, releasing both signals and later differentiating as somite mesoderm. Since both signals are needed to form OV vells, the somite territory will lie between the other two.

The O signal can be arbitrarily said to spread more rapidly in a lateral direction than the V signal, as both signals spread in the animal direction, so that OV cells will come to occupy two wing-shaped territories separating the O and V territories, and the O territory will have widened slightly. This pattern is a reasonable approximation of the fate map for mesoderm cell types. Thus, the organizer may serve as the source of a signal that acts on animal cap cells in conjunction with the V signal to form somites (dorsalization) or that acts on animal cap cells alone to form more notochord (mesodermalization). While the point is not made use of here, it is possible that late-induced cells (those at the top of the marginal zone) differ from those induced early, by virtue of age from the time of fertilization, thus giving an anterior–posterior difference to all territories.

Now we come to neural induction, the organizer's second function. As shown in the bottom two panels of Figure 1, we use a new rule of interaction to increase further the signaling complexity of the marginal zone: O cells in time convert spontaneously to ON cells, which release a neuralizing inductor onto cells of the overlying ectoderm of the animal cap. (During the previous period of dorsalization and mesodermalization, the marginal zone has moved underneath the animal cap.) However, OV cells produce an inhibitor that blocks this conversion, keeping O cells in their dorsalizing–mesodermalizing state. Since there are many more OV cells in the later-formed part of the marginal zone, they prevent the conversion and thereby reduce the release of neuralizing signals from that region. Only the early-formed part of the O territory becomes highly neuralizing, and the induction strength grades off in the animal direction. OV cells do not become neuralizing since they inhibit their own conversion.

This ad hoc model is infinitely pliable since rules can be added whenever new cell behaviors are needed in specific regions of the marginal zone. Recent work on the anteroposterior patterning of *Drosophila* embryos indicates that rules of interaction operate at all steps of a long series of patterning events.

What experimental results can be explained by this outline for amphibian organizer function?

1. The initial dorsalizing–mesodermalizing organizer: In several of the classical models (reviewed by Toivonen, 1978), this function is not discussed whereas neuralization is emphasized (see Yamada, 1950, as an exception, though). However, Hama, Kaneda, Suzuki, and others have obtained data (mostly from urodele amphibians) over the past 30 years supporting the idea that the organizer starts gastrulation with no neuralizing activity but with a strong mesodermalizing activity in its leading (most vegetal) part and with no inductive activity at all in its posterior (most animal) part (see Kaneda, 1981; Kaneda and Suzuki, 1983; Suzuki et al., 1984). This information has been incorporated into the model of Figure 1. The initial organizer concerns itself with the patterning the mesoderm first: it ''dorsalizes'' more lateral mesoderm to become somites, and mesodermalizes the posterior part of the organizer to gain immediate signaling activity and the ability to differentiate eventually to notochord. This dorsalizing–mesodermalizing function is expected to propagate by cell to cell signaling in the plane of the marginal zone tissue (so-called homeogenetic induction). This function continues for at least a quarter of the gastrulation period, while the leading part of the organizer involutes beneath the posterior part. Thereafter the leading part of the organizer comes to underlie prospective neural cells and the anterior organizer might be expected to lose dorsalizing mesodermalizing activity and gain neuralizing activity. The rate of this change will be discussed later.

2. The final neuralizing organizer: By 1950–1960 it became accepted that by the end of gastrulation the organizer might have *two* spatially arrayed signals, a neuralizing signal emitted from the anterior extreme and a mesodermalizing signal from the posterior extreme, both acting on the overlying animal cap ectoderm (Toivonen, 1978). According to various proposals, these might be nonoverlapping (Nieuwkoop, 1985) or one or both graded into the other (Saxen and Toivonen, 1961; Hamburger, 1988, respectively). A pure neuralizing signal would induce the animal cap ectoderm to form anterior dorsal neural structures (i.e., the head) whereas a pure mesodermalizing signal would induce ectoderm to form posterior dorsal neural plate structures, namely the tail. In amphibians, the posterior part of the neural plate does indeed form mesodermal components of the tail. By these classical models, no trunk neural inducer is needed: trunk neural structures would result from the combined effect of both inducers, either as a steady-state mix or as a time-averaged exposure of ectoderm cells to the different inducers passing by in gastrulation. The model of Figure 1 accepts the notion of a two part signal for neuralization and generates such an arrangement of inducers. It does not exclude the possibility that some neuralizing signals pass in the plane of the ectoderm while others cross tissue layers.

3. The switch from dorsalizing–mesodermalizing signals to neuralizing signals: Hama, Kaneda, and Suzuki favor the possibility that leading organizer cell population switches abruptly to full neuralizing activity as it reaches the prospective neural tissue, to become a full fledged two-part organizer (neuralizing–dorsalizing) which would follow the operations proposed by Nieuwkoop (1985) for an ordered two-step inductive process: first, all prospective neural plate tissue must first be neuralized, and second, it may be mesodermalized in which case it forms trunk and tail parts of the plate in relation to increasing amounts of the signal. Only neuralized but *not* mesodermalized neural plate cells can form head structures.

However, it is not known how fast this switch of signals is made by the leading organizer, and whether there is an obligatory order in which signals must be received by the responding tissue. As done in Figure 1, one could postulate the opposite: as the leading organizer moves forward under the ectoderm's surface, it might *gradually* lose dorsalizing activity and gain neuralizing activity, slowly becoming trunk inductive (mixed neural and mesodermalizing signals) and only later becoming head inductive (strong neuralizing activity but no mesodermalizing activity), while more posterior parts of the organizer would have just become trunk inductive by the end of gastrulation. If a pure neuralizing source arises only at the end of gastrulation, the head could not be induced until the time when leading organizer cells reach the prospective head end of the neural plate. By this model of gradual change, each part of the neural plate is underlain in succession by passing inductive sources that are inductive of just that level, not of more anterior levels. Also, a mesodermalizing signal is allowed to precede a neuralizing signal in trunk-tail parts of the neural plate. It is not clear why the older models preferred to have the neuralizing signal received prior to the mesodermalizing one.

4. Wide organizers: When eggs are treated with D_2O before cortical rotation or with LiCl at the early blastula stage, they develop to cylincrically symmetric embryos with a radial sucker and a band of eye pigment, a large heart, and a large core of notochord. They lack not only ventral structures but also dorsal structures of the trunk and tail mesoderm and neural plate; they are "dorsoanteriorized," the opposite of the ventralized–posteriorized embryos described before. The organizer, which is 60° wide in a normal embryo, occupies the entire 360° of these treated embryos (Kao and Elinson, 1988). Vice versa, they contain no ventral marginal zone material. When there is no ventral marginal zone material for the organizer to act upon, somites are not formed. This is incorporated in the model by the dependence of somites on both O and V signals. And when OV cells are absent, all O cells convert to neuralizing sources, with the result that only anterior neural plate structures are induced.

As shown by Kao and Elinson (1988), when a 60° wide piece of organizer from a treated embryo is grafted to the ventral marginal zone of a normal embryo, it induces somites and a full anteroposterior axis. This behavior is allowed by the model of Figure 1: some O cells are prevented by OV cells from switching to ON cells, that is, part of the organizer is kept in a trunk-tail inductive state.

5. Narrow organizers: The organizer of the *Xenopus* late blastula is approximately 60° wide. When this is removed surgically at the late blastula stage, the embryo develops only ventral structures (e.g., red blood cells and coelomic cavities). This shows that the ventral marginal zone cells and the animal cap cells depend on signals from the organizer to develop as somite and neural plate, respectively.

We have recently prepared embryos with quarter and half width organizers to see the subsequent effect on embryonic organization (Stewart and Gerhart, 1990). An embryo with a quarter organizer (15° wide, taken from either edge or from the center) develops a bilateral tadpole in which, on average, the body axis is missing dorsal structures anterior to the ear while the trunk and tail are approximately normal. Ventral parts are overdeveloped. An embryo with half an organizer (30° width) gives a bilateral tadpole, which is more axially complete but it is still missing forebrain and some midbrain structures, on average. Thus, in the range from none to one-half an organizer, we find that the anteroposterior as well as dorsoventral completeness of the dorsal structures of the body axis depends on the mediolateral width (quantity) of the organizer, and we find that the organizer behaves differently according to its width, losing anterior inductive ability first as its size is reduced.

These results suggest that the organizer is not rigidly patterned at the start of gastrulation. The model incorporates this width dependence in the provision that the organizer is far from completely formed at the start of gastrulation. The initial O cells must induce animal cap cells to become additional O cells. If the initial population is too small, it might emit too little O signal and not induce a full organizer. During propagation of the signals, neighboring V cells might convert all prospective O cells to OV cells, terminating further organizer enlargement. The remaining O cells would be blocked in their conversion to ON cells, and so anterior neuralization would fail. Depending on the severity of these effects, the body axis would be truncated from the anterior end. The smaller the organizer, the less its ability to withstand the effects of V and OV cells.

These results with surgically reduced organizers and anterior truncation help to explain the effects of partial rotation in the first cell cycle and the effects of partial removal of DVR cells in the blastula period: these just lead to the formation of a reduced initial O cell population.

6. Partial gastrulation: Various agents, applied during gastrulation, inhibit morphogenetic movements and have the consequence that the embryo is finally missing anterior parts of the axis, just as if the organizer had been reduced in size (Gerhart et al., 1989). If the block is imposed early and completely, the only part of the body axis that develops is the most posterior part, and it

contains the leading part of the organizer. If the block is imposed midway in gastrulation, embryos develop a trunk and tail but no head. Only near the end of gastrulation do they become resistant to the inhibitors and form the entire body axis including the head. Lithium and D_2O-treated embryos, which would develop excessive dorsoanterior structure on their own, can be driven back to a normal anteroposterior pattern by exposure to these same inhibitors. These antigastrulation agents may interfere with the propagation of the O signal more than the V signal, and this would make the O cell population smaller *in effect* as a signaling source, with the same consequences as when O cells are surgically removed. The results also suggest that the O signal drives morphogenetic movements, especially those of convergent extension (Gerhart and Keller, 1985), but the connection between signaling and movement is completely obscure.

Regulation During Gastrulation?

Even at this important stage, development is not highly regulative. The organizer, ventral marginal zone, and animal cap seem to possess little pattern at the start of gastrulation, and so there is little pattern to restore when perturbed. Even with this simple pattern, the early gastrula can only regulate its development in the range where the marginal zone contains from one-half to twice the normal amount of organizer, and this deserves further study. Outside this range there is an unregulatable alteration of the embryonic anatomy in the dorsalized or ventralized direction.

In general, though, I would suggest that true pattern regulation is not a property of early amphibian development at times when complexity is increasing rapidly at each stage. The early embryologists' assumption that regulation exists may reflect a misperception of early developmental processes. Most of the perturbed embryo's ability to generate a normal final pattern may just reflect the absence of pattern at the experimentally perturbed stage. Still, at later stages, there are examples of true developmental regulation in vertebrates, as in the case of limb regeneration. And of course in other organisms such as slime molds and hydra, such regulation is well known to occur.

References

Gerhart J (1989): The primacy of cell interactions in development. Trends Genet 5:233–236.

Gerhart J, Danilchik M, Doniach T, Roberts S, Rowning B, Stewart R (1989): Cortical rotation of the *Xenopus* egg: Consequences for the anteroposterior pattern of embryonic dorsal development. Development 107 (Suppl):37–51.

Gerhart JC, Keller RE (1985): Region-specific cell activities in amphibian gastrulation. Annu Rev Cell Biol 2:201–229.

Gerhart JC, Pardee AB (1962): The enzymology of control by feedback inhibition. J Biol Chem 237:891–896.

Hamburger V (1988): The Heritage of Experimental Embryology, Hans Spemann and the Organizer. New York: Oxford University Press.
Hemmati-Brivanlou A, Harland RM (1989): Expression of an engrailed-related protein is induced in the anterior neural ectoderm of early *Xenopus* embryos. Development 106:611–617.
Kaneda T (1981): Studies of the formation and state of determination of the trunk organizer in the newt, *Cynops pyrrhogaster.* III. Tangential induction in the dorsal marginal zone. Dev Growth Differ 23:553–564.
Kaneda T, Suzuki AS (1983): Studies on the formation and state of determination of the trunk organizer in the newt, *Cynops pyrrhogaster.* IV. The association of neural inducing activity with the mesodermization of the trunk organizer. Roux's Arch Dev Biol 192:8–12.
Kao KR, Elinson RP (1988): The entire mesodermal mantle behaves as a Spemann's organizer in dorsoanterior enhanced *Xenopus laevis* embryos. Dev Biol 127:64–77.
Nieuwkoop PD (1985): Inductive interactions in early amphibian development and their general nature. J Embryol Exp Morphol 89 (Suppl):333–347.
Saxen L, Toivonen S (1961): The two gradient hypothesis in primary induction. The combined effect of two types of inductors in different ratios. J Embryol Exp Morphol 9:514–528.
Scharf SR, Rowning B, Wu M, Gerhart JC (1989): Hyperdorsoanterior embryos from *Xenopus* eggs treated with D_20. Dev Biol 134:175–188.
Spemann H (1938): Embryonic Development and Induction. New York: reprinted by Hafner Publishing Co., 1967.
Stewart RM, Gerhart JC (1990): The anterior extent of dorsal development of the *Xenopus* embryonic axis depends on the quantity of organizer in the late blastula. Development 109: 363–372.
Suzuki AS, Mifune Y, Kaneda T (1984): Germ layer interactions in pattern formation of amphibian mesoderm during primary induction. Dev Growth Differ 26:81–94.
Toivonen S (1978): Regionalization of the embryo. In O. Nakamura and S. Toivonen (eds): Organizer: A Milestone of a Half Century from Spemann. Amsterdam: Elsevier/North-Holland Biomedical Press, pp. 119–156.
Yamada T (1950): Dorsalization of the ventral marginal zone of the *Triturus* gastrula. I. Ammonia treatment of the medio-ventral marginal zone. Biol Bull 47:98–121.

Perspectives on Cellular Regulation:
From Bacteria to Cancer, pages 201–212

Regulation of Vascular Cell Proliferation

John J. Castellot, Jr.

Department of Anatomy and Cellular Biology, Tufts University Health Science Schools, Boston, Massachusetts 02111

Introduction

The proliferation of vascular endothelial and smooth muscle cells plays a crucial role in a wide variety of physiologic and pathologic processes. These include wound healing, neoplasia, diabetic retinopathy, rheumatoid arthritis, and arteriosclerosis. Although this statement is now widely accepted, the seminal observations and experimental paradigms underlying this concept were not established until the early- to mid-1970s. It was during this period (1974–1978) that I was a graduate student with Arthur Pardee, learning the tools of the mammalian cell cycle trade and fashioning a thesis in which I developed a permeabilized mammalian cell system to study some of the biochemical mechanisms implicated in regulating the G_0 to S phase transition. To put this period in a temporal perspective, I began working with Art just as Lorraine Gudas was finishing her Ph.D. thesis, and I defended my thesis the same month that Judy Campisi joined the laboratory as a postdoctoral fellow. This period included Art's last year at Princeton and his first three years at the Dana-Farber Cancer Institute and Harvard Medical School. Among the principal members of the laboratory during my time there were Joyce Hamlin, Rolf Kletzien, Michael Miller, Peter Rossow, and Van Cherington. It is a tribute to the intellectual excitment and comraderie generated in the laboratory that we have all kept in touch and remained friends since then. The influence of Art on my scientific development cannot be overestimated—I arrived in his laboratory having been rather classically trained in chemistry and biology with no prior exposure to basic research. With patience and insight he taught me how to think about science and to be an experimentalist. My interest in the regulation of vascular cell proliferation really began when Art returned from a meeting on this topic and predicted that this field would become a very important one. As usual, how right he was!

The overall goal of my current research is to elucidate the cellular and molecular mechanisms that control the proliferation of vascular cells. Within this broad context there are two major objectives:

1. How is the formation of new blood vessels (a process termed angiogenesis or neovascularization) regulated? This is basically a problem in endothelial cell biology, since new blood vessels arise from preexisting microvessels and these are comprised primarily of endothelial cells.
2. How is smooth muscle cell proliferation in large arteries controlled?

The remainder of this chapter will review my laboratory's past work in these two areas and will describe our current efforts to understand these fundamental problems in mammalian cell proliferation.

Angiogenesis

Angiogenesis is an integral part of many normal and pathological processes, including embryogenesis, inflammation, wound healing, neoplasia, diabetic retinopathy, and rheumatoid arthritis. New blood vessels arise from endothelial cells (EC) in preexisting microvascular beds. The process is complex and requires proteolytic degradation of basement membrane in the parent vessels, directed migration toward the target tissue, and proliferation to provide cells for the new capillaries. Angiogenesis factors are characterized by their ability to promote neovascularization in vivo and to stimulate at least one of the following EC functions in culture: protease production, motility, and mitogenesis. Many cells and tissues have been shown to possess angiogenesis-stimulating activity (reviewed by Folkman and Klagsbrun, 1987). Angiogenesis factors can be broadly grouped into two classes: polypeptides such as acidic and basic fibroblast growth factor, angiogenin, tumor necrosis factor, and the transforming growth factors; and nonpeptide molecules such as prostaglandins E_1 and E_2, hyaluronate fragments, and nicotinamide. Although it is clear that these molecules can stimulate neovascularization in bioassay systems, their developmental and physiologic relevance as angiogenesis stimulators remains unclear.

Neovascularization during normal tissue development is the most common form of angiogenesis. For example, angiogenesis is tightly coordinated with adipocyte differentiation during embryogenesis, and the newly forming adipose tissue depends on continued neovascularization for further development (Wassermann, 1965). Since new blood vessels are formed from EC, and since cultured mouse 3T3 adipocytes have most of the characteristics of fat cells in vivo (Green, 1978), the interaction between 3T3-adipocytes and cultured EC provides an excellent model system for the study of developmental angiogenesis. In collaboration with Dr. Bruce Spiegelman, we have shown that 3T3 adipocytes secrete factors that stimulate angiogenesis in vivo and stimulate EC protease production, motility, and proliferation in vitro (Castellot et al., 1980; Castellot et al., 1982; Castellot et al., 1986b). Production of these angiogenesis factors is differentiation dependent, and at least 10-fold more angiogenic activity is secreted by 3T3-adipocytes than by their undifferentiated precursors.

Purification of the Major Adipocyte-Derived Angiogenesis Factor

Much of our recent effort in understanding angiogenesis has centered on the purification and identification of the molecule(s) responsible for the 3T3-adipocyte-derived angiogenic activity (Dobson et al., 1990). Preliminary biochemical analysis suggested that the adipocyte-derived angiogenic factors were small lipophilic molecules. Based on this, we carried out reverse-phase chromatography of adipocyte-conditioned medium on C18 columns. Factors that stimulate angiogenesis in the chick chlorioallantoic membrane (CAM) assay system and EC motility are retained on the column and elute with 30–50% aqueous ethanol. In contrast, an endothelial cell mitogen does not bind to the C_{18} column and is recovered in the aqueous flowthrough.

The 30–50% ethanol fraction from the C_{18} column was applied to a silica gel TLC system. Two of eight fractions were found to have angiogenic activity; however, one of the fractions (TLC5) contained 5- to 8-fold more angiogenic activity than the other fraction (TLC4). Further experiments revealed that adipocyte TLC4 had only 50% more angiogenic activity than the comparable fraction from preadipocytes, whereas the activity in TLC5 was strongly differentiation dependent. We calculated that adipocyte TLC5 contained nearly 40-fold more angiogenic activity than preadipocyte TLC5.

Purifying the angiogenic factor(s) to homogeneity via bioassays appeared to be an onerous task. We therefore sought to identify molecules that had the characteristics of the partially purified active molecules: lipid-like, differentiation-dependent, and migration in TLC5 after ethanol elution from C_{18} columns. When [^{14}C]arachidonate was used to metabolically label cellular lipids, no bands were seen in the TLC5 region. A prominent band was seen in the TLC4 region, which we subsequently identified as PGE_2. Radioimmunoassay indicated that the levels of PGEs produced in adipocyte-conditioned medium were sufficient to account for the angiogenic activity in the TLC4 fraction. When [^{14}C]acetate was used to label cellular lipids, we observed a single band that migrated in the adipocyte TLC5 fraction; this band was absent in preadipocyte TLC5. The acetate-labeled band was excised directly and found to contain nearly 90% of the angiogenic activity in TLC5.

Further purification of the acetate-labeled band in TLC5 was carried out by reverse-phase C_{18} high-performance liquid chromatography (HPLC). A single, ^{14}C-labeled peak was eluted which contained approximately 90% of the total radioactivity in the sample. Gas chromatographic (GC) analysis of this peak showed that the most abundant molecular species eluted at 8.49 min. The increase in this molecular species during adipocyte differentiation was approximately 200-fold. When the 8.49 min GC peak was analyzed by electron impact mass spectrometry, the spectrum indicated that the compound is 1-butyryl-glycerol (monobutyrin). Chemically synthesized monobutyrin gave GC and mass spectrometry patterns identical to the adipocyte-derived molecule.

Final confirmation that monobutyrin was the major adipocyte-derived angiogenic molecule required an examination of the biological activities of the synthetic material. Monobutyrin was angiogenic in the CAM assay at concentrations as low as 20 pg per sample; maximum activity was achieved using 20 ng per sample. Microvascular EC motility was also stimulated at concentrations as low as 7×10^{-8} M; the ED_{50} was approximately 7×10^{-6} M. In contrast, monobutyrin had no effect on EC mitogenesis. Thus, synthetic monobutyrin had a biological activity profile that was identical to the adipocyte-derived angiogenesis factor: stimulation of angiogenesis and EC motility, but no direct stimulation of EC proliferation.

Current Studies

Monobutyrin is a monoacyglycerol that has not been previously reported to have angiogenic activity or any other pharmacological action. The relatively low levels at which monobutyrin stimulates angiogenesis and EC motility suggest that it works via an intracellular or cell surface receptor. Although a monobutyrin receptor has not been described, it is intriguing to speculate that monobutyrin might mimic diacylglycerols, which are known to bind to protein kinase C or diacylglycerol kinase (Bell, 1986).

Monobutyrin is not mitogenic for EC, yet angiogenesis requires an increase in cell number. An important question is how mitogenesis is stimulated. One possibility is that monobutyrin interacts with the as yet uncharacterized adipocyte-secreted mitogen, which is found in the aqueous flowthrough of the C_{18} column. Another possibility is that monobutyrin synergizes with endogenous mitogens present in the extracellular matrix. Preliminary evidence suggests that monobutyrin can act synergistically with basic fibroblast growth factor and with prostaglandin E_2 in both CAM assays for angiogenesis and in EC motility assays. There is some specificity for this synergy, since monobutyrin does not appear able to augment the angiogenic activity of tumor necrosis factor-α or transforming growth factor-β. These data support the notion that monobutyrin acts in concert with mitogens or other angiogenic factors to stimulate a strong angiogenic response.

A major challenge facing the angiogenesis field is to determine the biological relevance of molecules that stimulate neovascularization in bioassay systems. To date, this has not been rigorously accomplished for any of the known angiogenic factors. Our data suggest strongly that monobutyrin is a key regulator of the angiogenesis program during the development of adipose tissue in vivo. More convincing evidence for this notion will require the development of monobutyrin-neutralizing antibodies. The use of antibodies should permit a critical evaluation of the physiologic role of monobutyrin in angiogenesis.

Regulation of Smooth Muscle Cell Proliferation

The importance of growth control in the artery wall is clearly demonstrated by three observations. The first is that in the normal arterial wall, both the EC and smooth muscle cells (SMC) lining the lumen are in a quiescent growth state. The second observation is that SMC proliferation is a hallmark of the early atherosclerotic lesion. The third observation is that at least 20% of the over 500,000 vascular surgical procedures (e.g., vein grafts, coronary bypass, angioplasty, arteriovenous shunts for kidney dialysis) done each year in the United States fail within a few months, mainly due to SMC hyperplasia in the operated vessels.

Much attention has been focused on the cellular and molecular mechanisms that stimulate SMC proliferation (reviewed by Ross, 1986). Powerful mechanisms must exist for regulating the responsiveness of cells to mitogenic stimuli, since the proliferation of most cells is tightly controlled (Sager, 1986). In recent years, a number of molecules have been described that can inhibit the growth of various cell types. These include retinoic acid, interferon, transforming growth factor-β (TGF-β), and heparan sulfates.

Heparan sulfates are structurally complex glycosaminoglycans (GAGs) composed of repeating disaccharide units of alternating glucosamine and uronic acid sugars. These molecules inhibit the growth of SMC, both in vivo and in vitro (reviewed in Castellot et al., 1987). Taken together with results from other laboratories, the data clearly demonstrate that heparan sulfates (which include the highly sulfated, iduronic acid-rich heparin) have the potential to act as effective antiproliferative molecules.

In Vivo Studies

My interest in heparan sulfates arose from in vivo studies by Dr. Morris Karnovsky's laboratory on the response of SMC in the rat carotid artery to endothelial injury. In these studies, endothelial injury was produced by the air-dry method, which destroys all of the endothelium in a well-defined segment but does not damage the underlying media. Little SMC proliferation occurred at the edges of the injury, where reendothelialization of the damaged segment was rapid. However, in the middle portion of the injured segment, which was the last site to be reendothelialized, massive SMC hyperplasia occurred. Interestingly, SMC proliferation ceased when complete reendothelialization occurred. From these observations, we hypothesized that the overlying layer of EC might regulate SMC proliferation.

Because the clotting sequence is activated in vascular injury and thrombin is a mitogen for some cell types, the effect of heparin in the rat carotid after the air-dry injury was tested. Heparin almost completely abolished the SMC hyperplasia following EC injury. The separation of heparin into purified anticoagulant (AC) and nonanticoagulant (NAC) fractions, using antithrombin III affinity chromatography, allowed testing of the hypothesis that heparin suppressed SMC proliferation through its effects on blood coagulation. When the rat carotid injury experiments were repeated with AC and NAC heparin infusions, it was clear that NAC heparin was just as effective as AC material.

The animal studies described above clearly demonstrate that heparin blocks the proliferation of SMC in vivo, and that its antiproliferative activity is independent of its anticoagulant activity.

In Vitro Studies

To develop a model system for analyzing the antiproliferative effect of heparin, we tested its ability to inhibit the growth of cultured SMC (Castellot et al., 1981). SMC were growth-arrested at subconfluent densities by serum deprivation to mimic more closely the in vivo situation, in which SMC exist in a quiescent (G_0) growth state. The SMC were released from G_0 by replacing the low-serum medium with medium containing 20% fetal calf serum in the presence or absence of heparin. Heparin was very effective in suppressing the serum-stimulated growth of cultured SMC. The ED_{50} was 1–5 μg/ml, a level that can be readily achieved in blood. Using this assay, we compared the effects of heparin with other classes of GAGs. Both AC and NAC heparins were highly effective antiproliferative agents, whereas dermatan sulfate, chondroitin 4- and 6-sulfates, and hyaluronic acid were much less effective. SMC from a variety of vessels and sources have been examined, including rat, calf, and monkey aorta, rat mesenteric artery, and human iliac artery and umbilical vein; all are very sensitive to the growth inhibitory effect of heparin.

When we tested heparin on other cell types, we found that EC, many fibroblast cell lines, virally transformed cells, and Madin–Darby canine kidney epithelial cells, among others, were much less sensitive than SMC. These results initially suggested that the effect of heparin was relatively specific for SMC. However, we subsequently found that two other cell types are also sensitive to the antiproliferative effect of heparin. One is the glomerular mesangial cell (Castellot et al., 1985a), a close relative of the SMC; the other is the rat cervical epithelial cell (Wright et al., 1985). The finding that cervical epithelial cell growth can be inhibited by low concentrations of heparin suggests that GAGs may have a more general role in regulating cell proliferation.

Two other characteristics of the antiproliferative effect of heparin on SMC are worth noting: (1) SMC that are quiescent (i.e., growth arrested) before exposure to growth medium containing heparin are 50- to 100-fold more sensitive than SMC that are exponentially growing when first exposed to heparin, and (2) the growth inhibitory effect is reversible, that is, washing off the heparin-containing medium and replacing it with normal growth medium allows cells to resume their normal growth rate.

Structure–Function Studies

A series of structure–activity experiments have been carried out in collaboration with Drs. Morris Karnovsky, Robert Rosenberg, Thomas Wright, and Jean Choay (Castellot et al., 1984; Castellot et al., 1986a; Wright et al., 1989a). Only the briefest summary of these findings will be given here. The smallest heparin fragment that retained antiproliferative activity was a pentasaccharide; di- and tetrasaccharide fractions were inactive. The maximum antiproliferative activity was obtained with dodecamer and larger fragments that gave activities equivalent to native heparin. Using chemically modified heparins, we found that O-sulfation, or at least a negative charge at the O-position, is necessary for the antiproliferative activity. The situation at the N-position is somewhat different. When the N-sulfates were removed, the N-position became positively charged and the growth of SMC was not suppressed. Neutralization of the charge at the N-positions by acetylation restored the antiproliferative activity, suggesting that a negative charge at N-positions is not required.

The structure–function experiments have identified several heparin analogs that have lost AC activity but retain antiproliferative activity. These molecules may have important clinical applications, especially in treating the SMC hyperplasia that occurs so frequently following vascular surgery. The use of an antiproliferative, NAC heparin immediately following vascular surgery may prevent SMC proliferation, while avoiding serious hemorrhagic complications. Clinical trials are now underway to test this possibility.

Studies on the Antiproliferative Mechanism

Cell cycle studies. One important characteristic of a growth inhibitor is the point in the cell cycle at which it exerts its effect. We have taken several experimental approaches to answer this question (Castellot et al., 1989). All of the data indicate at least two heparin-sensitive points in the SMC cycle: a block in the competence phase (i.e., within the first 30 min after release from G_0), and a second block at or near the R point (i.e., 2–4 hr before S phase).

Binding and internalization of heparin. One key to understanding the mechanism responsible for heparin's antiproliferative effect is to determine whether the heparin added to cultured SMC actually becomes associated with the cells or remains in the medium. Heparan sulfate proteoglycans are present in a number of different tissues and appear to be preferentially located at the surface of cells. A number of extracellular matrix molecules including fibronectin and laminin have heparan sulfate binding sites.

Heparin has been shown to bind to a variety of cell types including endothelial cells, macrophages, Chinese hamster ovary cells, and hepatocytes. However, no studies had been performed on the binding and fate of heparin in cells that were sensitive to its growth inhibitory effects. Utilizing ^{125}I- and ^{3}H-labeled heparin, specific, saturable heparin-binding sites were found on the surface of SMC (Castellot et al., 1985b). The affinity ($K_d = 10^{-9}$ M) and the number of binding sites (100,000 per cell) were typical of the binding sites for other growth regulatory molecules. Interestingly, we showed that growth-arrested SMC bind eight times more heparin than exponentially growing cultures. This may account, in part, for the observation that growth arrested SMC are 50–100 times more sensitive to the antiproliferative effects of heparin than exponentially growing cells.

When the fate of the radiolabeled heparin bound to SMC was studied (Castellot et al., 1985b), we found that it was internalized with biphasic kinetics. There was an initial rapid entry of labeled heparin, followed by a slow rate of uptake that occurred over several days. The $t_{1/2}$ for the initial phase was approximately 15 min. The binding and internalization of fluorescently labeled heparin on SMC was also studied using fluorescence video image intensification microscopy. SMC bind heparin diffusely on the cell surface at 4°C. On warming to 37°C, the fluorescent heparin that remained associated with the cells formed surface clusters and became internalized in endocytic vesicles. These vesicles localized to the perinuclear region within 2 hr. The presence of high-affinity heparin-binding sites on the SMC surface, combined with the rapid initial internalization and the pattern of fluorescent-labeled heparin internalization suggest that heparin enters SMC via receptor-mediated endocytosis.

Endothelial Cell Regulation of Smooth Muscle Cell Proliferation

In the carotid artery injury model, the cessation of SMC hyperplasia in a particular region of the damaged segment was correlated with the reendothelialization of that area. We therefore thought it possible that EC could exert a regulatory influence on SMC proliferation, and undertook to examine this using cultured cells (Castellot et al., 1981). Conditioned medium from confluent EC cultures strongly inhibited the growth of SMC released from G_0. Interestingly, medium conditioned by exponentially growing EC stimulated SMC growth, suggesting that the cessation of EC growth after regeneration in vivo may be an important step in the regulation and inhibition of SMC proliferation. The mitogenic factors elaborated by EC have been explored in detail by others (Gajdusek et al., 1980; DiCorleto, 1984).

Biochemical characterization showed that the antiproliferative activity from EC was unaffected by heating and was insensitive to proteases, hyaluronidases, and chondroitin sulfate ABC lyase. However, a highly specific heparinase almost completely abolished the activity. This enzyme, which has no detectable activity against proteins or other glycosaminoglycans, degrades heparin into tetrasaccharide and smaller fragments, which lack inhibitory activity. In addition, a trichloroacetic acid-soluble, ethanol-insoluble glycosaminoglycan fraction prepared from EC-conditioned medium was highly antiproliferative. When this material was treated with the heparinase, the activity was abolished. We therefore concluded that the inhibitory activity produced by EC was a heparan sulfate.

Subsequent work by Rosenberg and his colleagues suggests that confluent, but not exponentially growing, SMC synthesize a highly antiproliferative heparin species (Fritze et al., 1985). This raises the interesting possibility of a heparin-mediated, autocrine growth regulatory mechanism operating in SMC. It also suggests a functional homology between heparin and TGF-β and interferon, since all of these molecules inhibit the growth of cells that produce them. The roles of EC- and SMC-derived heparin in the growth dynamics of the arterial wall await elucidation; an intriguing possibility is that EC-derived heparin acts as a "trigger" for SMC to begin synthesizing their own heparin.

Role of Extracellular Matrix

There is mounting evidence that the extracellular matrix may play an important role in the regulation of many cell processes, including proliferation. Support for the ability of extracellular matrix to modulate the antiproliferative effect of heparin comes from recent experiments done in collaboration with Dr. Ira Herman, in which SMC were grown on different extracellular matrices and tested for their sensitivity to heparin (Herman and Castellot, 1987). The substrate on which SMC were most sensitive to heparin was EC-derived matrix prepared by EGTA treatment. Interestingly, the substrate on which SMC were the least sensitive was EC-derived matrix prepared by deoxycholate extraction. As discussed above, EC are capable of secreting antiproliferative heparan sulfate species. When grown on EGTA-prepared EC matrix, EC secreted a 10-fold more potent SMC inhibitory activity than EC cultured on deoxycholate-prepared EC matrix. When the matrix effects on secretion of antiproliferative heparan sulfates by EC and sensitivity of SMC to heparin were combined, a 30- to 40-fold difference between the best and worst combinations was observed. These observations indicate that extracellular matrix can strongly influence the growth of vascular cells. It is especially intriguing that the method of preparing the EC matrix (EGTA vs. deoxycholate) yielded vastly different growth properties.

Current Studies

Signal transduction pathways. Our recent efforts have focused on the biochemical and molecular mechanisms involved in the antiproliferative activity of heparin. The mitogenic response of SMC, like that of most mesenchymal cells, appears to involve at least two postreceptor signal transduction pathways. One pathway is mediated by protein kinase C (PKC) and is activated by PDGF, bombesin, and phorbol esters. The other pathway is dependent on cAMP and calcium, and is activated by EGF. Using calf aortic SMC, we demonstrated that heparin strongly inhibited phorbol ester-stimulated growth, but was much less effective in blocking EGF-stimulated mitogenesis. This result suggests that heparin selectively inhibits the PKC-dependent mitogenic pathway in these cells. Further support for this idea is provided by our observation (done in collaboration with Dr. J. Campisi) that heparin inhibits phorbol ester-stimulated induction of c-*fos* and c-*myc* mRNA in SMC and BALB/c 3T3 fibroblasts (Wright et al., 1989b; Castellot et al., 1988), but does not inhibit EGF-stimulated expression of these protooncogene mRNAs. Additional corroboration comes from recent experiments using heparin-resistant cultures of SMC obtained through classical drug-selection procedures. These cells are at least 40-fold less sensitive to heparin than their normal counterparts. No decrease in c-*fos* mRNA induction was observed in these cells, even at high heparin doses (200 μg/ml). We have also taken advantage of the availability of heparin analogs devoid of antiproliferative activity. Inactive heparin species were unable to block protooncogene expression in sensitive SMC. These results suggest that one site of action of heparin is located at or past the stimulation of PKC, and at or before the induction of protooncogene mRNAs. Homing in on the specific step within this part of the PKC-mediated pathway will be an important objective of my laboratory.

Heparin receptors. The binding and internalization studies described earlier strongly suggest the presence of specific heparin receptors on the SMC surface. If one prepares SMC membrane proteins and applies them to a heparin-Sepharose column, several dozen proteins bind to the column. Determining which, if any, of these many heparin-binding proteins might mediate the antiproliferative effect could be difficult. However, we have compared membrane proteins prepared from heparin-sensitive and heparin-resistant SMC. In these studies, done in collaboration with Dr. Chitra Biswas, we have found only a few consistent differences in the profile of heparin-binding membrane proteins. The most striking difference is the almost complete absence of a pair of proteins in the 31–33 kDa range from the membranes of resistant cells. We will soon have sufficient quantities of both proteins to allow amino acid sequencing and antibody preparation. If antibodies to these proteins abrogate the antiproliferative effect of heparin, then this would provide strong evidence for a role in the growth inhibitory mechanism. Clearly, biochemical and molecular analysis of the precise role of the 31–33 kDa heparin-binding proteins would then be warranted.

Conclusion

This chapter has described two projects that, in the broadest sense, are designed to elucidate the cellular and molecular mechanisms that regulate the growth of vascular cells in both large and small vessels. Their biological and clinical implications for cardiovascular function and pathobiology are quite different, yet a number of experimental and conceptual commonalities exist between them. Both projects will require a 1990s approach to understanding growth regulation: receptor biology and biochemistry, signal transduction mechanisms, and gene regulation studies. All of these approaches, however, are firmly rooted in the fundamental training given to a scientifically naive graduate student over 15 years ago by Arthur Pardee.

References

Bell RM (1986): Protein kinase C activation by diacylglycerol second messengers. Cell 45:631–32.

Castellot JJ, Addonizio ML, Rosenberg RD, Karnovsky MJ (1981): Cultured endothelial cells produce a heparin-like inhibitor of smooth muscle cell growth. J Cell Biol 90:372–379.

Castellot JJ, Beeler DL, Rosenberg RD, Karnovsky MJ (1984): Structural determinants of the capacity of heparin to inhibit the proliferation of vascular smooth muscle cells. J Cell Physiol 120:315–320.

Castellot JJ, Choay J, Lormeau J-C, Petitou M, Sache E, Karnovsky MJ (1986a): Structural determinants of the capacity of heparin to inhibit the proliferation of vascular smooth muscle cells. II. Evidence for a pentasaccharide sequence that contains a 3-O-sulfate group. J Cell Biol 102:1979–1984.

Castellot JJ, Hoover RL, Harper PA, Karnovsky MJ (1985a): Heparin and epithelial cell-secreted heparin-like species inhibit mesangial cell proliferation. Am J Pathol 120:427–435.

Castellot JJ, Kambe AM, Dobson DE, Spiegelman BM (1986b): Heparin potentiation of 3T3-adipocyte stimulated angiogenesis: Mechanisms of action on endothelial cells. J Cell Physiol 127:323–329.

Castellot JJ, Karnovsky MJ, Spiegelman BM (1980): Potent stimulation of vascular endothelial cell growth by differentiated 3T3 adipocytes. Proc Natl Acad Sci USA 77:6007–6011.

Castellot JJ, Karnovsky MJ, Spiegelman BM (1982): Differentiation-dependent stimulation of neovascularization and endothelial cell chemotaxis by 3T3 adipocytes. Proc Natl Acad Sci USA 79:5597–5601.

Castellot JJ, Pukac L, Wright TC, Karnovsky MJ (1988): Heparin alters cell cycle progression, the response to mitogens, and gene expression in vascular smooth muscle cells. FASEB J 2:A418.

Castellot JJ, Pukac LA, Caleb BL, Wright TC, Karnovsky MJ (1989): Heparin selectively inhibits a protein kinase C-dependent mechanism of cell cycle progression in calf aortic smooth muscle cells. J Cell Biol 109:3147–3155.

Castellot JJ, Wong K, Herman B, Hoover RL, Albertini DF, Wright TC, Caleb BL, Karnovsky MJ (1985b): Binding and internalization of heparin by vascular smooth muscle cells. J Cell Physiol 124:13–20.

Castellot JJ, Wright TC, Karnovsky MJ (1987): Regulation of vascular smooth muscle cell growth by heparin and heparan sulfates. Sem Thromb Hemostasis 13:489–503.

DiCorleto PE (1984): Cultured endothelial cells produce multiple growth factors for connective tissue cells. Exp Cell Res 153:167–72.

Dobson DE, Kambe AM, Block E, Dion T, Lu H, Castellot JJ, Spiegelman BM (1990): 1-Butyryl-glycerol: A novel angiogenesis factor secreted by differentiating adipocytes. Cell 61:223–230.

Folkman J, Klagsbrun M (1987): Angiogenic factors. Science 235:442–447.

Fritze LMS, Reilly CF, Rosenberg RD (1985): An antiproliferative heparan sulfate species produced by postconfluent smooth muscle cells. J Cell Biol 100:1041–1049.

Gajdusek C, DiCorleto P, Ross R, Schwartz SM (1980): An endothelial cell-derived growth factor. J Cell Biol 85:467–472.

Green H (1978): The adipose conversion of 3T3 cells. In F. Ahmad, J. Schultz, T.R. Russell, and R. Werner (eds): 10th Miami Symposium on Differentiation and Development. New York: Academic Press, p. 13.

Herman IH, Castellot JJ (1987): Regulation of vascular smooth muscle cell growth by endothelial-synthesized extracellular matrices. Arteriosclerosis 7:463–469.

Ross R (1986): The pathogenesis of atherosclerosis—an update. N Engl J Med 314:488–500.

Sager R (1986): Genetic suppression of tumor formation: A new frontier in cancer research. Cancer Res 46:1573–1580.

Wassermann F (1965): The development of adipose tissue. Handbook of Physiology Vol. 5. Washington, DC: American Physiological Society, pp. 87–100.

Wright TC, Castellot JJ, Petitou M, Lormeau J-C, Choay J, Karnovsky MJ (1989a): Structural determinants of heparin's growth inhibitory activity: Interdependence of oligosaccharide size and charge. J Biol Chem 264:1534–1542.

Wright TC, Johnstone TV, Castellot JJ, Karnovsky MJ (1985): Inhibition of rat cervical epithelial cell growth by heparin and its reversal by EGF. J Cell Physiol 125:499–506.
Wright TC, Pukac LA, Castellot JJ, Karnovsky MJ, Levine RA, Kim-Park H-Y, Campisi J (1989b): Heparin suppresses the induction of c-fos and c-myc mRNA in murine fibroblasts by selective inhibition of a protein kinase C-dependent pathway. Proc Natl Acad Sci USA 86:3199–3203.

Perspectives on Cellular Regulation:
From Bacteria to Cancer, pages 213–223

A Model for the Study of Cellular Heterogeneity in Human Tumors

Estela E. Medrano

Department of Dermatology, University of Cincinnati College of Medicine, Cincinnati, Ohio 45267

Introduction

It is with great emotion that I dedicate this paper to Arthur B. Pardee in honor of his 70th birthday. I arrived to his laboratory at the Dana-Farber Cancer Institute in the spring of 1979, naive and scared, a biochemist with the intention to increase my knowledge in cellular biology. In Art's lab I obtained expertise from his constant dedication and advice, but also the intangible, fragile, and always precious freedom to work and to create. When I left two years later, I was no longer a biochemist. Since then, tumor biology has occupied and fascinated me every day.

Tumors may be considered cell-renewal systems that differ from their normal counterparts in that the controls that regulate the steady state size of the tissue have been lost. Moreover, tumors can be viewed as a stem cell system. Tumor cells are capable of self-renewal, and, in many cases, of generating clonal hierarchies that present differentiated features and a limited growth capacity (Buick and Pollack, 1984). It may be considered also that cellular heterogeneity within a tumor is the consequence of the existence of multiple neoplastic subpopulations, each capable of maintaining their genetic identity under different environmental conditions for long periods of time (Heppner, 1984). Differences in the microenvironment of a tumor, e.g., oxygen gradients, pH, nutritional and growth factors, cellular waste products, and interaction with diverse extracellular matrixes, also play an important role in the observed heterogeneity within the tumor mass. Heterogeneity is not a property unique to human tumors and it has been documented in many normal tissues, most thoroughly in the hemopoietic system, but also in the epidermis, endometrium, and in the embryonic cells that differentiate to generate tissues (Nicolson, 1987).

Tumor heterogeneity is also an inevitable outcome of tumor progression, during which cells become genetically unstable through successive mutations

(Nowell, 1986; Hart and Fidler, 1981; Hart et al., 1989). This changing genetic background includes the activation of oncogenes and inactivation of suppressor genes (Sager, 1988). Thus, the recognition of tumor heterogeneity has become essential to any theory of neoplastic development. I will not go through the extensive examples on the subject, since it has been covered in excellent reviews (references already mentioned; Houghton et al., 1987). Rather, I will try to convince the reader with a few examples of experiments performed in my laboratory that "a particular isolated subpopulation of tumor cells is unimportant except as a reminder of the diversity of the cell society from which it came" (Heppner, 1984).

Phenotypic Heterogeneity in a Human Breast Tumor Cell Line: A Unidirectional Differentiation Process

Using the uncloned human breast tumor cell line MCF-7, I and my colleagues have shown that the cells can be fractionated in Percoll gradients to yield several subpopulations (Resnicoff et al., 1987). Figure 1 shows the distribution of exponentially growing MCF-7 cells among the subpopulations obtained by Percoll density gradient centrifugation. The distribution is highly heterogeneous; subpopulation C is the most abundant, whereas subpopulations D and E contain the fewest cells. The inset shows that a single peak is ob-

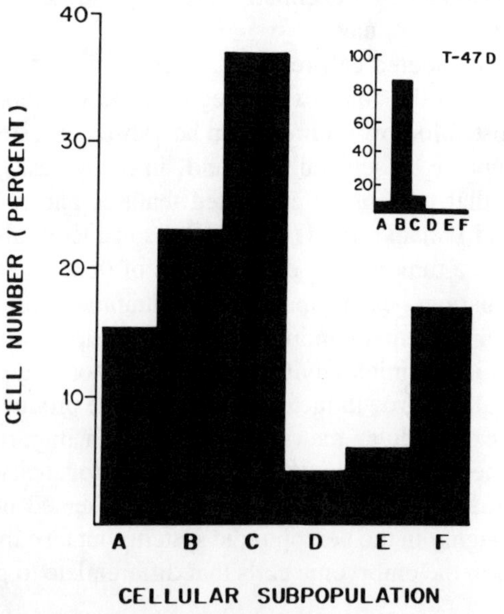

Fig. 1. Distribution of the human breast tumor cells MCF-7 and T-47D in Percoll density gradients. From Resnicoff et al. (1987).

tained when clone 11 of the T-47D breast tumor line is sedimented under identical conditions. This result demonstrates that the density heterogeneity of the MCF-7 cells is not an artifact of the Percoll gradient, but an intrinsic property of these cells. The question arises: Does the existence of these subpopulations support the clonal or the stem cell model for tumor heterogeneity? The reader will find that the second alternative is applicable in this case.

If the different subpopulations isolated from the gradient derive from different steps in the cell hierarchies originating from a stem cell, then only the stem cell fraction should be able to generate all the other subpopulations. To determine which, if any, of the subpopulations contained stem cells, the individual subpopulations were cultured for various times and again submitted to Percoll density gradient centrifugation. In Figure 2, it is shown that cells in subpopulation E (E cells) were able to generate all the other subpopulations, thus reproducing the pattern of the starting MCF-7 cell population. The A

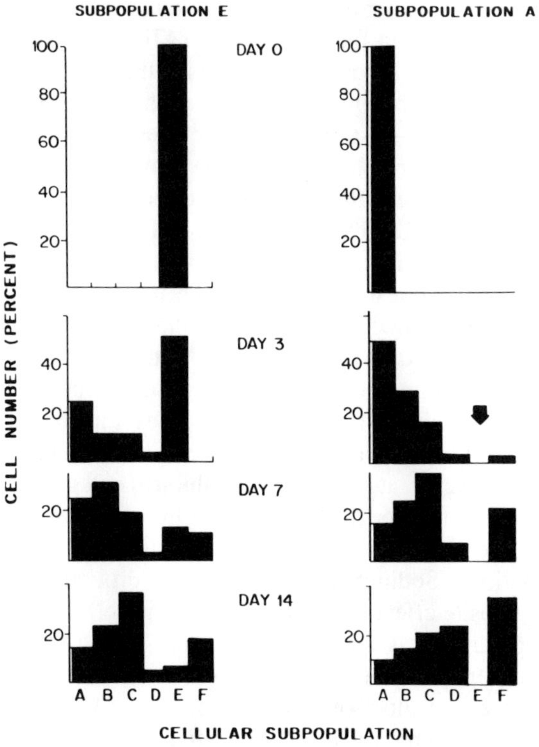

Fig. 2. Differing ability of subpopulations E and A to regenerate other subpopulations. Subpopulations E and A were isolated by density gradient centrifugation and subcultured. At the indicated times they were again separated by Percoll gradients and their distribution was estimated. From Resnicoff et al. (1987).

cells gave rise to fractions of incresing density (B, C, D, and F), but did not generate E cells. B cells gave rise to C, D, and F cells; C cells generated D and F cells; D cells generated F cells; and F cells remained unchanged for up to 19 days in culture. The data shown by Figure 2 are consistent with the sequence E → A → B → C → D → F.

Tumor Subpopulations Are Differentially Regulated by Growth Factors and Hormones

Very little is known about the growth factors and hormones that affect stem cell renewal and differentiation. The reason for this is that it has been difficult to obtain enough stem cells from solid human tumors. With the Percoll technique we can now study not only the stem cells, but also their differentiated progeny in culture.

In previous work, I and my colleagues have provided evidence for a role for several growth factors in the growth and differentiated behavior of human mammary tumor cells (Medrano et al., 1987). We found that the serine protease thrombin is a potent mitogen for MCF-7 and T47D cells, either alone or in combination with epidermal growth factor and insulin. We therefore analyzed the response of MCF-7 subpopulations to insulin, thrombin, prostaglandin $F_{2\alpha}$ ($PGF_{2\alpha}$), estradiol, and 13-*cis*-retinal (Resnicoff and Medrano, 1989). Table 1 shows that insulin and thrombin promote the growth of the most differentiated subpopulations, whereas $PGF_{2\alpha}$ has a dual effect: it was growth stimulatory for subpopulations C and D at 600 ng/ml and strongly growth inhibitory for the stem cells (E cells) at doses from 200–400 ng/ml. Estradiol had little effect over the 24- to 48-h test period. However, it played a major role in the anchorage-independent growth of E cells when the cells were treated with this hormone for more than 10 days (Resnicoff and Medrano, 1989).

The effect of the vitamin A derivative 13-*cis*-retinal was particularly interesting because we have shown that at very low doses it increased proliferation and the accumulation of nuclear proteins in breast tumor cells (Resnicoff and Medrano, 1987). However, at high doses, it inhibited cell growth and, thus, is considered a potential therapeutic agent. From the results shown in Figure 3 (upper panel) it is clear that 13-*cis*-retinal prevents maturation of the A, B, and C subpopulations. Sodium butyrate, a known differentiation promoting agent, had the opposite effect. In Figure 3 (lower panel) the effect of 13-*cis*-retinal on subpopulation A is analyzed. More than 70% of A cells were unable to proceed through the sequence A → B → C → D → F (Resnicoff et al., 1987). Based on these results, we have proposed that 13-*cis*-retinal is a negative modulator of differentiation for MCF-7 cells.

Our results show that each subpopulation has its own characteristics with respect to their response to growth factors and hormones. More interestingly, the isolated subpopulations show a transient response to growth factors and hormones as cells proceed to more mature phenotypes.

TABLE 1. Dose–Response to Insulin, $PGF_{2\alpha}$, Thrombin, and Estradiol in the Isolated Subpopulations of MCF Cells[a]

Incorporation of [^{3}H]thymidine (cpm/10^5 of growth factor treated cells/cpm/10^5 of control cells ± SE)

Subpopulations	Insulin 0.5 μg/ml	Insulin 1 μg/ml	Insulin 1.5 μg/ml	Insulin 2 μg/ml	Insulin 5 μg/ml	Insulin 10 μg/ml	$PGF_{2\alpha}$ 100 ng/ml	$PGF_{2\alpha}$ 300 ng/ml	$PGF_{2\alpha}$ 600 ng/ml
A	1.0 ± 0.1	1.0 ± 0.1	N.D.	1.0 ± 0.1	1.0 ± 0.1	1.0 ± 0.1	1.26 ± 0.12	1.24 ± 0.16	1.1 ± 0.1
B	1.5 ± 0.1	1.7 ± 0.4	N.D.	1.6 ± 0.1	1.4 ± 0.2	1.4 ± 0.3	1.1 ± 0.1	0.7 ± 0.1	1.8 ± 0.1
C	1.2 ± 0.1	1.8 ± 0.1	2.0 ± 0.3	2.2 ± 0.1	2.2 ± 0.3	3.5 ± 0.2	2.6 ± 0.1	2.3 ± 0.2	4.2 ± 0.8
D	1.6 ± 0.5	2.2 ± 0.2	9.0 ± 1.0	28.0 ± 4.0	30.0 ± 3.7	42.0 ± 2.0	2.0 ± 0.2	2.0 ± 0.1	7.8 ± 2.2
E	1.0 ± 0.1	1.0 ± 0.1	N.D.	1.0 ± 0.1	1.0 ± 0.1	1.0 ± 0.1	1.0 ± 0.1	0.3 ± 0.1[a]	1.3 ± 0.3
F	1.4 ± 0.1	3.5 ± 0.5	6.4 ± 0.2	8.0 ± 0.2	10.0 ± 0.6	22.0 ± 3.0	1.0 ± 0.1	0.7 ± 0.1	0.75 ± 0.05

	Thrombin 0.25 U/ml	Thrombin 0.5 U/ml	Thrombin 0.75 U/ml	Thrombin 1 U/ml	Thrombin 1.5 U/ml	Thrombin 2 U/ml	Estradiol 10^{-10} M	Estradiol 10^{-9} M
A	0.8 ± 0.2	0.8 ± 0.4	N.D.	0.75 ± 0.10	0.7 ± 0.1	0.7 ± 0.1	0.7 ± 0.1	1.0 ± 0.3
B	0.9 ± 0.1	0.9 ± 0.1	N.D.	1.25 ± 0.05	1.0 ± 0.1	0.7 ± 0.1	1.2 ± 0.1	1.25 ± 0.10
C	0.9 ± 0.1	0.9 ± 0.1	N.D.	1.5 ± 0.1	1.6 ± 0.1	1.7 ± 0.2	1.4 ± 0.6	1.6 ± 0.2
D	10.0 ± 1.1	14.0 ± 3.0	N.D.	13.0 ± 3.2	13.0 ± 1.4	17.0 ± 2.3	0.7 ± 0.1	0.76 ± 0.20
E	1.0 ± 0.1	1.9 ± 0.1	N.D.	2.2 ± 0.2	2.0 ± 0.1	3.0 ± 0.2	0.8 ± 0.1	0.85 ± 0.10
F	1.0 ± 0.1	1.0 ± 0.2	8.9 ± 2.1	39.0 ± 2.1	28.0 ± 3.0	20.0 ± 2.3	0.5 ± .1	0.6 ± 0.2

[a]Because 300 ng/ml is so inhibitory for fraction E, we performed experiments using 200 and 400 ng/ml $PGF_{2\alpha}$, which gave results of 0.6 ± 0.1 and 1.4 ± 0.1, respectively. Results are the means of two different experiments, each performed in triplicate. N.D., not determined. SE, standard error. From Resnicoff and Medrano (1989). Reproduced by permission of Academic Press.

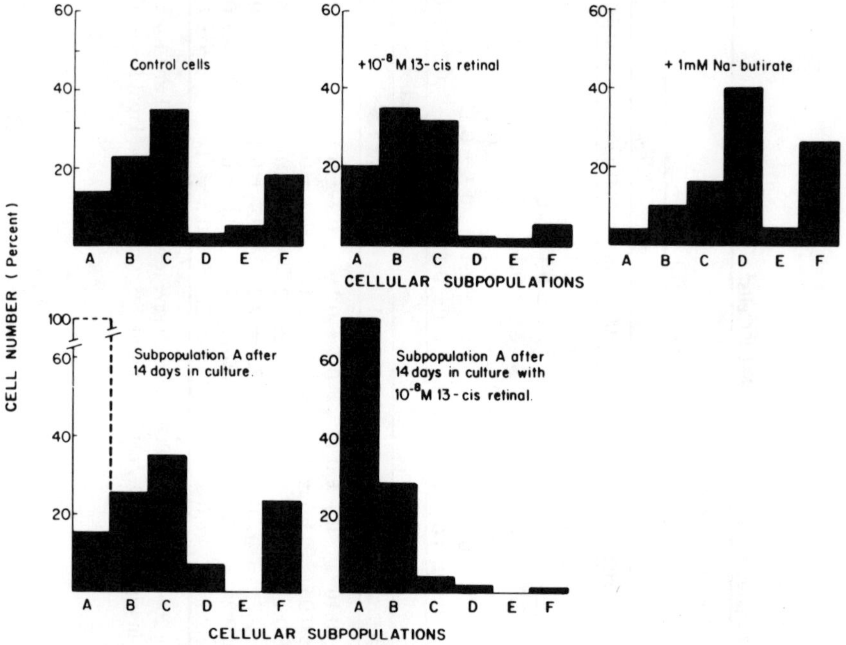

Fig. 3. Effect of 13-*cis*-retinal and sodium butyrate on the Percoll gradient profile of the total or subpopulation A of MCF-7 cells. Total MCF-7 cells after 5 days of 10^{-8} M 13-*cis*-retinal or 1 mM sodium butyrate treatment (upper panel). Subpopulation A was isolated also by density gradient centrifugation and subcultured for 14 days untreated (left panel) or treated with 10^{-8} M 13-*cis*-retinal (right panel) and again centrifuged in density gradients. From Resnicoff and Medrano (1989). Reproduced by permission from Academic Press.

Cellular Interactions

Tumors are cell societies that change many of their characteristics when the representation of one component changes at the expense of another (Heppner, 1984). We have used our model to study the ability of MCF-7 subpopulations to influence each other's growth. First, we reconstituted the entire population of cells by mixing the isolated subpopulations in relative proportion to their original representation (Fig. 4). The reconstituted population, called MCF-7 "R," grow in culture similarly to the unfractionated parental cells. The remarkable influence of the subpopulation E was shown when E cells were not included in the MCF-7 "R" cells. MCF-7 "R" (−E) cells grow much slower than the unfractionated parental population, and after 16–20 days in culture they ceased growth. By Percoll separation, we found that more than 85% of the cells were in the densest part of the gradient (data not shown). The addition of conditioned medium from the E fraction allowed MCF-7 "R" (−E) to grow for 3–10 days at a rate similar to that of the parental cells (Fig. 4).

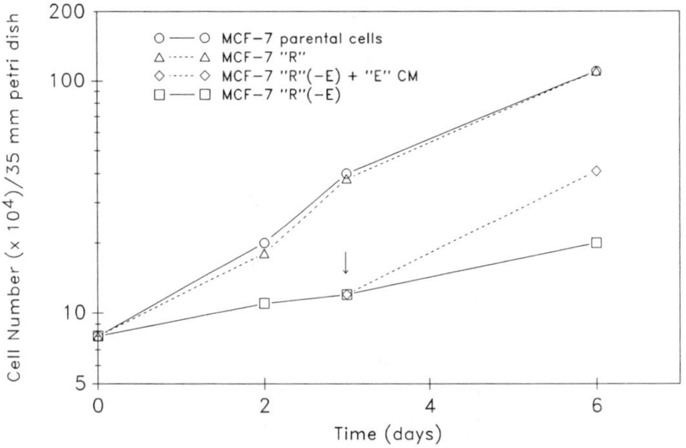

Fig. 4. Growth curves of the parental MCF-7, MCF-7 "R," and MCF-7 "R"(−E) cells. MCF-7 "R" were obtained by mixing the isolated subpopulations in the proportions originally found in the parental cells (see Fig. 1). MCF-7 "R"(−E) were obtained by mixing all but subpopulation E. The arrow indicates the day at which conditioned medium (CM) from subpopulation E was added.

The strong cellular interactions that coexist and regulate tumor growth may derive in part from the putative positive and negative growth factors released to the culture medium by the different subpopulations. We tested the ability of medium conditioned by different subpopulations to affect the growth of the stem cells. Figure 5 shows that medium from fraction F had a strong inhibitory effect on E cells. In addition, medium conditioned by the stem cells had a profound stimulatory effect on the stem cells themselves. These results are probably related to the ability of MCF-7 cells to produce positive and negative growth factors (TGF-α and TGF-β). We postulate that growth is the final result of the exquisite balance of autocrine and paracrine factors released not only by the stem cells, but also by their more differentiated descendants. The phenotypic heterogeneity in the uncloned MCF-7 cell line is a clear example that, in some cases, tumor cells retain some of the characteristics of the differentiation pathway followed by the normal tissue from which they originated.

Melanoma and Precursor Cells: Multiple (Independent?) Subpopulations That Correspond to Various Stages of Differentiation

Melanoma, a tumor originating from pigmented cells, has been the subject of extensive studies related to its characteristic heterogeneity presented in vivo and in vitro (Houghton et al., 1987). Tissue sections of melanoma lesions

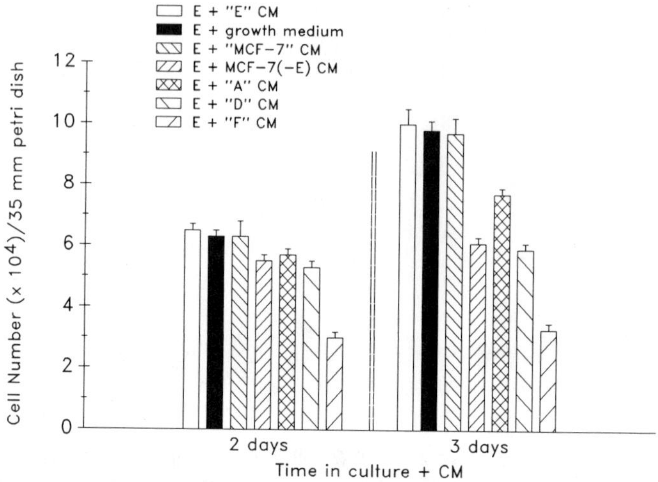

Fig. 5. Effect of conditioned media from several subpopulations on the stem cells (''E'' fraction). Cells were seeded at 2.0 × 10^4 cells/35-mm petri dish in the presence of conditioned media (CM were tested at 50% concentration). E + ''E'' CM: subpopulation ''E''+CM obtained from separate cultures of the same fraction. E + growth medium: subpopulation ''E'' with fresh growth medium. ''A,'' ''D,'' and ''F'' represent CM obtained from these isolated subpopulations.

show substantial variability in morphology: epithelioid, spindle, dendritic, giant, and unclassified cells, all positive for the neural crest-derived antigen S-100 and other markers for melanomas. Variability is also seen in the degree of pigmentation within the tumor mass and in the variety of antigens expressed by different clones derived in vitro from parental cultures. Observations made by us and other investigators led us to consider that subpopulations of cells at different stages of differentiation may coexist in melanoma and cultures derived from them. To address the possibility, we recently derived a highly heterogeneous cell line, IIB-Mel-J, which may be helpful for the study of cellular differentiation and interactions between subpopulations of melanoma cells (Guerra et al., 1989). The cellular heterogeneity shown by IIB-Mel-J cells is morphologically related to the three stages of the melanocyte differentiation and is reflected in the different subpopulations obtained by Percoll fractionation and by differences in ganglioside expression and labeling index (Table 2). Preliminary studies on the isolated subpopulations showed that they have different growth rates. After Percoll gradient separation, each fraction was enriched for cells having a particular morphology, i.e., dendritic cells in A, epithelioid and round cells in B and C. However after weeks in culture, the subpopulations were once again morphologically heterogeneous. It remains

TABLE 2. Expression of Gangliosides GD_2 and GD_3 and Thymidine Labeling Index (TLI) in Subpopulations of IIB-Mel-J Cells[a]

Peak	Density (g/ml)	GD_2 (%, mean ± SD)	GD_3 (%, mean ± SD)	TLI
A	1.050	17.0 ± 0.3	1.9 ± 0.9	10.9
B	1.060	78.0 ± 2.6	80.8 ± 3.5	24.5
C	1.070	89.6 ± 0.3	75.0 ± 0.5	27.0
Unfractionated cells		72.4 ± 0.3	75.0 ± 0.5	34.2

[a]IIB-Mel-J cells were cultured and submitted to a Percoll gradient centrifugation. The cells recovered in fractions A, B, and C were 19, 52, and 27%, respectively. GD_2 and GD_3 were determined by immunocytochemistry. For the TLI determination, cells were labeled with [^{3}H]thymidine for 18 hr, washed, and submitted to density gradient centrifugation. After recovering the different fractions, they were autoradiographed with Kodak NTB-2 emulsion. Reproduced from Guerra et al. (1989), with permission.

to be determined if this heterogeneity is due to the presence of several clones, each of which may also be able to generate cells with several differentiation states. Interestingly, we found that dysplastic nevi, a class of pigment lesions with potential for malignant transformation, also have extensive heterogeneity. Differences in pigment production and in the morphology of cells (small bipolar, dendritic and multinucleated cells) may account for the observed heterogeneity. When the growth rates of the isolated subpopulations were studied (Table 3), we found that one of the slowest subpopulations became, after few weeks in culture, a fast growing one. These results suggest strong cellular interactions among dysplastic nevi cells.

Thus, melanoma and putative precursor lesions are another example that shows that only by considering all subpopulations within a tumor will we eventually be able to elaborate a meaningful conclusion regarding tumor growth and more completely understand the action of cancer therapeutic agents. Also,

TABLE 3. Growth in Culture of Subpopulations of Dysplastic Melanocytes[a]

Peak	Days in culture after isolation	Generation time (hr)
A	4	24–36
B	4	70–80
C	4	>120
C	30	60–70
Total population	—	72–96

[a]Dysplastic nevi melanocytes were exponentially growing (V passage) at the moment they were fractionated by Percoll density gradients. Subpopulation C remained quiescent for more than a week. Thereafter, cells began growing with a generation time slightly faster than the parental cells. We discarded the possibility of selection of a transformed clone of C cells since both "C" and total cells reached senescence after more than 100 doublings. Growth conditions will be explained elsewhere (Medrano et al., in press).

the possibility of regulating tumor cell growth by controlling the finely tuned gene expression that is regulated not only by the environment of the tumor (immune cells, organ of nesting, extracellular matrix) but also by the tumor's own cells may become a reality in the future.

Conclusion

Phenotypic heterogeneity is a common feature of the majority of human tumors and is now recognized as one of the reasons for the failure of chemotherapy and radiation therapy. By using a very simple technique of cellular fractionation, we evaluated the response of tumor subpopulations to growth factors and hormones, not as isolated clones but as components of a cell society from which they came. We also have begun to study the very complex interactions that coexist within a tumor. We proposed that in some cases tumor societies mimic in part the "Yin-Yang" regulation which has been proposed as a central feature of normal growth control (Sager, 1988). However, I want to emphasize that ours is also a very simplistic model for the analysis of cellular interactions. In vivo, tumor cells interact with immune cells as well as the tumor microenvironment, particularly during the metastatic nesting of cells. Only new advances in understanding cellular interactions at the molecular level will clarify some of the ideas we have on tumor growth and dissemination.

Acknowledgments

The author is indebted to several colleagues in her laboratory who carried out the studies described in this paper. In particular, Mariana Resnicoff who played an essential role in the laborious experiments with the breast tumor cells. The author also gratefully acknowledges the influence of Dr. Jose Mordoh and colleagues (IIB F. Campomar) on the development of many of the ideas summarized in this work, and Dr. James J. Nordlund, Chairman of the Department of Dermatology, University of Cincinnati College of Medicine, for introducing her to the fascinating world of pigment cell research. These studies were supported by grants from the Consejo Nacional de Investigaciones Cientificas y Técnicas de Argentina (CONICET), the Fundación Maria Calderón de la Barca, the University of Buenos Aires, and a research grant from the American Dermatology Foundation (U.S.A.). The author wishes to thank Joan Griggs for her expertise in the preparation of this manuscript and Drs. Raymond Boissy and Zalfa Abdel-Malek for helpful discussions.

References

Buick RN, Pollack MN (1984). Perspectives on clonogenic tumor cells, stem cells and oncogenes. Cancer Res 44:4909–4918.

Guerra L, Mordoh J, Slavitsky I, Larripa I, Medrano EE (1989). Characterization of IIB-Mel-J: A new and highly heterogeneous human melanoma cell line. Pigment Cell Res 2:504–509.
Hart IR, Fidler IJ (1981). The implications of tumor heterogeneity for studies on the biology and therapy of cancer metastasis. Biochim Biophys Acta 651:37–50.
Hart IR, Goode NT, Nilson RE (1989). Molecular aspects of the metastatic cascade. Biochim Biophys Acta 989:65–84.
Heppner GH (1984). Tumor heterogeneity. Cancer Res 44:2259–2265.
Heppner GH, Miller BE (1989). Therapeutic implications of tumor heterogeneity. Sem Oncol 16(2):91–105.
Houghton AN, Real XF, Davis LJ, Cordon-Cardo C, Old LJ (1987). Phenotypic heterogeneity of melanoma. J Exp Med 164:812–829.
Medrano EE, Cafferata EGA, Larcher F (1987). Role of thrombin in the proliferative response of T-47D mammary tumor cells. Exp Cell Res 172:354–364.
Nicholson GL (1987). Tumor cell instability, diversification and progression to the metastatic phenotype: From oncogene to oncofetal expression. Cancer Res 47:1473–1487.
Nowell PC (1986). Mechanism of tumor progression. Cancer Res 46:2203–2207.
Resnicoff M, Medrano EE (1987). 13-cis-Retinal stimulates proliferation and induces intranuclear protein accumulation in the human mammary tumor cells MCF-7. Biochim Biophys Acta 143:309–315.
Resnicoff M, Medrano EE, Podhajcer OL, Bravo AI, Bover L, Mordoh J (1987). Subpopulations of MCF-7 cells separated by Percoll gradient centrifugation: A model to analyze the heterogeneity of human breast cancer. Proc Natl Acad Sci USA 84:7295–7299.
Resnicoff M, Medrano EE (1989). Growth factors and hormones which affect survival, growth and differentiation of the MCF-7 stem cells and their descendants. Exp Cell Res 181:116–125.
Sager R (1988). Tumor suppressor genes. In N. Colburn et al. (eds): UCLA Symposia on Molecular and Cellular Biology, Vol 58. New York: Alan R. Liss, pp 353–357.

Perspectives on Cellular Regulation:
From Bacteria to Cancer, pages 225–234

The Biology of Human Mammary Epithelium in Culture: The Path From Viral Transformation to Human Cancer

Helene S. Smith

Geraldine Brush Cancer Research Institute, Pacific Presbyterian Medical Center, San Francisco, California 94115

Introduction

At a recent meeting, Art confided his disappointment at not getting one of his grant awards funded. His career, he felt, spoke for itself, but the most frustrating part to him was that he believed the work he was currently proposing was the best of his career so far, yet his peers seemed to miss the point entirely. I mentioned that his comment reminded me of an article by Gunther Stent (Stent, 1972) describing the premature discovery as one made by an intuitive leap. Since it is far ahead of its time, colleagues cannot connect it to their own work; hence, it is generally ignored. Most of science proceeds in increments where details are filled in and the major challenge is to be the first one to complete the next obvious step. Art listened without comment, but the next day, he hurried up to tell me that he never wanted to do the kind of science that filled in small gaps. The challenge for him was to take the big leap that pointed the way for the future. This summarizes my perception of the quintessential Art Pardee—a scientific aristocrat.

It also summarizes my experiences as his postdoctoral fellow. Although the research direction that I subsequently pursued did not directly stem from my studies in his laboratory, his approach to research profoundly influenced me. I became acutely aware of the need to trust my own judgment and to ask what I believed to be the most important questions, regardless of the scientific fashion of the moment. That goal led me by a circuitous route to the biology of human mammary epithelium in culture.

After studying with Art, I went to NIH where I worked with George Todaro on viral transformation. There, together with Charles Scher, I discovered abortive transformation by simian virus 40 (SV40) (Smith et al., 1971). Subsequently I discovered that latent viral DNA was present in the abortive transformants

(Smith et al., 1972). I then asked whether abortive transformants were tumorigenic in animals (Wright et al., 1973). From these studies, I realized that the commonly used rodent culture systems were inadequate models of malignancy. The "normal" 3T3 negative controls made tumors as frequently as did the SV40 transformed positive controls, albeit requiring more time. These disturbing results led me to start reading about human cancer. I soon became aware of the fact, obvious to clinicians but not to most molecular virologists at the time, that 90% of human malignancies are of epithelial origin. Unfortunately, all of the cell culture systems then available for studying transformation were of fibroblastic origin. Thus began my fascination with developing epithelial cell culture systems for studying malignancy, first using mouse liver and mammary gland and subsequently with human tissues (for reviews see Smith et al., 1981, 1984; Hackett and Smith, 1986; Smith et al., 1987a). My purpose in developing these systems was to provide the means for asking how advances in molecular and cell biology using animal models could be applied to understanding human cancer.

Early Studies on Mouse Epithelium

After NIH, I moved to the School of Public Health of the University of California, Berkeley, where I began a long and fruitful collaboration with Dr. Adeline Hackett and the late Mr. Robert Owens. Robert Owens had been trying to culture epithelial cells from various mouse organs. He was using a collagenase digestion technique developed by Lasfargues (Lasfargues and Moore, 1971) to degrade the stromal matrix and basement membrane. After digestion, the standard approach was to mechanically aspirate the digest, obtaining a single cell suspension for subsequent culture. Residual clumps were routinely discarded. Mr. Owens and I hypothesized that the key to success was to plate the clumps rather than the single cells. Since epithelial cells are characterized by tight junctional complexes that are not destroyed by collagenase, we reasoned that the shear stress necessary to obtain a single epithelial cell suspension might kill most of the epithelial cells. If, however, we allowed the clumps of epithelium to attach to the culture vessel, we found that the epithelial cells migrated out of the clump and readily proliferated. Using this approach, we were able to develop numerous epithelial cell lines from various normal and malignant murine tissues (Owens et al., 1974). Subsequently, I was able to characterize the in vitro biology and viral transformation of the cell lines originating from mammary gland and bile duct (Anderson and Smith, 1978, 1980; Dollbaum et al., 1980).

During the first few passages in culture, the normal epithelial cell cultures behaved predictably. They had cuboidal morphology, formed desmosomes and other tight junctions, and formed fluid-filled secretory domes at confluence. We were able to inject them subcutaneously into mice and show that they organized into fluid-filled, benign cysts, characteristic of the tissue of origin (Owens et al., 1976).

Yet, as I delved more deeply into the murine epithelial systems, I became as dissatisfied with them as with their fibroblastic counterparts. After approximately 15–20 passages, despite the fact that the cells exhibited no obvious change in culture, they began to form carcinomas when they were reinoculated into mice. Thus, the cultures were inadequate for understanding malignant transformation since, by the time we had isolated sufficient transformed cells to study, the untreated controls also had become malignant.

It was known that human fibroblasts did not spontaneously transform in culture. The fact that murine epithelial cells and fibroblasts both readily immortalized and became malignant suggested that there were some fundamental differences in cellular growth controls between humans and rodents. These concerns led me to face the seemingly insurmountable logistical problems inherent with working on human systems.

I hypothesized that the biochemical behavior of different organs would be as variable as their function, hence the task would be much too complicated if more than one organ was chosen. After some initial studies using any type of available human tissue, we decided to concentrate on the mammary gland. The mammary gland was chosen because it had a relatively simple organizational structure. Furthermore, it was one of the only organs where large amounts of normal tissue were readily available from plastic surgery (reduction mammaplasties) and also from lacteal secretions. For other organs, the only nonmalignant tissue available for study came from nonmalignant regions of tissues removed during surgery for cancer. It was possible that a ''field'' effect might be associated with carcinomas, hence, such tissues might not be really normal.

Normal Mammary Epithelium in Culture

We initially utilized milk as the source of normal mammary cells. A postdoctoral fellow at the laboratory, Gertrude Buehring, discovered that large numbers of epithelial cells could be obtained from milk, particularly during early stages of weaning, when mammary ducts and alveoli involute and slough into the luminal contents (Buehring, 1972). These milk-derived cells had the advantage of being free of fibroblast contamination; however, they grew poorly in whatever media formulations were given to them.

While we were focusing on milk cells, Dr. Richard Hallows, at the Imperial Cancer Research Fund, had modified our technique for digesting rodent epithelial tissues to isolate epithelial cells from the massive amounts of connective tissue and fat comprising reduction mammoplasties. He grossly dissected the epithelial tissue and enzymatically digested it with collagenase and hyaluronidase to degrade the stromal matrix and basement membranes. As with the rodent tissues, the epithelial cells remained as clumps while the stromal fibroblasts and connective tissues were dissociated to single cells. The epithelial cells were then easily isolated free of fibroblasts by filtration through nylon mesh filters. When Richard Hallows came for a sabbatical, he brought this technique for handling reduction mammaplasties to the laboratory.

At the same time, Dr. Martha Stampfer started working at the laboratory. She developed MM medium, a formulation using much less serum (1% fetal calf serum), which allowed considerably more proliferation than previously described media (Stampfer et al., 1980). The major components of this medium are conditioned media from three specific human cell lines previously developed by Robert Owens and me (Owens et al., 1976; Hackett et al., 1977). One difficulty with using these cell lines to provide conditioned media is that they grow poorly and have a finite life span. [Circumventing this problem, I have recently found that conditioned media from short-term endothelial cultures of human umbilical chord or bovine aorta are equally effective at providing the conditioning factors needed by mammary epithelial cells (unpublished observation).] In addition to conditioned media factors, MM contains estradiol, insulin, hydrocortisone, and triiodothyronine. When cholera toxin was added to it and the cells sparsely plated on irradiated fibroblast feed cells, Dr. Sam Lan (a visiting Scientist from Einstein Medical Center, NY) and I found that single mammary epithelial cells proliferate rapidly, resulting in a highly efficient clonogenic assay (Smith et al., 1981).

Martha Stampfer then continued to modify and improve the culture medium focusing on in vitro transformation, particularly with chemical carcinogens. I turned toward studies to characterize breast cancers in culture.

Breast Cancer in Culture

Rationale

As my own thinking evolved, an additional paradox emerged. My initial objective was to develop more relevant normal cell substrates for studying transformation. I wanted to make the normal cells become malignant using external agents that could be manipulated in a reproducible manner. Initially, I assumed that the criteria for selecting transformants would be the same as for rodent fibroblasts. However, it became increasingly clear that the criteria for defining transformation of human mammary epithelium might be very different from those of rodent cells. For example, by transforming normal cells with carcinogens so that they become tumorigenic in nude mice, I could define only one pathway by which the cells could possibly become cancerous. However, these nude mouse studies would tell me nothing about which pathways are commonly used by human mammary cells when they become cancerous, since breast cancers rarely form tumors in nude mice.

The only way to know what constitutes an appropriate assay for malignant transformation would be to characterize the phenotypic differences between breast cancers and normal mammary cells. To do this, it became necessary to culture breast cancers as well as normal mammary epithelium. Cell culture is necessary because breast carcinoma cells in vivo are in various proliferative states, with some cells dividing while others are viable but quiescent, and still others are in various stages of cell death. In contrast, normal mammary epithelium is quiescent, viable, and has a much more uniform and appropriate blood supply. When normal and malignant tissues are compared, differences in their physiologic states may mask underlying differences associated with malignancy. Placing cells in culture would allow both the normal and malignant cells to be in the same proliferative and physiologic state, thereby highlighting consistent differences associated with malignant transformation.

Cell Lines

When I started working in this field, there were a number of permanently established breast cancer cell lines. Although these lines provide substrates for many important studies, I found that they had limited utility for defining transformation criteria.

One problem with cell lines is that breast cancers are heterogeneous, but only a small subpopulation within the cancerous tissue develops into a cell line. To develop a cell line, usually the tumor cells are held in a maintenance state for prolonged periods of time, sometimes months, before a cell population emerges that is able to grow continuously in culture. During the initial period, most of the carcinoma cells obtained from the malignant tissue proliferate a few times and then undergo a phenomenon, termed ''crisis'' in which most of them stop proliferating, deteriorate, and disappear from the culture vessel. The cell line subsequently emerges from a subpopulation of the remaining cells.

I also realized that breast cancer cell lines are strongly selected to represent only late stages of malignant progression. Only a small number of breast cancer specimens contain a subpopulation capable of proliferating subsequent to crisis and most of these are derived from metastatic lesions. Even among effusion metastases, the most widely studied type of metastatic lesion, less than 10% actually develop into lines. Among primary breast cancers, we found that only 1 of more than 200 specimens developed into a cell line. The line that we established was from a carcinosarcoma, a very unusual histopathologic type of breast cancer, that does not necessarily reflect the biology of much more common infiltrating breast carcinomas (Hackett et al., 1977).

The ability to survive crisis and become an immortalized cell line is not random, either in relation to culture technique or to tumor progression. We examined the properties in culture of three breast cancer effusion metastases, obtained over approximately 2 years from the same patient. Despite repeated attempts with cryopreserved cells, only the last specimen reproducibly exhibited immortality in culture; the first two specimens grew initially but failed to survive the crisis. Each specimen was unique in morphology, growth properties, and oncogene aberrations (Smith et al., 1987b; Liu et al., 1988), although karyotypic markers indicated a common origin. The observation that the last effusion metastasis could develop reproducibly into a cell line when prior malignant effusions from the same patient could not suggests that the capacity for infinite life in culture depends on inherent change(s) in the biological phenotype of the tumor rather than on irreproducible vagaries of cell culture. This study together with numerous observations that metastatic specimens much more commonly develop into cell lines than primary breast cancers indicate that the capacity for infinite life in vitro results from a phenotype that is usually acquired by breast cancer cells at a late stage of malignant progression.

Short-Term Cultures

Utilizing the MM medium developed by Martha Stampfer, we attempted to culture primary breast carcinomas (Smith et al., 1981), and in the vast majority of cases, cultures were successfully established. They tended to grow more slowly and clone with lower efficiency than mammary cells either from reduction mammoplasties or from nonmalignant mastectomy tissue peripheral to carcinomas (for review see Smith et al., 1987a).

The carcinoma-derived cultures were very heterogeneous, containing populations with varying growth potential. Although even the most rapidly proliferating cells in tumor cultures proliferated more slowly than those from nonmalignant specimens, still certain tumor-derived subpopulations tended to overgrow the initial cultures. These cells had a near-diploid karyotype (Wolman et al., 1985; Smith et al., 1985b). They also tended to be morphologically similar to normal cells in culture (Smith et al., 1981).

Despite their slower growth rate, the similarities in karyotype and morphology between tumor-derived and nonmalignant cells raised the question of whether the carcinoma-derived cultures contained bona fide tumor cells or nonmalignant cells originating from tissue peripheral to the malignancy. There are a number of reasons why it is unlikely that the predominantly diploid carcinoma-derived cells originate from nonmalignant tissue. First, the cellular clumps dissociated from tumors tend to be much smaller and less structured than clumps digested from nomalignant tissue. Second, it is misleading to argue that the morphologic similarities between cultured nonmalignant and malignant breast epithelium indicate that only nonmalignant cells are being cultured. We found that normal mammary cells themselves acquire many morphologic features associated with malignancy in vivo, such as oncogene overexpression (Benz et al., 1989) and high nuclear:cytoplasmic ratio, and irregular chromatin (Smith et al., 1987a). The fact that the normal cells in culture acquire these changes suggests that many of the criteria commonly used to characterize malignancy may be related to proliferation or loss of appropriate three-dimensional cellular architecture rather to malignancy per se.

Further substantiating the suggestion that carcinoma-derived cultures are not derived from normal tissues were additional reports describing consistent differences between cultures derived from malignant and nonmalignant tissues. Asaga et al. (1983) found a significant increase in multinucleated cells after incubation of human mammary carcinoma cultures with cytochalasin when compared with cultures derived from various benign tissues. Similar results were reported for cultured rodent mammary tissues (Medina et al., 1980). Carcinoma-derived cultures also showed increased variability in surface antigen expression when compared to nonmalignant tissue from the same donor (Ceriani et al., 1978). When malignant and nonmalignant cultures were compared using antiserum to a 19.5-kDa glycoprotein, all tumor-derived cultures were positive while all nonmalignant cultures were negative (Stampfer et al., 1982). Additionally, in an in vitro assay for invasion utilizing denuded human amnions, we found that the tumor-derived cells were capable of invasive growth while nonmalignant cells were not (Smith et al., 1985a).

Recently we compared tumor and normal cells for sensitivity to tumor necrosis factor (TNF) (Dollbaum et al., 1988). Although the response of the tumors varied from specimen to specimen, the vast majority were very sensitive to TNF. In contrast, all of the specimens derived from nonmalignant breast epithelium were very resistant to TNF. This differential sensitivity to TNF was even seen in nonmalignant and malignant mammary epithelium from the same patient.

Whatever the origin of the diploid carcinoma-derived cells, it is clear that there are additional tumor-derived populations that do not proliferate readily in culture. In approximately 20% of cultures from primary carcinomas, another cell type with abnormal morphology was also observed. These abnormal cells, designated E′ cells, were thought to resemble cells from some metastatic lesions. However, unlike the cells from metastases, they were unable to grow in culture (Taylor-Papadimitrious et al., 1980).

Additional studies verify the observations that very aberrant subpopulations that grow poorly in culture are found in primary breast cancers. A subpopulation appearing in about 50% of primary carcinomas in vivo has been defined by a cytochemical reaction for reduced nicotinamide adenine dinucleotide phosphate neotetrazolium (NADPH-NT) reductase (Peterson and van Deurs,1987). In primary culture of NADPH-NT reductase positive tumors, positive cell islets were seen, some of which were highly aneuploid. The population doubling times of these NADPH-NT reductase positive tumor cells ranged from 5 days to infinite. In another study, a subpopulation of cells was seen in primary breast cancers that adhered poorly, if at all, to collagen substrates. In most cases, these nonadherent cells proliferated very poorly, and were morphologically very aberrant (Rudland et al., 1985).

Thus the biology of malignant mammary epithelium in culture is very complicated. The commonly held belief that malignant cells grow better than normal ones is incorrect. Although occasional breast carcinoma specimens develop into permanent cell lines, most carcinoma-derived cells grow very poorly in culture. In most cases, only less deviant, predominantly diploid carcinoma-derived subpopulations found in primary breast carcinomas and hypodermal metastases grow to any significant extent.

I have hypothesized that the diploid carcinoma cells represent early stages of malignant progression (Smith et al., 1988). As such they may provide important insights into fundamental differences between normal and malignant mammary epithelium because they lack secondary but irrelevant changes. Furthermore, I have found them to be useful for predicting patient response to chemotherapeutic drugs (Smith et al., 1985b, 1990). However, a fundamental characteristic of breast cancers is their increased aggressiveness as a function of malignant progression. To understand the nature of this progression, it will be necessary to study the later, more aggressive forms of the disease. For such studies, the ability to readily culture the more deviant tumor subpopulations will be necessary. To reach this goal, it is clear that new culture procedures and media formulations will be needed.

Summary

The goal of all these studies has been to develop ways to ask how advanced in molecular and cell biology using animal models could be applied to understanding human cancer. Recently, new molecular techniques have permitted exciting advances by directly studying breast cancer tissues. However, these molecular advances raise some of the same questions that have fascinated me for so long. For example, to characterize a putative breast cancer suppressor gene, it will be necessary to put the gene into tumor cells and show that the tumor cells regain a normal phenotype. But what criteria will be used to define normality? Certainly, dysregulated growth is one critical aspect of malignancy. Putting a suppressor gene into tumor cell lines may inhibit proliferation and, therefore, prevent tumorigenicity in nude mice. But loss of other key functions also may be critical for the acquisition of malignancy. Such changes may not be measurable as loss of tumorigenicity in nude mice by breast cancer lines, as discussed above. Thus cell lines may be an artificially selected system that will miss important markers. Further studies on short-term cultures of primary breast may be necessary to answer some of the critical questions facing us in the next decade.

Acknowledgments

Supported in part by Grant P01 CA 44768 from the National Cancer Institute, National Institutes of Health.

References

Anderson LW, Smith HS (1979): Premalignancy in vitro: Progression of an initially benign epithelial cell line to malignancy. Br J Exp Pathol 60:575–581.

Anderson LW, Smith HS (1980): Simian virus 40 and Moloney-murine sarcoma virus infection of bona fide mouse epithelium. J Gen Virol 49:443–446.

Asaga T, Suzuki K, Takemiya S, Okamoto T, Tamura N, Umeda M (1983): Difference in polynucleation of cultured cells from human mammary tumors and normal mammary glands on treatment with cytochalasin. B. Gann 74:95–99.

Benz C, Scott GK, Santos GF, Smith HS (1989): Expression of cmyc, c-Ha-ras1, and c-erbB-2 proto-oncogenes in normal and malignant human breast epithelial cells. J Natl Cancer Inst 81:1704–1709.

Buehring GC (1972) Brief communication: Culture of human mammary epithelial cells: Keeping abreast with a new method. J Natl Cancer Inst 49:1433–1434.

Ceriani RL, Peterson JA, Blank EW (1978): Variability in surface antigen expression of human breast epithelial cells cultured from normal breast, normal tissue peripheral to breast carcinomas and breast carcinomas. Cancer Res 44:3033–3039.

Dollbaum C, Creasey A, Dairkee S, Hiller AJ, Rudolph AR, Lin L, Vitt C, Smith HS (1988): Specificity of TNF toxicity for human mammary carcinomas relative to normal mammary epithelium and correlation with response to doxorubicin. Proc Natl Acad Sci USA 85:4740–4744.

Dollbaum CM, Anderson LW, Smith HS (1980): An in vitro model of neoplastic progression in murine epithelial cells. Cancer Lett 11:121–127. Hackett AJ, Smith HS (1986): In Vitro Models for Cancer Research,'' Vol. III. Boca Raton, FL: Webber and Sekely, pp. 31–49.

Hackett AJ, Smith HS (1986): In Vitro Models for Cancer Research,'' Vol. III. Boca Raton, FL: Webber and Sekely, pp. 31–49.

Hackett AJ, Smith HS, Springer EL, Owens RB, Nelson-Rees WA, Riggs JL, Gardner M (1977): Two synergistic cell lines from human breast tissue: The aneuploid mammary epithelial (Hs578T) and the diploid nyopithelial (Hs578Bst) cell lines. J Natl Cancer Inst 58:1795–1806.

Lasfargues EY, Moore DH (1971): A method for the continuous cultivation of mammary epithelium. In Vitro 7:21–25.

Liu E, Dollbaum C, Rochlitz C, Benz C, Smith HS (1988): Molecular lesions involved in the progression of human breast cancer. Oncogene 3:323–327.

Medina D, Osborn CJ, Ash BB (1980): Distinction between preneoplastic and neoplastic mammary cell populations in vitro by cytochalasin B-induced multinucleation. Cancer Res 40:329–333.

Owens R, Smith HS, Hackett AJ (1974): Epithelial cell culture from normal glandular tissue of mice. J Natl Cancer Inst 53:261–269.

Owens RS, Smith HS, Nelson-Rees WA, Springer EL (1976): Epithelial cell cultures from normal and cancerous human tissues. J Natl Cancer Inst 56:8430–8439.

Peterson OW, van Deurs B (1987): Preservation of defined phenotypic traits in short-term cultured human breast carcinoma derived epithelial cells. Cancer Res 47:856–866.

Rudland PS, Hallowes RC, Cox SA, Ormerod EJ, Warburton MJ (1985): Loss of production of myoepithelial cells and basement membrane proteins but retention of response to certain growth factors and hormones by a new malignant human breast cancer cell strain. Cancer Res 45:3864–3877.

Stampfer MR, Hallowes RC, Hackett AJ (1980): Growth of normal human mammary cells in culture. In Vitro 16:415–425.

Stampfer MR, Hackett AJ, Hancock M, Leung JP, Edgington TS, Smith HS (1982): Cold Spring Harbor Symp Cell Prolif 9:819–829.

Smith HS, Dollbaum CM (1981): Growth of human tumors in culture In R. Baserga (ed): Handbook of Experimental Pharmacology: Tissue Growth Factors.

Smith HS, Gelb LD, Martin MA (1972): Detection and quantitation of SV40 genetic material in abortively transformed Balb/3T3. Proc Natl Acad Sci USA 69:152.

Smith HS, Scher CD, Todaro GJ (1971): Induction of cell division in medium lacking serum growth factor by SV40. Virology 44:359.

Smith HS, Lan S, Ceriani R, Hackett AJ, Stampfer MR (1981): Clonal proliferation of cultured nonmalignant and malignant breast epithelia. Cancer Res 41:4637–4643.

Smith HS, Wolman SR, Hackett AJ (1984): The biology of breast cancer at the cellular level. Biochim Biophys Acta Rev Cancer 738:103–123.

Smith HS, Liotta LA, Hancock MC, Wolman SR (1985a): Invasiveness and ploidy of human mammary carcinomas in short-term culture. Proc Natl Acad Sci USA 82:1805–1809.

Smith HS, Lippman ME, Hiller AJ, Stampfer MR, Hackett AJ (1985b): Response to doxorubicin of cultured normal and malignant mammary epithelial cells. J Natl Cancer Inst 74:341–348.

Smith HS, Dairkee SH, Ljung B-M, Mayall B, Sylvester SS, Hackett AJ (1987a): Cellular and Molecular Biology of Experimental Mammary Cancer. New York: Medina, Kidwell, Heppner, pp. 437–452.

Smith HS, Wolman SR, Dairkee SH, Hancock MC, Lippman M, Leff A, Hackett AJ (1987b): Immortalization in culture: Occurrence at a late stage in progression of breast cancer. J Natl Cancer Inst 78(4):611–615.

Smith HS, Dollbaum CM, Ljung BM, Mayall B, Hackett AJ (1988): Tumor Progression and Metastasis. New York: Alan R. Liss, pp. 143–150.

Smith HS, Zoli W, Volpi A, Hiller A, Lippman M, Swain S, Mayall AB, Dollbaum C, Hackett AJ, Amadori D (1990): Preliminary correlations of clinical outcome with in vitro chemosensitivity of second passage human breast cancer cells. Cancer Res, in press.

Stent GS (1972): Prematurity and uniqueness in scientific discovery. Sci Am 227:84–93.

Taylor-Papadimitriou J, Fentiman IS, Burchell J (1980): Problems and Directions in Cell Biology of Breast Cancer. New York: McGrath, Brennan, and Rich, pp. 347–362.

Wolman SR, Smith HS, Stampfer MR, Hackett AJ (1985): Growth of diploid cells from breast cancers. Cancer Gent Cytogenet 16:49–64.

Wright P, Smith HS, McCoy J (1973): Tumorigenicity and antigenicity of SV40 infected cells. J Natl Cancer Inst 51:951.

Perspectives on Cellular Regulation:
From Bacteria to Cancer, pages 235–249

Mitosis: Normal Control Mechanisms and Consequences of Aberrant Regulation in Mammalian Cells

Robert Schlegel and Ruth W. Craig

Laboratory of Toxicology, Harvard School of Public Health, Boston, Massachusetts 02115 (R.S.), and Department of Physiology, Johns Hopkins University School of Medicine, Baltimore, Maryland 21218 (R.W.C.)

Introduction

Cell cycle events are often thought to be regulated by a series of interdependent biochemical steps, with the initiation of late events requiring the successful completion of those preceding them. The loss of this ordered progression can lead to a decrease in cell viability and may be an important contributor to neoplastic transformation.

In reality, cell cycle control is not quite so straightforward. It appears that some cellular events, while having the illusion of being dependent on the completion of earlier steps, are, or can be manipulated to be, quite independent. It may, therefore, be more accurate to think of cell cycle control as a combination of interdependent, as well as autonomous, but synchronized, cellular programs.

How normal cell cycle events are controlled and how this regulation is altered by DNA damaging agents, DNA repair modifiers, oncogenes, and neoplastic transformation are some of the questions to which Dr. Arthur Pardee has applied his talents. This chapter addresses these questions as they pertain to the regulation of mitosis in mammalian cells. My studies on mitosis began when I was a postdoctoral fellow in Art's lab. The area of mitotic control has blossomed in recent years, and its development owes much of its conceptual foundation to the pioneering work of this renowned scientist.

Normal Control of Mitotic Onset

The control of meiosis and mitosis has received much attention during the last few years, and considerable progress has been made in understanding some of the important regulatory pathways. Much of this attention can be attributed to the discovery by Lee and Nurse (1987) that human DNA could functionally complement a cell division cycle mutant (*cdc2*) known to be defective in the transition from interphase to mitosis in the yeast *Schizosaccharomyces pombe*. Previous searches for human genes corresponding to those of yeast *cdc* mutants had used techniques based on sequence homology, rather than functional complementation. These attempts were to a large extent unsuccessful, presumably due to extensive evolutionary divergence in nucleotide and amino acid sequence. The *cdc2* gene encodes a serine–threonine protein kinase with a molecular weight of 34,000 ($p34^{cdc2}$). Microinjection experiments using an antibody to $p34^{cdc2}$ (Riabowol et al., 1989) and cell cycle studies with a mouse cell line containing a temperature-sensitive *cdc2* gene product (Th'ng et al., 1990) have shown that this kinase is essential for entry of mammalian cells into mitosis.

The cyclins, first discovered in sea urchin and clam embryos, are proteins that progressively increase in concentration as cells approach mitosis. These proteins are degraded rapidly during the transition from metaphase to anaphase (Evans et al., 1983). A role for cyclins in the regulation of meiosis and mitosis was established by the experiments of Swenson et al. (1986). They discovered that microinjection of cyclin A (one of several cyclin forms currently known) into *Xenopus* oocytes induced meiotic maturation. Cyclins have now been identified in cells of numerous species, including humans (Pines and Hunter, 1989). Cyclins are complexed with $p34^{cdc2}$ during mitosis, coincident with increased activity of the kinase, while cyclin degradation is associated with exit from mitosis and decreased kinase activity.

The importance of protein phosphorylation in the regulation of mitosis has become well established. The activity of the $p34^{cdc2}$ protein kinase is regulated by phorylation, and the phosphorylation state of this enzyme is cell cycle regulated. Dephosphorylation of $p34^{cdc2}$ increases its kinase activity during mitosis (Morla et al., 1989). The activity of other protein kinases, including $pp60^{c\text{-}src}$ (Chackalaparampil and Shalloway, 1988), is elevated during mitosis as well. The cyclins undergo extensive phosphorylation at mitosis (Meijer et al., 1989), and peptide mapping has identified at least nine sites of phosphorylation (Pines and Hunter, 1989). In addition, proteins such as lamins, histones, high mobility group proteins, nuclear matrix proteins, intermediate filaments, and nonhistone proteins become phosphorylated during mitosis (Adlakha et al., 1985 for review).

Dephosphorylation events have often been neglected in discussions of mitotic regulatory mechanisms. Other than the previously mentioned work by Morla et al. (1989), demonstrating the activation of $p34^{cdc2}$ by dephosphorylation, and the finding by Schlegel et al. (1990), that the protein kinase inhibitors 2-aminopurine and 6-dimethylaminopurine can induce premature mitosis, little research has focused on the role of protein dephosphorylation in controlling mitotic onset in mammalian cells. Earlier studies using microinjection of

Xenopus oocytes found that the regulatory subunit of cAMP-dependent protein kinase induced meiotic maturation (Maller and Krebs, 1977), whereas an inhibitor of phosphoprotein phosphatase-1 (PP-1) suppressed maturation (Huchon et al., 1981). More recent work showed that an inhibitor of PP-1 delayed posttranslational activation of maturation promoting factor in vitro (Cyert and Kirschner, 1988). It is clear that future research in this field must be concerned with both phosphorylation and dephosphorylation events. Work in the immediate future will most certainly focus on identifying the kinases and phosphatases regulating the phosphorylation state of $p34^{cdc2}$, and on the in vivo protein substrates for the $p34^{cdc2}$ kinase.

A temperature-sensitive mutant of Syrian hamster (BHK) fibroblasts, referred to as tsBN2, was initially isolated as a DNA synthesis mutant (Nishimoto et al., 1978). This mutant was shown to undergo premature mitosis at the restrictive temperature. The hamster and human genes associated with this defect have been cloned and sequenced and show no resemblance to known protein kinases or phosphatases (Uchida et al., 1990). These findings indicate that additional control mechanisms that we currently do not understand play important roles in mitotic regulation and that considerable work will be required before these processes are well characterized.

Aberrant Regulation of Mitotic Onset

Chemically Induced Alterations

The timing of mitosis is responsive to stresses induced by chemical and physical agents. Aside from global RNA and protein synthesis inhibition, DNA damage has long been known to delay mitotic onset (Whitmore et al., 1961). DNA damage causes a delay in the G_2 phase of the cell cycle that can persist for many hours. The length of this delay is directly related to the extent of DNA damage. This response is important for cell survival because DNA repair occurs during the G_2 delay, making it possible to repair damaged DNA before the onset of chromosome condensation and segregation. Although much of this work was conducted using ionizing radiation, the same principles apply to chemically induced damage.

Exposure of cells to methylxanthines, and caffeine in particular, can suppress the G_2 delay induced by DNA damage (Walters et al., 1974; Lau and Pardee, 1982). Methylxanthines also advance slightly the onset of mitosis in undamaged cells (Scaife, 1971). Moreover, caffeine can induce the onset of mitosis before DNA replication is completed, thereby uncoupling a normally dependent relationship between these two cell cycle events (Schlegel and Pardee, 1986).

Synchronized Syrian hamster (BHK) fibroblasts that were arrested in early S phase with DNA synthesis inhibitors underwent premature chromosome condensation (PCC), nuclear envelope breakdown, morphological "rounding up," and mitosis-specific phosphoprotein synthesis following exposure to caffeine

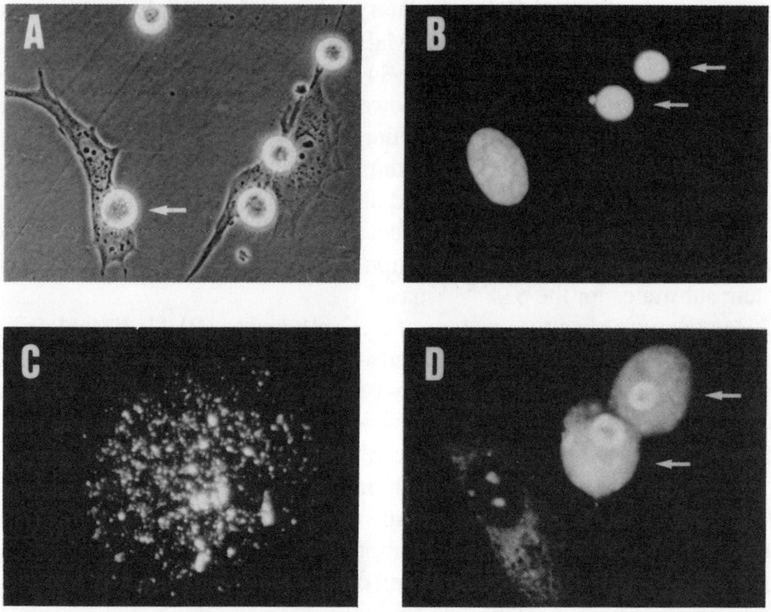

Fig. 1. Cytoplasmic, chromosomal, and phosphoprotein synthesis changes in PCC-containing BHK cells (indicated by arrows). Cells were synchronized and arrested in early S phase and then treated with caffeine (5 mM) and colcemid (0.3 μg/ml) as described previously (Schlegel and Pardee, 1986). **(A)** Phase contrast photograph (× 1000) of "rounded up" PCC-containing cells 4 hr after the addition of caffeine. No "rounded up" cells were seen in untreated controls. **(B)** Condensed chromatin of cells 8 hr after the addition of caffeine. Cells were fixed in absolute methanol and stained for 10 min with Hoechst 33242 (1 μg/ml) (× 2500). **(C)** Chromosome preparation stained with Hoechst 33242 showing S phase PCC 4 hr after the addition of caffeine (× 2500). **(D)** Indirect immunofluorescence of mitosis-specific phosphoproteins present in both the cytoplasm and chromatin of the same PCC-containing cells shown in **B**. Monoclonal antibody MPM-2 (Davis et al., 1983) and rhodamine-conjugated, goat-anti-mouse secondary antibody were used as described previously (Schlegel and Pardee, 1986). From Schlegel and Pardee (1986). Reprinted with permission from *Science*.

(Fig. 1). Caffeine concentrations of 200 μM or greater were required to induce these premature mitotic events. Figure 1A is a phase contrast photograph showing the rounded mitotic morphology of cells undergoing premature mitosis following exposure to 5 mM caffeine for 4 hr. The condensed state of the chromatin can be seen easily in these cells following fixation and staining with the DNA-specific stain Hoechst 33242 (Fig. 1B). Chromosome preparations of these premature mitotic cells demonstrate the pulverized appearance of the chromatin (Fig. 1C). This pulverized appearance is identical to that seen when S phase cells are forced to undergo premature chromosome condensation by fusion with mitotic cells (Johnson and Rao, 1970). Figure 1D illustrates that the cells undergoing premature mitosis in Figure 1B were also synthesizing mitosis-specific phosphoproteins, as determined by indirect immunofluorescence with a monoclonal antibody that recognizes a family of such proteins (Davis et al., 1983).

New protein, but not new RNA, synthesis was required for the caffeine-induced premature mitosis, indicating that the RNA required for mitotic onset was already present in cells arrested in S phase. Caffeine may, therefore, enhance the translation of this preexisting RNA or stabilize the corresponding proteins or posttranslational modifications. Experiments that will be discussed later have indicated that alterations in the stability of mitosis-related RNA and protein(s) have an important role in premature mitotic onset (Schlegel et al., 1987, 1990). Caffeine is known to inhibit the enzyme cyclic AMP phosphodiesterase, thereby increasing the intracellular levels of cyclic AMP. Phosphodiesterase inhibition cannot be solely responsible for these effects, however, because the potent phosphodiesterase inhibitor 3-isobutyl-1-methylxanthine was inactive at doses up to 5 mM.

Time-lapse videomicroscopy revealed that caffeine could induce multiple entries into and exits from mitosis, while the cells remained blocked in S phase (Schlegel and Pardee, 1987), thereby creating a mitotic "subcycle" that is independent of the G_1 and S phase events of the normal mammalian cell cycle (Fig. 2). The levels of mitosis-specific phosphoproteins, detected by indirect immunofluorescence, increased during periods of chromatin condensation and decreased following chromatin decondensation (Fig. 3). These periodic mitotic events occurred even under conditions that normally make BHK cells quiescent. Exposing the cells to low concentrations of serum (0.5%) or cycloheximide (0.1 μg/ml) had little or no effect, indicating that caffeine was able to override the normal control mechanisms that are sensitive to environmental conditions. An identical, independent mitotic cycle was seen in the BHK temperature-sensitive mutant tsBN2. As described above, this mutant undergoes premature mitosis when raised to the restrictive temperature. These results indicate that certain chemicals and genetic mutations can override the normal cell cycle control mechanisms governing the activation and inactivation of mitotic factors, thereby creating a mitotic oscillation that is independent of environmental and cell cycle controls.

The protein kinase inhibitors 2-aminopurine and 6-dimethylaminopurine have recently been shown to induce premature mitosis in hamster fibroblasts in a manner similar to that reported for caffeine (Schlegel et al., 1990). These kinase inhibitors caused changes in cell morphology, premature chromosome condensation, and nuclear envelope breakdown in cells arrested during DNA synthesis. As with caffeine, the activity of these compounds was dependent on new protein synthesis, but not new RNA synthesis. 2-Aminopurine and 6-dimethylaminopurine acted cooperatively with each other and with caffeine, suggesting a common mechanism of action that involves protein dephosphorylation. This hypothesis is supported by the fact that all of these compounds

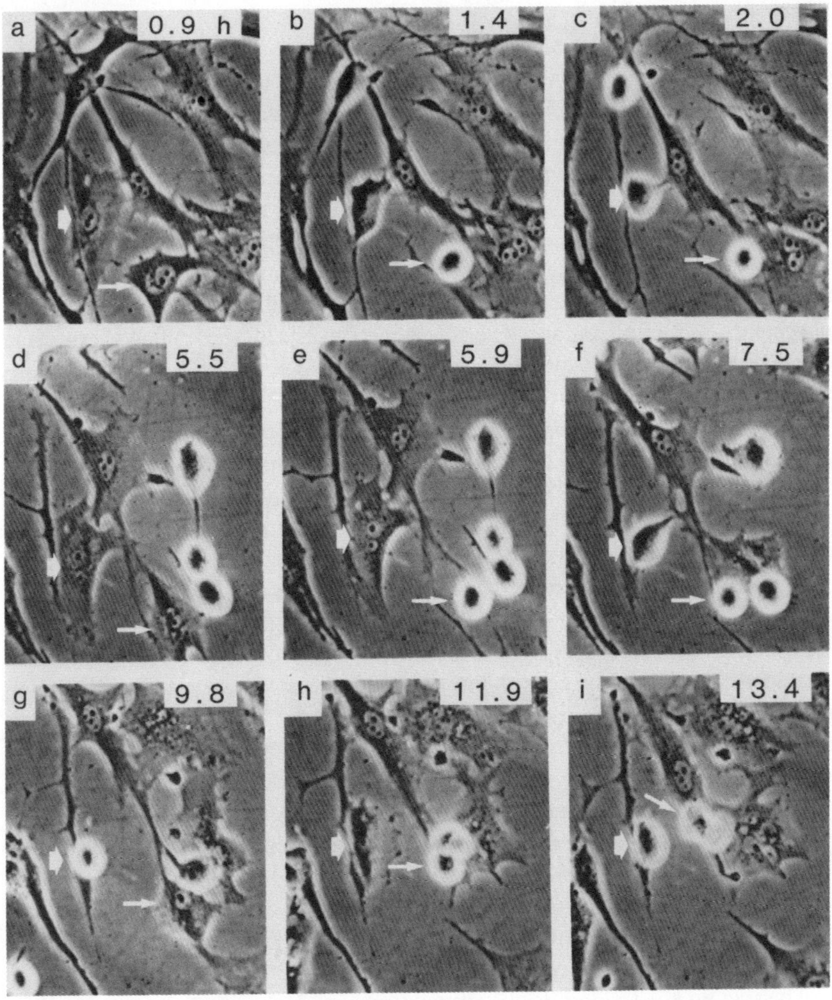

Fig. 2. Time-lapse videomicroscopy of caffeine-induced mitotic events in BHK cells arrested in early S phase. Cells were plated and synchronized in early S phase as described previously (Schlegel and Pardee, 1987). Caffeine (5 mM) was then added ($t = 0$) and videomicroscopy was conducted at $\times 320$ magnification. Photographs were taken from the video monitor at times (hours) after caffeine addition shown in the upper right corner. Arrows show the progression of two different cells, each undergoing three mitotic cycles while remaining arrested in S phase. From Schlegel and Pardee (1987).

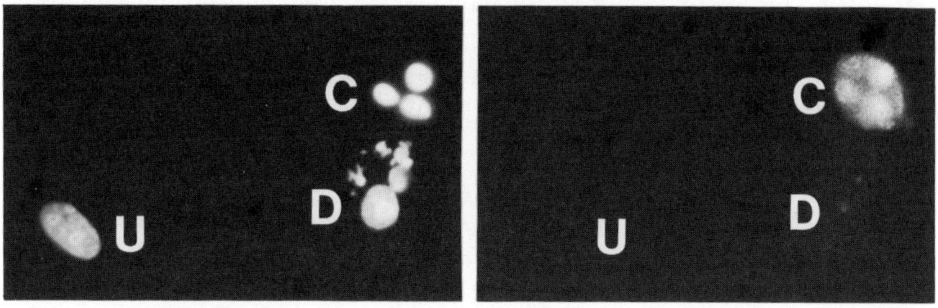

Fig. 3. Chromatin condensation and expression of mitosis-specific phosphoproteins during caffeine-induced mitotic cycles. BHK cells were synchronized and arrested in early S phase as described previously (Schlegel and Pardee, 1987). Eight hours after caffeine (5 mM) addition, cells were fixed in absolute methanol. (Left) Hoechst 33242 staining shows the uncondensed (U), condensed and fragmented (C), and decondensed and micronucleated (D) chromatin states ($\times 3300$). (Right) Indirect immunofluorescence with monoclonal antibody MPM-2, which recognizes a family of mitosis-specific phosphoproteins (Davis et al., 1983). Cells in left and right are the same. From Schlegel and Pardee (1987).

are known to inhibit protein kinase activity (e.g., Legon et al., 1974; Farrell et al., 1977).

Effect of Cell Cycle Perturbations

The study of cell cycle events often involves the use of cell synchrony techniques. It is common practice to perturb normal cell cycle transit to achieve this synchrony. Reversible arrest of cells in G_0 or G_1 by serum or nutrient deprivation, or by confluence arrest, is frequently used to study early cell cycle events, whereas reversible arrest near the G_1/S boundary with DNA synthesis inhibitors is preferred for examining later events. These experimental manipulations have been shown to affect the timing of subsequent mitoses.

Walters et al. (1974) reported that synchronization of Chinese hamster ovary cells in G_1 by isoleucine deprivation resulted in a reduced mitotic delay when the cells were subsequently released from G_1 arrest and irradiated. Similar results were reported when cells were reversibly arrested near the G_1/S boundary by isoleucine deprivation followed by release into hydroxyurea. We have noticed similar effects in other cell lines and have found that synchronization at the G_1/S boundary can lead to an acceleration of the later phases of the cell cycle in unirradiated cells (unpublished results). These results suggest that factors required for later cell cycle events may accumulate prematurely in cells arrested at the G_1/S boundary.

Our work concerning chemically induced premature mitosis supports the concept that cell cycle arrest does not prevent, and can actually stimulate, biochemical events associated with later cell cycle stages. Caffeine, 2-aminopurine, and 6-dimethylaminopurine induce premature mitosis only when cells are arrested in S or G_2. Premature mitosis was not detected when exponentially growing cultures were exposed to these compounds. Experiments conducted with caffeine indicated that mRNA needed for mitosis was synthesized during arrest in S phase, and was unstable when DNA replication was resumed (Schlegel et al., 1987). Earlier studies using RNA and protein synthesis inhibitors indicated that normal cycling cells synthesize these mRNA species and their protein products during G_2 (Kishimoto and Liberman, 1964).

The increased stability of mitosis-related mRNA following the completion of DNA replication may be important for the normal regulation of mitotic onset. DNA synthesis inhibitors may mimic the postreplicative state of G_2 cells and thereby lead to a premature accumulation of mRNA essential for mitosis. The mechanism by which cells regulate histone mRNA levels during the cell cycle provides an example of how mRNA stability can be coupled to the replicative state of the cell. The abundance of histone mRNA is regulated at both the transcriptional and posttranscriptional levels (Stein et al., 1984 for review). Histone mRNA is destabilized when DNA replication is completed or when it is suppressed with DNA synthesis inhibitors. This coupling of stability with DNA replication has been linked to both the 3′ and 5′ sequences of these genes (e.g., Morris et al., 1986).

Analysis Using Interspecies Cell Hybrids

Caffeine induces premature mitosis in Syrian and Chinese hamster cell lines (e.g., BALB/c 3T3, BHK, CHEF/18, CHO). Of the 12 different human cell types we examined, which included finite lifespan and immortal cells, as well as cells of fibroblastic and epithelial origin, none displayed caffeine-induced premature mitosis. The cells tested included normal human foreskin fibroblasts, FS-2; bladder, breast, and cervical transformed epithelioid cells, T-24, BT-20, HeLa; SV40 and *ras* oncogene-infected osteosarcoma and foreskin fibroblast cells, K-HOS, FSVK. Since caffeine permeability and metabolism are essentially the same in all mammalian cells examined, and caffeine, rather than its metabolites, is the most active agent (Timson, 1977; Pardee et al., 1987), it seemed likely that genetic differences in the regulation of mitosis were responsible for this species-specific response. To investigate this possibility, hamster–human whole cell hybrids were created to determine whether hamster chromosomes transferred mitosis-inducing activity to human cells, or whether human chromosomes suppressed premature mitotic activity in hamster cells.

Approximately 50–70% of Chinese hamster cells (CHEF/18 or CHO) underwent premature mitosis within 8 hr after caffeine addition, while human foreskin fibroblast and human fibrosarcoma cells showed less than a 1% response (Table 1). Original whole cell hybrids of CHEF/18 and human diploid foreskin fibroblasts (NLN1) or CHO and human fibrosarcoma (HT1080-6TG)

TABLE 1. Premature Chromosome Condensation (PCC) in Hamster, Human, and Hybrid Cells[a]

Cell type	% PCC
Parents	
Human foreskin fibroblast (NLN1)	<1
Human fibrosarcoma (HT1080-6TG)	<1
Chinese hamster embryo fibroblast (CHEF/18)	48
Chinese hamster ovary (CHO)	69
Original hybrids	
NLN1-CHEF/18 # 4	11
NLN1-CHEF/18 # 5	9
NLN1-CHEF/18 # 10	7
(HT1080-6TG)-CHO #3	7
(HT1080-6TG)-CHO #4	10
Hybrid subclones	
NLN1-CHEF/18 # 1	16
NLN1-CHEF/18 # 7	36
NLN1-CHEF/18 # 14	45

[a]Hybrids were selected with G418 and either ouabain or HAT medium. Exponentially growing cells were exposed to 2.5 mM hydroxyurea for 5 hr and then treated with 5 mM caffeine and 0.3 μg/ml colcemid for 8 hr. Premature chromosome condensation was assayed by fixing cells in absolute methanol and staining for 10 min with Hoechst 33258 (1 μg/ml). At least 300 cells were examined at ×400 by fluorescence microscopy to determine the percentage of cells containing PCC.

cells showed an approximately 10% response. The positive response of hamster–human hybrids was not surprising since interspecies fusions of mitotic and interphase cells are known to induce premature chromosome condensation in the interphase cell (Johnson and Rao, 1970). The reduction in frequency of caffeine-induced mitosis in the hybrids (10% versus 50–70% in hamster parents) could be the result of several factors. Mitosis-promoting factors produced by the hamster genomes could be diluted in the hybrid cells, and thus less able to condense approximately twice the amount of chromatin. An analogous condition was reported when single mitotic cells were fused to either one or two interphase cells wherein the frequency of premature chromosome condensation in the interphase cells decreased from 50% in binucleate fusions to 12% in trinucleate fusions (Rao, 1982). Alternatively, the reduced frequency of caffeine-induced mitosis in hamster–human hybrids could be due to the presence of mitosis-inhibitory factors in human cells. This explanation is supported by our finding that subclones of original NLN1–CHEF/18 hybrids had responses ranging from 16 to 45%. With time, these hybrids typically lose human and retain hamster chromosomes, suggesting that the frequency of premature mitosis increases as certain human chromosomes are lost. Since the subclones displaying the highest and lowest responses had roughly equivalent amounts of DNA, as determined by flow cytometry, the variability is most likely caused by a loss of different human chromosomes from the various subclones, rather than by a general decrease in human DNA content. The results from these initial fusion studies strongly suggest that hamster chromosomes can transfer mitosis-inducing activity to human cells, and that certain human chromosomes can inhibit premature mitosis in hamster cells. Identification of the chromosomes and genes responsible for these activities awaits further research.

Consequences of Aberrant Regulation

Enhanced Cytogenetic Damage and Cell Lethality

An important cellular response to DNA damage is a delay in the onset of mitosis. This delay in the G_2 phase of the cell cycle allows additional time for the repair of DNA before the initiation of cell division. Prevention of G_2 delay by chemicals or genetic mutants results in increased cytogenetic damage and cell lethality.

Exposure of cells to methylxanthines, and caffeine in particular, potentiates chromosome aberrations, nuclear fragmentation, and cytotoxicity following DNA damage induced by radiation and chemicals (e.g., Busse et al., 1978; Lau and Pardee, 1982). Caffeine has been reported to enhance the cytotoxicity of DNA-damaging agents by as much as 10 to 100-fold. Inhibition of new protein synthesis, using cycloheximide, following caffeine addition abolished these effects (Lau and Pardee, 1982). These results suggest that caffeine promotes the accumulation of proteins that accelerate the G_2-to-mitosis transition. It also appears that mitosis can occur prematurely in certain damaged, but otherwise normal, cells without subsequent chemical treatment. Sognier and Hittleman (1986) reported that mitomycin C-treated HeLa cells could undergo mitosis even though 10–20% of the DNA remained unreplicated. They proposed that the chemically induced chromatid breaks seen in these cells were a consequence of this incomplete replication. Similar results have been reported with the hamster mutant tsBN2. If, after DNA damage, these cells are forced to initiate premature mitosis by incubating them for a short period of time at the restrictive temperature, cell survival is greatly reduced (Sasaki and Nishimoto, 1987).

At present, only one gene has been cloned that is known to regulate G_2 delay following DNA damage. The *rad*9 mutant of *Saccharomyces cerevisiae,* which was initially isolated due to its sensitivity to ionizing radiation, has subsequently been found to be essential for radiation-induced mitotic delay (Weinert and Hartwell, 1988). As one would suspect, the absence of G_2 delay leads to increased cytotoxicity following irradiation. The corresponding genes in mammalian cells have yet to be identified.

Premature mitosis alone, without additional DNA damage, has also been shown to have profound deleterious effects. When cells are arrested during DNA replication and exposed to certain methylxanthines or protein kinase inhibitors, premature mitosis is followed by a return to interphase morphology, with the nuclear envelope reassembled around decondensed and fragmented chromatin to form numerous micronuclei (Schlegel and Pardee, 1987; Schlegel et al., 1990) (Figs. 2 and 3). A nearly identical response is seen when tsBN2 cells are arrested in S phase and undergo premature mitosis at the restrictive temperature (Nishimoto et al., 1981). The cytogenetic damage produced is at least in part due to improper segregation of the chromosomes. Brinkley et al. (1988) reported that caffeine-induced premature mitosis in S phase-arrested cells is characterized by essentially normal mitotic spindle formation and function. The spindle forces generated during this process, however, physically separate the kinetochores from the unreplicated chromosomes, preventing normal segregation of chromosomal DNA and producing numerous DNA strand breaks.

Mitotic Control and Neoplastic Transformation

The protein products of the c-*mos*, Ha-*ras*, Ki-*ras*, and adenovirus E1A oncogenes have been shown to have important roles in the regulation of meiosis and in the progression of cells from G_2 to mitosis. Deletion analysis has revealed that the transformation-related domain of the E1A protein is the same region as that which is essential for the G_2-to-mitosis transition, but is distinct from that which stimulates cells to initiate DNA replication (Zerler et al., 1987). In *Xenopus* oocytes, microinjection of the human Ha-*ras* protein can induce meiosis (Birchmeier et al., 1985), whereas microinjection of an antibody to *ras* proteins can inhibit insulin-induced meiosis (Deshpande and Kung, 1987). In quiescent rodent cells incubated in serum-free medium, the Ki-*ras* gene product (p21) can stimulate reentry into the replication cycle. If a temperature-sensitive p21 is inactivated when restimulated cells reach mid-S phase, however, cells become arrested in G_2 (Durkin and Whitfield, 1987). Arrested cells can resume G_2 transit and complete cell division when p21 is reactivated at the permissive temperature.

The c-*mos* protooncogene has recently been shown to be a cytostatic factor responsible for meiotic arrest in vertebrate eggs (Sagata et al., 1989). Cytostatic factor is required for stabilizing maturation-promoting factor (MPF) in germ cells and appears to be essential for the accumulation of active MPF and subsequent germinal vesical breakdown. Microinjection of c-*mos* antisense oligonucleotides blocks progesterone-induced maturation of *Xenopus* (Sagata et al., 1988) and mouse (Paules et al., 1989) oocytes, whereas microinjection of in vitro transcribed c-*mos* RNA induces meiosis in *Xenopus* oocytes (Freeman et al., 1989). Antisense c-*mos* oligonucleotides also inhibit Ha-*ras*-induced maturation of *Xenopus* oocytes (Barrett et al., 1990). Expression of c-*mos* appears to be restricted to male and female germ cells. Although little or no expression is seen in normal mammalian somatic cells, c-*mos* readily transforms such cells. It remains to be determined whether the resulting neoplastic transformation is due to aberrant mitotic control.

As mentioned previously, meiosis can be initiated in *Xenopus* oocytes by microinjection of cyclin A mRNA (Swenson et al., 1986). Recent findings now suggest that abnormal regulation of cyclin may lead to a transformed phenotype. A human hepatocellular carcinoma was found to contain hepatitis B virus integrated into the cyclin A gene (Wang et al., 1990), and the 60-kDa cellular protein known to associate with the E1A protein in adenovirus-infected cells has been shown to be cyclin A (Pines and Hunter, 1989). Since these preliminary experiments did not determine whether viral integration into the gene or complex formation with the protein altered cyclin A synthesis or function, association of aberrant cyclin regulation with a transformed phenotype remains tenuous.

Provocative associations have also been found between certain oncogene products and the serine/threonine protein kinase $p34^{cdc2}$. Both simian virus 40 large tumor antigen (McVey et al., 1989) and $pp60^{c\text{-}src}$ (Shenoy et al., 1989; Morgan et al., 1989) are in vitro substrates for this kinase. The sites of phosphorylation correspond to those seen in vivo and lead to activation of T antigen-dependent DNA replication and enhanced tyrosine kinase activity of $pp60^{c\text{-}src}$. One must be careful with the interpretation of such findings, however. Numerous other kinases could be responsible for these phosphorylation events in vivo, and it is currently very difficult to design experiments in mammalian cells that can establish an essential role for $p34^{cdc2}$. Taken together, however, these findings suggest that the loss of normal control of mitotic onset may be an important factor in the neoplastic transformation of cells by certain oncogenes and viruses.

Acknowledgments

We thank Elizabeth McGurty for her help in the preparation of this manuscript. This work was supported in part by a Mellon Foundation Junior Faculty Development Award and a grant from the National Institutes of Health (CA49749) to R.S., and by a grant from the American Cancer Society (CD-434) to R.W.C.

References

Adlakha RC, Davis FM, Rao PN (1985): Role of phosphorylation of nonhistone proteins in the regulation of mitosis. In A.L. Boynton and H.L. Leffert (eds): Control of Animal Cell Proliferation. New York: Academic Press, pp 485–513.

Barrett CB, Schroetke RM, Van der Hoorn FA, Nordeen SK, Maller JL (1990): Ha-$ras^{Val\text{-}12,Thr\text{-}59}$ activates S6 kinase and $p34^{cdc2}$ kinase in *Xenopus* oocytes: Evidence for c-mos^{xe}-dependent and -independent pathways. Mol Cell Biol 10:310–315.

Birchmeier C, Broek D, Wigler M (1985): *Ras* proteins can induce meiosis in Xenopus oocytes. Cell 43:615–621.

Brinkley BR, Zinkowski RP, Mollon WL, Davis FM, Pisegna MA, Pershouse M, Rao PN (1988): Movement and segregation of kinetochores experimentally detached from mammalian chromosomes. Nature (London) 336:251–254.

Busse PM, Bose SK, Jones RW, Tolmach LJ (1978): The action of caffeine on X-irradiated HeLa cells: Enhancement of x-ray induced killing during G_2 arrest. Radiat Res 76:292–307.

Chackalaparampil I, Shalloway D (1988): Altered phosphorylation and activation of $pp60^{c\text{-}src}$ during fibroblast mitosis. Cell 52:801–810.

Cyert MS, Kirschner MW (1988): Regulation of MPF activity *in vitro*. Cell 53:185–195.

Davis FM, Tsao TY, Fowler SK, Rao PN (1983): Monoclonal antibodies to mitotic cells. Proc Natl Acad Sci USA 80:2926–2930.

Deshpande AK, Kung H-F (1987): Insulin induction of *Xenopus laevis* oocyte maturation is inhibited by monoclonal antibody against p21 *ras* proteins. Mol Cell Biol 7:1285–1288.

Durkin JP, Whitfield JF (1987): The viral Ki-*ras* gene must be expressed in the G_2 phase if *ts* Kirsten sarcoma virus-infected NRK cells are to proliferate in serum-free medium. Mol Cell Biol 7:444–449.

Evans T, Rosenthal ET, Youngbloom J, Distel D, Hunt T (1983): Cyclin: A protein specified by maternal mRNA in sea urchin eggs that is destroyed at each cleavage division. Cell 33:389–396.

Farrell PJ, Balkow K, Hunt T, Jackson RJ (1977): Phosphorylation of initiation factor eIF-2 and the control of reticulocyte protein synthesis. Cell 11:187–200.

Freeman RS, Pickham KM, Kanki JP, Lee BA, Pena SV, Donoghue DJ (1989): *Xenopus* homolog of the *mos* protooncogene transforms mammalian fibroblasts and induces maturation of *Xenopus* oocytes. Proc Natl Acad Sci USA 86:5805–5809.

Huchon D, Ozon R, Demaille JG (1981): Protein phosphatase-1 is involved in *Xenopus* oocyte maturation. Nature (London) 294:358–359.

Johnson RT, Rao PN (1970): Mammalian cell fusion: Induction of premature chromosome condensation in interphase nuclei. Nature (London) 226:717–722.

Kishimoto S, Lieberman I (1964): Synthesis of RNA and protein required for the mitosis of mammalian cells. Exp Cell Res 36:92–101.

Lau CC, Pardee AB (1982): Mechanism by which caffeine potentiates lethality of nitrogen mustard. Proc Natl Acad Sci USA 79:2942–2946.

Lee MG, Nurse P (1987): Complementation used to clone a human homologue of the fission yeast cell cycle control gene *cdc*2. Nature (London) 327:31–35.

Legon S, Brayley A, Hunt T, Jackson RJ (1974): The effect of cyclic AMP and related compounds on the control of protein synthesis in reticulocyte lysates. Biochem Biophys Res Commun 46:745–752.

Maller JL, Krebs EG (1977): Progesterone-stimulated meiotic cell division in *Xenopus* oocytes. J Biol Chem 252:1712–1718.

McVey D, Brizuela L, Mohr I, Marshak DR, Gluzman Y, Beach D (1989): Phosphorylation of large tumour antigen by cdc2 stimulates SV40 DNA replication. Nature (London) 341: 503–507.

Meijer L, Arion D, Golsteyn R, Pines J, Brizuela L, Hunt T, Beach D (1989): Cyclin is a subunit of the sea urchin M-phase specific histone H1 kinase. EMBO J 8:2275–2282.

Morgan DO, Kaplan JM, Bishop JM, Varmus HE (1989): Mitosis-specific phosphorylation of $pp60^{c\text{-}src}$ by $p34^{cdc2}$-associated protein kinase. Cell 57:775–786.

Morla AO, Draetta G, Beach D, Wang JYJ (1989): Reversible tyrosine phosphorylation of cdc2: Dephosphorylation accompanies activation during entry into mitosis. Cell 58:193–203.

Morris T, Marashi F, Weber L, Hickey E, Greenspan D, Bonner J, Stein J, Stein G (1986): Involvement of the 5′-leader sequence in coupling the stability of a human H3 histone mRNA with DNA replication. Proc Natl Acad Sci USA 83:981–985.

Nishimoto T, Eilen E, Basilico C (1978): Premature chromosome condensation in a ts DNA^- mutant of BHK cells. Cell 15:475–483.

Nishimoto T, Ishida R, Ajiro K, Yamamoto S, Takahashi T (1981): The synthesis of protein(s) for chromosome condensation may be regulated by a post-transcriptional mechanism. J Cell Physiol 109:299–308.

Pardee AB, Schlegel R, Boothman DA (1987): Pharmacological interference with DNA repair. In P.A. Cerutti, O.F. Nygaard, and M.G. Simic (eds): Anticarcinogenesis and Radiation Protection. New York: Plenum Press, pp 431–436.

Paules RS, Buccione R, Moschel RC, Vande Woude GF (1989): Mouse *mos* protooncogene product is present and functions during oogenesis. Proc Natl Acad Sci USA 86:5395–5399.

Pines J, Hunter T (1989): Isolation of a human cyclin cDNA: Evidence for cyclin mRNA and protein regulation in the cell cycle and for interaction with $p34c^{dc2}$. Cell 58:833–846.

Rao, PN (1982): The phenomenon of premature chromosome condensation. In P.N. Rao, R.T. Johnson, and K. Sperling (eds): Premature Chromosome Condensation. New York: Academic Press, pp 1–14.

Riabowol K, Draetta G, Brizuela L, Vandre D, Beach D (1989): The cdc2 kinase is a nuclear protein that is essential for mitosis in mammalian cells. Cell 57:393–401.

Sagata N, Oskarsson M, Copeland T, Brumbaugh J, Vande Woude GF (1988): Function of c-*mos* proto-oncogene product in meiotic maturation in *Xenopus* oocytes. Nature (London) 335: 519–525.

Sagata N, Watanabe N, Vande Woude GF, Ikawa Y (1989): The c-*mos* proto-oncogene product is a cytostatic factor responsible for meiotic arrest in vertebrate eggs. Nature (London) 342:512–518.

Sasaki H, Nishimoto T (1987): chromosome condensation may enhance x-ray-related cell lethality in a temperature-sensitive mutant (tsBN2) of baby hamster kidney cells (BHK21). Radiat Res 109:407–418.

Scaife JF (1971): Cyclic 3′-5′ adenosine monophosphate: its possible role in mammalian cell mitosis and radiation-induced mitotic G_2-delay. Int J Radiat Biol 19:191–195.

Schlegel R, Pardee AB (1986): Caffeine-induced uncoupling of mitosis from the completion of DNA replication in mammalian cells. Science 232:1264–1266.

Schlegel R, Pardee AB (1987): Periodic mitotic events induced in the absence of DNA replication. Proc Natl Acad Sci USA 84:9025–9029.

Schlegel R, Croy RG, Pardee AB (1987): Exposure to caffeine and suppression of DNA replication combine to stabilize the proteins and RNA required for premature mitotic events. J Cell Physiol 131:85–91.

Schlegel R, Belinsky GS, Harris MO (1990): Premature mitosis induced in mammalian cells by the protein kinase inhibitors 2-aminopurine and 6-dimethylaminopurine. Cell Growth Differ 1:171–178.

Shenoy S, Choi J-K, Bagrodia S, Copeland TD, Maller JL, Shalloway D (1989): Purified maturation promoting factor phosphorylates $pp60c^{-src}$ at the sites phosphorylated during fibroblast mitosis. Cell 57:763–774.

Sognier MA, Hittelman WN (1986): Mitomycin-induced chromatid breaks in HeLa cells: A consequence of incomplete DNA replication. Cancer Res 46:4032–4040.

Stein GS, Sierra F, Stein JL, Plumb M, Marashi F, Carozzi N, Prokopp K, Baumbach L (1984): Organization and expression of human histone genes. In G.S. Stein, J.L. Stein, and W.F. Marzluff (eds): Histone Genes: Structure, Organization and Regulation. New York: John Wiley, pp 397–455.

Swenson KI, Farrell KM, Ruderman JV (1986): The clam embryo protein cyclin A induces entry into M phase and the resumption of meiosis in *Xenopus* oocytes. Cell 47:861–870.

Th'ng JPH, Wright PS, Hamaguchi J, Lee MG, Norbury CJ, Nurse P, Bradbury EM (1990): The FT210 cell line is a mouse G2 phase mutant with a temperature-sensitive *cdc2* gene product. Cell 63:313–324.

Timson J (1977): Caffeine. Mutat Res 47:1–52.

Uchida S, Sekiguchi T, Nishitani H, Miyauchi K, Ohtsubo M, Nishimoto T (1990): Premature chromosome condensation is induced by a point mutation in the hamster RCC1 gene. Mol Cell Biol 10:577–584.

Walters RA, Gurley LR, Tobey RA (1974): Effects of caffeine on radiation-induced phenomena associated with cell-cycle traverse of mammalian cells. Biophys J 14:99–118.

Wang J, Chenivesse X, Henglein B, Brechot C (1990): Hepatitis B virus integration in a cyclin A gene in a hepatocellular carcinoma. Nature (London) 343:555–557.

Weinert TA, Hartwell LH (1988): The *rad*9 gene controls the cell cycle response to DNA damage in *Saccharomyces cerevisiae*. Science 241:317–322.

Whitmore GF, Stanners CP, Till JE, Gulyas S (1961): Nucleic acid synthesis and the division cycle in X-irradiated L-strain mouse cells. Biochim Biophys Acta 47:66–77.

Zerler B, Roberts RJ, Mathews MB, Moran E (1987): Different functional domains of the Adenovirus E1A gene are involved in regulation of host cell cycle products. Mol Cell Biol 7:821–829.

Perspectives on Cellular Regulation:
From Bacteria to Cancer, pages 251–263

Methylxanthines: From Cell Biology to Clinical Oncology

Bruce J. Dezube, Howard J. Fingert, and Ching C. Lau

Division of Cell Growth and Regulation, Department of Biological Chemistry and Molecular Pharmacology and Department of Medicine, Dana-Farber Cancer Institute, and Division of Medical Oncology, Department of Medicine, Beth Israel Hospital, Boston, Massachusetts 02215 (B.J.D.); Division of Hematology and Oncology, St. Elizabeth's Hospital, Boston, Massachusetts 02135 (H.J.F.); and Laboratory of Gynecologic Oncology, Brigham and Women's Hospital, Boston, Massachusetts 02115 (C.C.L.)

Introduction

Many of the drugs used in cancer chemotherapy act by damaging DNA. The lethality of the antineoplastic agents depend not only on the amount of DNA damage they cause, but also on the efficiency with which the damaged cells repair such lesions. DNA repair can diminish the lethality of an agent by several orders of magnitude. This has been shown with cells genetically defective in a repair process, of which xeroderma pigmentosum provides the best example (Setlow, 1978), or by using compounds that inhibit a repair process (Roberts, 1978). Such repair processes may be very important in relation to the therapeutic index, a measure of the relative lethality of a drug toward tumor cells versus normal cells. Cells of different tissues and of different tumors have been reported to have different repair capacities (Goth and Rajewsky, 1974).

Pharmacological control of DNA repair during chemotherapy would appear to be useful, since increased lethality leads to a smaller fraction of surviving tumor cells. Moreover, DNA repair is an important mechanism of chemotherapy resistance in human tumors (Harris, 1985). Thus DNA repair inhibitors represent a potential solution to two of the fundamental problems in cancer chemotherapy: how to maximize tumor cell kill and how to prevent the emergence of resistant tumor cells. Obviously the toxicity to the normal cells, such as those in the bone marrow and the gastrointestinal tract, would have to be low to provide a better therapeutic index when a combination of antineoplastic drug and DNA repair inhibitor is administered.

A number of DNA repair inhibitors have been studied, of which caffeine is by far the best known (Kihlman, 1977). The earlier literature contains a few rather preliminary reports on attempts to modulate chemotherapy with caffeine in humans (Cohen et al., 1980), in mice (Cohen, 1975; Gaudin and Yielding, 1969; Rose et al., 1978), and much more frequently in tissue culture systems (Roberts, 1978). The studies with humans and with animals have led to mixed success. Many in vitro experiments have, however, shown greatly enhanced lethality of antineoplastic agents toward tumor cells when repair is inhibited by caffeine. With regard to the animal work, it is fair to say that the cell biology and biochemistry of DNA repair and its modification were not sufficiently understood at that time to allow a rational choice of conditions such as dosage and scheduling.

During the last decade, Dr. Arthur Pardee has directed the attention of several investigators in his laboratory to some of the issues regarding the cellular and molecular mechanisms of DNA repair. The combined efforts of a number of basic scientists and clinical oncologists assembled in Dr. Pardee's group have contributed to understanding the complexities of the DNA repair process. This chapter highlights some of the important findings of this group regarding the mechanisms of action of caffeine and other methylxanthines as well as their applications in the clinical setting.

In Vitro Studies of Cytotoxic, Cell Cycle, and Chromosomal Effects of Caffeine

Caffeine potentiates the lethal effects of DNA damaging agents including X-rays, ultraviolet light, and many chemotherapeutic drugs. The mechanism(s) by which caffeine enhances the lethality of genotoxic agents is obfuscated by the multitude of molecular effects attributed to caffeine. For example, by inhibiting phosphodiesterase and increasing cyclic AMP levels, caffeine can arrest cells in G_1 phase and thereby decrease DNA synthesis in culture (Pardee and James, 1975; Walters et al., 1974). In addition, it can increase the rate of DNA synthesis, possibly by initiating new replicons (Robinson and Harris, 1979; Tatsumi and Strauss, 1979). Furthermore, caffeine and other methylxanthines can inhibit the phosphorylation of H1 histones during the G_1 phase (Dolby et al., 1981) and can increase H1 phosphorylation during G_2 (Walters et al., 1974). These results show that caffeine can have opposite effects in different phases of the cell cycle. The use of synchronized cells is thus essential to unravel the cell cycle-specific effects of caffeine.

Most early reports attributed the enhancement of the cytotoxicity of DNA damaging agents by caffeine to its effects in S phase (For review and references, see Roberts, 1978). It has long been recognized that one of the immediate consequences of genotoxic damage by chemical and physical agents is the inhibition of DNA replication (Roberts et al., 1971). It has also been shown that this early inhibition of DNA synthesis is frequently due to inhibiton of replicon initiation. The subsequent addition of caffeine to the damaged cell reverses the progressive inhibition in the rate of DNA synthesis, probably by reversing the depression of replicon initiation in cells with DNA damage. In addition, caffeine can also inhibit elongation of nascent DNA in cells pretreated with alkylating agents. As a consequence of this inhibition, low-molecular-weight DNA accumulates when caffeine is added to damaged cells (Van den Berg and Roberts, 1976).

Although the effects of caffeine on DNA replication in damaged cells are dramatic, later work showed that they may not account for the observed enhancement of lethality by caffeine (Cleaver, 1978). The results of our studies with alkylating agents, and those of others working with radiation, established clearly that the effects of caffeine during G_2 phase (Busse et al., 1978; Lau and Pardee, 1982; Lücke-Huhle, 1982) could account for its interference with DNA repair, and its enhancement of the lethal and DNA-damaging effects of alkylating agents.

We have demonstrated (Lau and Pardee, 1982) that following treatment with low doses of nitrogen mustard (75–90% survival), baby hamster kidney (BHK) cells were delayed in G_2. DNA repair (unscheduled DNA synthesis) was detected during this G_2 delay, and when it was completed, the cells proceeded into mitosis. Both the length of the G_2 delay and the amount of DNA repair synthesis were proportional to the concentration of nitrogen mustard used. Caffeine increased the lethality of nitrogen mustard 5- to 10-fold. It prevented the G_2 delay and enhanced chromosome aberrations and nuclear fragmentation. Caffeine did not prevent nitrogen mustard-treated cells from entering S phase, but rather allowed the G_2-arrested cells to divide prematurely without finishing the repair process. The extent of nuclear fragmentation was proportional to the degree of enhanced lethality over a wide range of nitrogen mustard concentrations.

All the above caffeine-induced effects, abolition of G_2 delay, nuclear fragmentation, and potentiation of lethality could be abolished by a low dose of cycloheximide (0.2 μg/ml) that partially inhibited protein synthesis (Das et al., 1982). This concentration did not block the ability of cells to enter mitosis after the transient G_2 delay. Caffeine, therefore, did not act directly to potentiate lethality in damaged cells, but rather acted in conjunction with a protein or proteins that needed to be newly synthesized. Protection from lethality by cycloheximide was not due to the inhibition of a caffeine-sensitive repair process. Nitrogen mustard-treated cells were no longer sensitive to caffeine when cycloheximide was removed after 24-hr treatment with caffeine and cycloheximide, indicating that repair continued even in the presence of low doses of cycloheximide.

Further experiments showed that the effects of both caffeine and cycloheximide were G_2 specific. By pulsing synchronized, nitrogen mustard-treated cells with caffeine and cycloheximide for short intervals corresponding to various phases of the cell cycle, the G_2 delay was shown to be abolished by caffeine when it was added after damaged cells had entered G_2 phase. The presence of caffeine during the G_2 arrest, but not during S phase, was sufficient to cause nuclear fragmentation and to enhance lethality. Cycloheximide reversed all the effects of caffeine, even when it was present only during the G_2 delay period. Further characterization of the effects of caffeine on BHK cells showed that they are not mediated by cyclic AMP or by adenosine receptors (Lau, 1982).

Based on these results, we proposed the following mechanism of caffeine-enhanced lethality (Fig. 1). When cells suffer DNA damage, they suppress the synthesis of proteins required for mitosis. Damaged cells are thereby delayed in G_2 to allow time for repair. This repair process may be part of a previously proposed surveillance mechanism operating in G_2 (Tobey, 1975). When DNA repair is completed, cells proceed to mitosis and divide. Caffeine may override the inhibitory signal(s) so that damaged cells enter mitosis before DNA repair is completed. This premature mitosis leads to pulverized chromosomes, fragmented nuclei, and cell death. Cycloheximide protects damaged cells from caffeine lethality by preventing the premature synthesis of mitosis-inducing proteins. Later studies demonstrated that caffeine can also induce the onset of mitosis before DNA replication is completed in undamaged cells (Schlegel and Pardee, 1986). This effect of caffeine also requires protein synthesis, consistent with the idea that caffeine works by enhancing either the synthesis or stability of mitosis-related proteins (Schlegel and Craig, 1991).

Our conclusions are significantly different from those derived from earlier studies which focused primarily on the effects of caffeine on DNA synthesis in damaged cells, especially those associated with replicon initiation and DNA elongation. It is important to point out that the effects of caffeine described in our work were studied with cells after *low* levels of DNA damage. At the

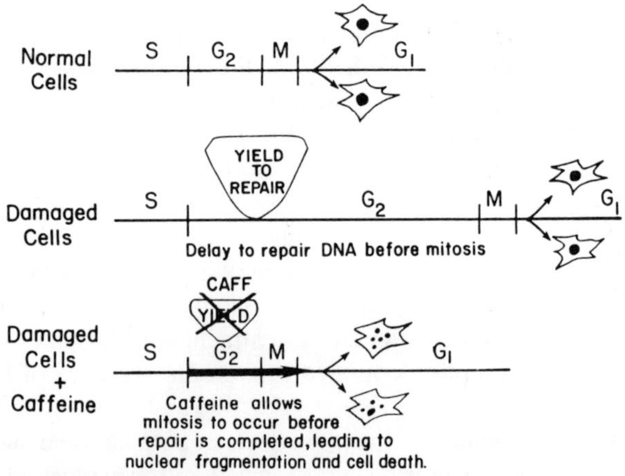

Fig. 1. Schematic representation of the proposed mechanism by which caffeine enhances the lethality of alkylating agents. Reproduced from Lau (1989).

doses of nitrogen mustard used, DNA synthesis was neither significantly suppressed nor modified by caffeine, and survival was maintained at 75–90%. Using a computer model that calculates the number of cross–links caused by bifunctional alkylating agents such as nitrogen mustard (Kohn et al., 1981), we estimated that with the low dose of nitrogen mustard we used (0.5 μM), there is 1 cross-link per 10^6 base pairs. This low level of crosslinking apparently allows most of the DNA to be replicated, since cells proceed to G_2 without any significant delay. This amount of damage is also efficiently repaired by BHK cells as indicated by the disappearance of most of the cross-links within 8 hr. Therefore, under these conditions, caffeine is acting on a recovery process which normally allows cells to survive DNA damage. *Higher* concentrations of nitrogen mustard, on the other hand, arrest cells in S phase (Murnane, et al., 1980) rather than in G_2 phase. Many of these S phase-arrested cells do not subsequently recover—they are unable to replicate their DNA. The G_2 effect of caffeine will not be apparent under such conditions.

The fragmentation of nuclei by caffeine in cells containing damaged DNA has led to a rapid genotoxicity test (Lau, 1989). The purpose of using caffeine to potentiate DNA damage in a screening test is analogous to the use of repair deficient mutants to enhance mutational sensitivities in the Ames test. Compounds tested with the nuclear fragmentation assay included alkylating agents, antimetabolites, DNA strand breakers, and cytotoxic antibiotics. All these agents caused nuclear fragmentation with caffeine whereas treatment with these agents alone had little or no effect (Table 1). Liver microsomal fractions can be used in this assay to detect genotoxicity of compounds that require metabolic activation. In addition, we demonstrated previously that DNA repair inhibitors, such as 3-aminobenzamide, can act synergistically with caffeine in inducing nuclear fragmentation (Das et al., 1984). Thus the nuclear fragmentation assay has promise for screening other agents that interfere with DNA repair.

Animal Models for Enhancement of Chemotherapy by Methylxanthines

In animal systems caffeine has been shown to enhance the antitumor activity of many chemotherapeutic agents. For example, caffeine substantially enhanced the antitumor effects of bleomycins and phleomycins on rat Walker 256 carcinosarcoma and murine Ehrlich ascites tumor (Allen et al., 1985). Nitrogen mustard-resistant plasmacytomas implanted in hamsters had an increased response to mechlorethamine when the animals were given caffeine in their drinking water (Gaudin and Yielding, 1969). In a model system using human pancreatic adenocarcinoma grown in nude mice, caffeine enhanced the antitumor effects of the combination of arabinoside-C and cisplatin (Kyriazis, et al., 1985).

We have chosen the methylxanthine pentoxifylline {1-(5′-oxohexyl)-3,7-dimethylxanthine (Fig. 2); Trental, Hoechst-Roussel Pharmaceuticals, Som-

TABLE 1. DNA Damaging Agents Tested in Nuclear Fragmentation Assay[a]

Types of DNA damaging agents	Concentration (μM)	Percent fragmented nuclei −CAFF	+CAFF	Lowest detectable concentration (μM)
Alkylating agents				
Monofunctional				
MMS	500	1	59	70
MNNG	3.4	2	61	0.2
Bifunctional				
HN2	0.5	1	73	0.05
BCNU	3	0	48	0.3
Mitomycin C	3	0	38	n.t.
Cisplatinum	33	3	38	1.7
Strand breaking agents				
Neocarzinostatin	0.1	1	45	0.003
Antimetabolites				
6-Thioguanine	3	1	42	1
ara-C	3	3	47	0.5
Intercalating agents				
Adriamycin	0.1	1	51	0.03
Isoleucine starvation (48 hr)	—	1	26	—
UV, 2 J/m^2	—	0	55	n.t.
No treatment	—	0	0	—

[a]Cells (8×10^3) in 1 ml medium were plated per well of 24 plates. Twenty-four hours after plating, test substances were added at varying doses and were allowed to remain in the medium for varying times. After that, cells were rinsed once with fresh medium and incubated with or without caffeine (2 mM) for 24 hr. At the end of the incubation period, cells were rinsed twice with phosphate-buffered saline, fixed with methanol/acetic acid (3:1 v/v) for 15 min at room temperature, and subsequently air dried. Cells were then stained with 4% Giemsa in phosphate/citrate buffer (pH 6.7). Percentage of nuclear fragmentation was determined by microscopic examination of a minimum of 300 cells for each determination. Length of treatments for each drug: adriamycin 30 min; all alkylating agents 1 hr except cisplatinum, which was 2 hr; neocarzinostatin 90 min; ara-C 4 hr; and 6-thioguanine, 6 hr. n.t., not tested. Reproduced from Lau (1989).

erville, NJ} for further animal studies because it is less cardiotoxic and neurotoxic than caffeine (Dalvi, 1986; Physicians' Desk Reference, 1990). Pentoxifylline is approved by the Food and Drug Administration for the treatment of patients with intermittent claudication from chronic occlusive arterial disease. Our earlier work established that pentoxifylline has similar effects as caffeine in vitro (Fingert et al., 1986). With the anticipation of testing clinically the modulating effects of pentoxifylline in cancer chemotherapy, we and others first established its efficacy in a number of animal systems, including the subrenal capsule, tumor growth delay, and tumor excision assays.

Subrenal Capsule Assay

Caffeine had been proposed to enhance the lethality of DNA damaging agents to rodent cells, but not to human cells (Roberts and Ward, 1973). In addition, earlier clinical studies using the combination of methyl-CCNU and caffeine

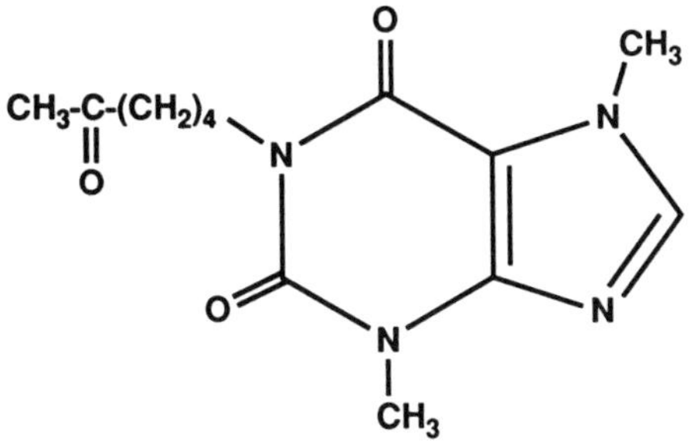

Fig. 2. Pentoxifylline—chemical structure.

failed to show any beneficial effect of caffeine (Cohen et al., 1980). Thus, it was important to demonstrate that pentoxifylline affects human tumor cells to a similar extent as rodent cells. To accomplish this, we used the subrenal capsule assay, which permits the testing of antineoplastic drugs on human tumors in mice (Bogden et al., 1978). In this assay, either tumor xenografts or cultured tumor cells embedded in a fibrin clot matrix are implanted under the subrenal capsule of mice. Tumor dimensions are measured in situ before and after various treatments.

The antitumor effects of thiotepa with and without pentoxifylline were evaluated on T-24 human bladder cancer cells and MX-1 human breast tumor xenografts (Fingert et al., 1988). The antitumor effect of the combination of thiotepa and pentoxifylline was greater than that of thiotepa alone in both instances ($P < 0.005$) (Fig. 3). Furthermore, the combination was not more toxic than thiotepa alone as determined by weight gain, alterations in histopathology of normal bladder cells, and survival of both groups of treated mice.

Tumor Growth Delay Experiments, Tumor Excision Assay

The ability of pentoxifylline to potentiate chemotherapy was confirmed in two additional model systems (Beverly Teicher, Dana–Farber Cancer Institute, unpublished data). In the tumor growth delay experiment, EMT6 mouse mammary carcinoma cells were implanted into the legs of mice. In the animals that received pentoxifylline in combination with an alkylating agent (thiotepa, cyclophosphamide, or carboplatin), it took 50–60% more time for tumors to reach 500 mm^3 in size than in animals that received an alkylating agent alone.

In the tumor excision assay, FSaII mouse fibrosarcoma cells were implanted in the legs of mice. The animals were treated with chemotherapy with and without pentoxifylline and sacrificed 24 hr later. The tumors were excised, made into single cell suspensions, and then plated at low cell densities to assess the ability of individual cells to give rise to multi-cellular colonies. This assay therefore allowed quantitative assessment of the impact of in vivo chemotherapy on clonogenic cell survival. The addition of pentoxifylline to che-

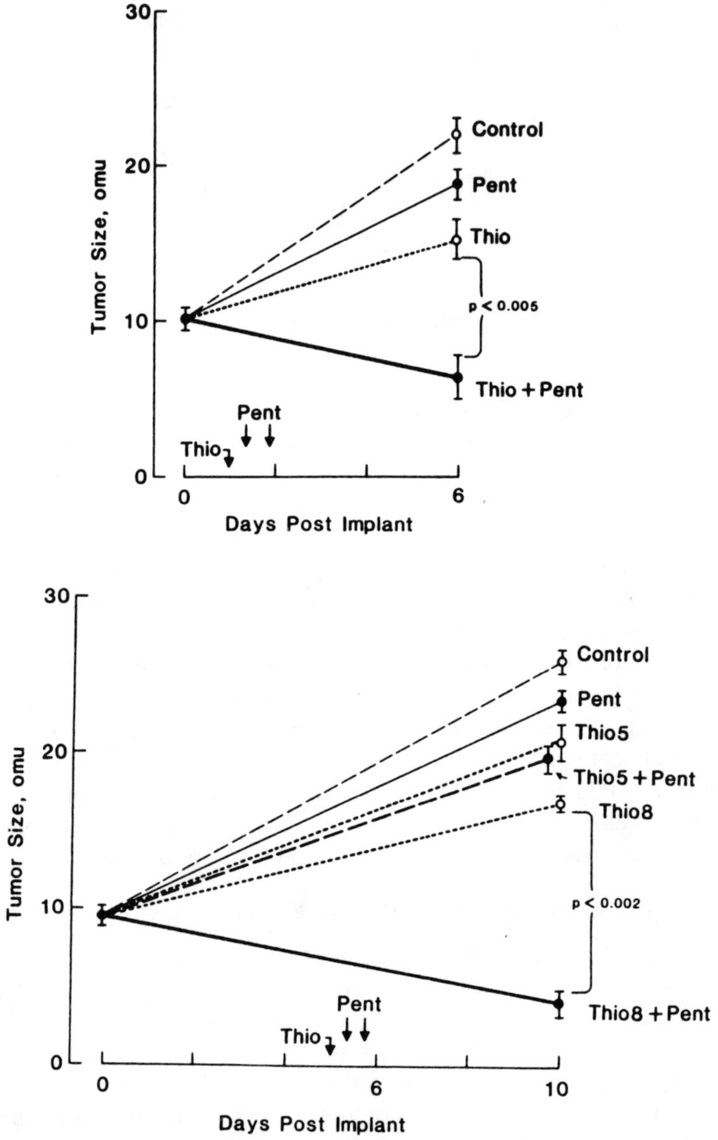

Fig. 3. Enhanced in vivo activity of thiotepa (Thio) combined with pentoxifylline (Pent) in T24 human bladder cancer cells (upper graph) and in MX-1 human breast tumor grafts (lower graph) in the subrenal capsule assay. Thiotepa doses used were 5 mg/kg intraperitoneally in the T24 experiment and 5 mg/kg or 8 mg/kg in the MX-1 experiment. Pentoxifylline dose was 257 mg/kg subcutaneously in both. Tumor size is expressed in ocular micrometer units (omu), where 10 omu equals 1 mm. Reproduced from Fingert et al. (1988).

motherapy decreased cell survival by up to one order of magnitude, compared to that seen with chemotherapy alone.

In a modified tumor excision assay, mice were injected with Hoechst 33342 prior to being sacrificed. Oxic tumor cells stain brighter than hypoxic cells, due to differential availability of the dye. After the tumor was excised, the cells were suspended and sorted on a flow cytometer into a bright population (enriched in oxic cells) and a dim population (enriched in hypoxic cells). The most striking finding of Teicher's work was that pentoxifylline selectively enhanced the killing of dim (hypoxic) cells by chemotherapy. This finding is important because hypoxia is one of the well-recognized causes of drug resistance. Pentoxifylline enhanced chemotherapy at doses as low as 50 mg/kg intraperitoneally. This dose will become important when we compare it to that given to patients as part of our human studies.

Human Studies of the Modulating Effects of Pentoxifylline During Chemotherapy

The ability of pentoxifylline to potentiate alkylator treatment served as the rationale for a Phase I trial at the Dana-Farber Cancer Institute. The purpose of this trial was to determine the maximum tolerated dose of pentoxifylline administered orally in combination with the alkylating agent thiotepa and to estimate whether this dose approaches that which is necessary to enhance chemotherapy, as predicted by the mouse models. The maximum tolerated dose of pentoxifylline was 1600 mg when given three times daily for 5 days. This dose is four times that approved for intermittent claudication. Nausea and vomiting precluded further escalation (Dezube et al., 1990a).

For pentoxifylline to have significant application in cancer chemotherapy, the maximum tolerated dose in humans should be greater than the dose required for an antitumor effect in mice. Using an interspecies comparison based upon surface area, 1600 mg in a typical 70 kg patient, or 23 mg/kg, is equivalent to 276 mg/kg in a mouse (Freireich et al., 1966). Even after taking into account the 20% bioavailability of pentoxifylline in humans (Ward and Clissold, 1987), this latter dose exceeds the intraperitoneal 50 mg/kg dose that delayed tumor growth in mice. These calculations suggest that antitumor effects of pentoxifylline can be expected in humans. To test this prediction, a Phase II trial to determine response rate in metastatic breast cancer patients has been initiated.

Pentoxifylline as a Down-Regulator of Tumor Necrosis Factor

Several patients who were given pentoxifylline in the Phase I trial experienced an increased sense of well-being even though no tumor shrinkage was detectable. One mediator of the wasting syndrome (cachexia) that adversely affects the lives of many terminal patients is tumor necrosis factor (Balkwill et al., 1987; Beutler and Cerami, 1987). Elevated tumor necrosis factor levels have been correlated with cytotoxicity in tissue culture and tissue injury in animals. Nude mice inoculated with tumor necrosis factor-secreting tumor cells became progressively wasted and died more quickly than mice inoculated with tumor cells that did not secrete tumor necrosis factor (Oliff et al., 1987). Furthermore, tumor necrosis factor mediates endotoxin shock, bone resorption via osteoclast-activating factor activity, and the hypercoagulable state of malignancy (Bauer et al., 1989; Beutler and Cerami, 1987).

Pentoxifylline blocks endotoxin-stimulated elevation of tumor necrosis factor levels in normal human volunteers (Zabel et al., 1989). This is consistent with the ability of pentoxifylline to suppress in vitro lipopolysaccharide-induced increases in murine macrophage-derived tumor necrosis factor at both the transcriptional and translational levels. Of note, the in vitro suppression was reported at micromolar concentrations, levels easily achieved in patients (Strieter, et al., 1988).

We have demonstrated on three independent occasions that pentoxifylline decreased the steady-state level of tumor necrosis factor mRNA in mononuclear cells from cancer patients. This suppression of TNF mRNA was accomplished by a striking improvement in the quality of life (Fig. 4) (Dezube et

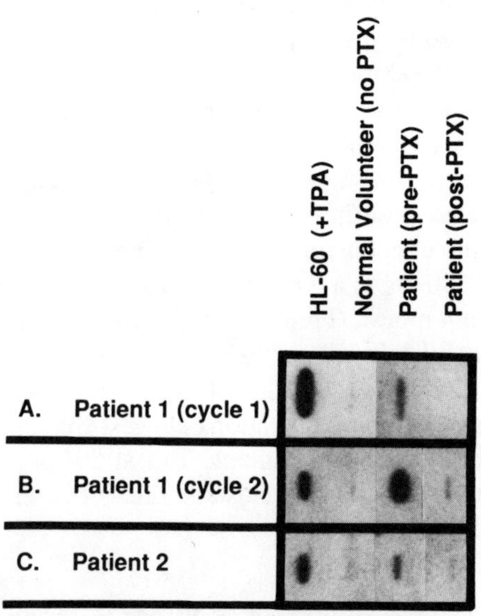

Fig. 4. Slot-blot analysis of tumor necrosis factor mRNA in peripheral mononuclear cells from cancer patients. Samples from patients on pentoxifylline (PTX) were tested in three independent experiments (**A, B, C**), positive (HL-60) and negative (normal volunteer) controls being included. Reproduced from Dezube et al. (1990b).

al., 1990b). These observations have led to several ongoing trials at the Dana-Farber Cancer Institute exploring the ability of pentoxifylline to decrease tumor necrosis factor levels and to improve the quality of life of cancer patients.

Acknowledgments

We thank Robert Schlegel, Ph.D., and Judith L. Fridovich-Keil, Ph.D., for critically reading the manuscript, William J. Novick, Jr., Ph.D., and Jane Chiurco, M.D., for encouragement, and Beverly A. Teicher, Ph.D., for allowing us to quote her unpublished data. The work described in this chapter was supported by grants from Aid for Cancer Research, Boston, Massachusetts; American Cancer Society PRTF-95 and RD-303; NIH 22427 and HL07516 (Hematology Career Training Grant); U.S. Public Health Service CA 22427; and Hoechst-Roussel Pharmaceuticals, Inc.

References

Allen TE, Aliano NA, Cowan RJ, Grigg GW, Hart NK, Lamberton JA, Lane A, (1985): Amplification of the antitumor activity of phleomycins and bleomycins in rats and mice by caffeine. Cancer Res 45:2516–2521.

Balkwill F, Burke F, Talbot D, Tavernier J, Osborne R, Naylor S, Durbin H, Fiers W, (1987): Evidence for tumor necrosis factor/cachectin production in cancer. Lancet ii:1229–1232.

Bauer KA, ten Cate H, Barzegar S, Spriggs DR, Sherman ML, Rosenberg RD, (1989): Tumor necrosis factor infusions have a procoagulant effect on the hemostatic mechanism of humans. Blood 74:165–172.

Beutler B, Cerami A, (1987): Cachectin: More than a tumor necrosis factor. N Engl J Med 316:379–385.

Bogden AE, Kelton DE, Cobb WR, Esber HJ, (1978): A rapid screening method for testing chemotherapeutic agents against human tumor xenografts. In D.P. Houchens and A.A. Ovejera (eds): Proceedings of the Symposium on the Use of Athymic (nude) Mice in Cancer Research. New York: Gustav Fischer, pp 231–250.

Busse PM, Bose SK, Jones RW, Tolmach LJ (1978): The action of caffeine on X-irradiated HeLa cells: III. Enhancement of X-ray induced killing during G2 arrest. Radiat Res 76:292–307.

Cleaver JE, (1978): DNA repair and its coupling to DNA replication in eukaryotic cells. Biochim Biophys Acta 516:489–516.

Cohen MH (1975): Enhancement of the antitumor effect of 1,3-bis(2-chloroethyl)-1-nitrosourea by various psychotropic drugs in combination with caffeine. J Pharm Exp Therap 194: 475–479.

Cohen MH, Schoenfeld D, Wolter J, (1980): Randomized trial of chlorpromazine, caffeine, and methyl-CCNU in disseminated melanoma. Cancer Treat Rep 64:151–153.

Dalvi RR, (1986): Acute and chronic toxicity of caffeine: A review. Vet Hum Toxicol 28:144–150.

Das SK, Lau CC, Pardee AB, (1982): Abolition by cycloheximide of caffeine-enhanced lethality of alkylating agents in hamster cells. Cancer Res 42:4499–4504.

Das SK, Lau CC, Pardee AB (1984): Comparative analysis of caffeine and 3-aminobenzamide as DNA repair inhibitors in Syrian baby hamster kidney cells. Mutat Res 131:71–79.

Dezube BJ, Eder JP, Pardee AB (1990a): Phase I trial of escalating pentoxifylline dose with constant dose thiotepa. Cancer Res 50:6806–6810.

Dezube BJ, Fridovich-Keil JL, Bouvard I, Lange RF, Pardee AB, (1990b): Pentoxifylline and wellbeing in patients with cancer. Lancet 335:662.

Dolby TW, Belmont A, Borun TW, Nicolini C, (1981): DNA replication, chromatin structure, and histone phosphorylation altered by theophylline in synchronized HeLa S3 cells. J Cell Biol 89:78–85.

Fingert HJ, Chang JD, Pardee AB, (1986): Cytotoxic, cell cycle, and chromosomal effects of methylxanthines in human tumor cells treated with alkylating agents. Cancer Res 46: 2463–2467.

Fingert HJ, Pu AT, Chen Z, Googe PB, Alley MC, Pardee AB, (1988): In vivo and in vitro enhanced antitumor effects by pentoxifylline in human cancer cells treated with thiotepa. Cancer Res 48:4375–4381.

Freireich EJ, Gehan EA, Rall DP, Schmidt LH, Skipper HE, (1966): Quantitative comparison of toxicity of anticancer agents in mouse, rat, hamster, dog, monkey, and man. Cancer Chemother Rep 50:219–244.

Gaudin D, Yielding KL, (1969): Response of a ''resistant'' plasmacytoma to alkylating agents and x-ray in combination with the ''excision repair inhibitors'' caffeine and chloroquine. Proc Soc Exp Biol Med 131:1413–1416.

Goth R, Rajewsky MF, (1974): Persistence of O^6-ethylguanine in rat brain DNA: Correlation with nervous system-specific carcinogenesis by ethylnitrosourea. Proc Natl Acad Sci USA 71:639–643.

Harris AL, (1985): DNA repair and resistance to chemotherapy. Cancer Surveys 4:601–624.

Kihlman BA (1977): Caffeine and Chromosomes. Amsterdam: Elsevier Science Publishers B.V.

Kohn KW, Ewig RAG, Erickson LC, Zwelling LA, (1981): Measurement of strand breaks and cross-links by alkaline elution. In E.C. Friedberg and P.C. Hanawalt (eds): DNA Repair: A Laboratory Manual of Research Procedures, Part 1. New York: Dekker, pp 379–402.

Kyriazis AP, Kyriazis AA, Yagoda A, (1985): Enhanced therapeutic effect of cis-diamminedichloroplatinum (II) against nude mouse grown human pancreatic adenocarcinoma when combined with 1-D-arabinofuranosylcytosine and caffeine. Cancer Res 45:6083–6087.

Lau CC, (1982): Mechanism by which caffeine enhances lethality of alkylating agents in mammalian cells. Ph.D. Thesis, Harvard University.

Lau CC, Pardee AB, (1982): Mechanism by which caffeine potentiates lethality of nitrogen mustard. Proc Natl Acad Sci USA 79:2942–2946.

Lau CC, (1989): A rapid assay for detecting clastogenic agents. In M.J.B. Marques, J.C. Alves DaCorte, G.N.M.P. DosSantos, and I.C.R. Lencastre (eds): Public Health and Protection of Population. Amsterdam: Elsevier Science Publishers B.V., pp 293–300.

Lücke-Huhle C, (1982): Alpha-irradiation-induced G2 delay: A period of cell recovery. Radiat Res 89:298–308.

Murnane JP, Byfield JE, Ward JF, Calabro-Jones P, (1980): Effects of methylated xanthines on mammalian cells treated with bifunctional alkylating agents. Nature (London) 285:326–329.

Oliff A, Defeo-Jones D, Boyer M, Martinez D, Kiefer D, Vuocolo G, Wolfe A, Socher SH, (1987): Tumors secreting human TNF/cachectin induce cachexia in mice. Cell 50:555–563.

Pardee AB, James LJ, (1975): Selective killing of transformed baby hamster kidney (BHK) cells. Proc Natl Acad Sci USA 72:4994–4998.

Physicians' Desk Reference: Trental (pentoxifylline). (1990): New Jersey: Medical Economics Company, pp. 1051–1052.

Roberts JJ, (1978): The repair of DNA modified by cytotoxic, mutagenic, and carcinogenic chemicals. Adv Radiat Biol 7:211–436.

Roberts JJ, Brent TP, Crathorn AR, (1971): Evidence for the inactivation and repair of the mammalian DNA temlate after alkylation by mustard gas and half mustard gas. Eur J Cancer 7:515–524.

Roberts JJ, Ward KN, (1973): Inhibitionof postreplication repair of alkylated DNA by caffeine in Chinese hamster cells but not HeLa cells. Chem Biol Interact 7:241–264.

Robinson AC, Harris WJ, (1979): The effect of caffeine on deoxyribonucleic acid synthesis in baby-hamster kidney cells (BHK-21/C13). Biochem Soc Trans 7:1299–1300.

Rose WC, Trader MW, Dykes DJ, Laster WR Jr, Schabel FM Jr, (1978): Therapeutic potentiation of nitrosoureas using chlorpromazine and caffeine in the treatment of murine tumors. Cancer Treat Rep 62:2085–2093.

Schlegel R, Pardee AB, (1986): Caffeine-induced uncoupling of mitosis from the completion of DNA replication in mammalian cells. Science 232:1264–1266.

Schlegel R, Craig RW, (1991): Mitosis: Normal control mechanisms and consequences of aberrant regulation in mammalian cells. In J. Campisi, D.D. Cunningham, M. Inouye, M. Riley (eds.): Perspectives on Cellular Regulation: From Bacteria to Cancer. New York: Wiley-Liss, pp 235–249.

Setlow RB, (1978): Repair deficient human disorders and cancer. Nature (London) 271:713–717.

Strieter RM, Remick DG, Ward PA, Spengler RN, Lynch JP III, Larrick J, Kunkel SL (1988): Cellular and molecular regulation of tumor necrosis factor alpha production by pentoxifylline. Biochem Biophys Res Commun 155:1230–1236.

Tatsumi K, Strauss BS, (1979): Accumulation of DNA growing points in caffeine-treated human lymphoblastoid cells. J Mol Biol 135:435–449.

Tobey RA, (1975): Different drugs arrest cells at a number of distinct states in G2. Nature (London) 254:245–247.

Van den Berg HW, Roberts JJ, (1976): Inhibition by caffeine of post-replication repair in Chinese hamster cells treated with cis platinum (II) diammine dichloride: The extent of platinum binding to template DNA in relation to the size of low molecular weight nascent DNA. Chem Biol Interact 12:375–390.

Walters RA, Gurley LR, Tobey RA, (1974): Effects of caffeine on radiation-induced phenomena associated with cell cycle traverse of mammalian cells. Biophys J 14:99–118.

Ward A, Clissold SP, (1987): Pentoxifylline: A review of its pharmacodynamic and pharmacokinetic properties, and its therapeutic efficacy. Drugs 34:50–97.

Zabel P, Schonharting MM, Wolter DT, Schade UF, (1989): Oxpentifylline in endotoxaemia. Lancet ii :1474–1477.

Perspectives on Cellular Regulation: From Bacteria to Cancer, pages 265–277

Regulation of Gene Expression in Late G_1: What Can We Learn From Thymidine Kinase?

Judith L. Fridovich-Keil, Jean M. Gudas, and Qing-Ping Dou

Department of Biological Chemistry and Molecular Pharmacology, Harvard Medical School, and Division of Cell Growth and Regulation, Dana-Farber Cancer Institute, Boston, Massachusetts 02115

Introduction

The proliferation of normal cells is a stringently regulated process that results in the production of pairs of daughter cells following cell division. A combination of exogenous growth factors and internal modulators controls the ordered transit of normal cells through the various phases of the cell cycle. This control is mediated by diverse sets of second messenger molecules that ultimately influence critical transcriptional, posttranscriptional, and biochemical events required for cells to duplicate their genetic material and subsequently divide (Pardee, 1989; Baserga, 1985).

Regulation of proliferation may be exerted at any of several distinct points in the cell cycle. During early embryogenesis, growth control occurs primarily at the boundary of the G_2/M phases and involves the cyclins, cdc2 and cdc25 proteins (Murray, 1989; Murray and Kirschner, 1989; O'Farrell et al., 1989, and Schlegel and Craig, this volume). In somatic cells, control of proliferation also occurs during the G_1 phase: first during emergence from quiescence, and later at the restriction point about 2 hr prior to the onset of DNA synthesis (Pardee, 1989). The biochemical and molecular events that transpire at these regulatory points have been the subject of intensive investigation for many years, largely due to the evidence that abnormal control at these points may result in the relaxed growth factor requirements characteristic of transformed and tumorigenic cells. Therefore, understanding these controls in normal cells is fundamental to our understanding of cancer. In the last decade, Art Pardee's laboratory has made a number of critical contributions towards understanding normal and abnormal growth control. Here, we summarize some of the most recent findings from his lab. These studies are ongoing—and we are excited to participate in their evolution.

Early G_1 Controls

Using a variety of molecular techniques, many genes have been identified that are rapidly induced by the addition of serum growth factors to quiescent cells. The expression of several of these early response or competence genes has been shown to be necessary, but not sufficient, for cells to leave the quiescent growth state, progress through G_1, and initiate DNA synthesis (Rollins and Stiles, 1989). The products of these genes include transcription factors, cytoplasmic proteins involved in signal transduction, nuclear proteins, and several structural secreted proteins. Convergent studies have corroborated the hypothesis that deregulation of the early response genes might be associated with transformation; several of these genes, including the c-*myc*, c-*fos*, c-*ras*, and c-*jun* protooncogenes, are deregulated in transformed and tumorigenic cells.

Many of the early response genes share common modes of regulation. Most are induced at the level of transcription initiation by processes that do not require de novo protein synthesis. In fact, the mRNAs for many early response gene are overexpressed in the presence of protein synthesis inhibitors. Many early response genes encode mRNAs that are very short-lived, so that their abundance may be regulated at the levels of both transcription initiation and mRNA stability. The cloning and characterization of early response genes are important steps toward understanding how their products influence subsequent events during the progression of cells from G_1 into S phase.

Late G_1 Controls

The Pardee laboratory has been interested for many years in elucidating the events that occur late in G_1 at the restriction or R point of the cell cycle (Pardee, 1974). The R point is defined as the time in late G_1 after which cells are no longer dependent on exogenous growth factors or de novo protein synthesis in order to enter the S phase. Previous studies from Art's laboratory have demonstrated that murine fibroblasts in G_1 require insulin-like growth factor-I (IGF-I) and efficient protein synthesis to reach the R point (Yang and Pardee, 1986). This requirement for rapid protein synthesis led to Art's proposal that a labile protein (or proteins) must accumulate to a critical threshold level in late G_1 before cells may proceed into S phase. One such protein that meets these criteria, p68, was identified by two-dimensional gel electrophoresis of proteins synthesized at various times during G_1 (Pardee, 1989). This 68-kDa protein is labile in nontransformed fibroblasts, but stable in transformed cells.

Other, more recent evidence suggests that the onset of S phase may also be controlled by negatively acting factors that prevent cells from initiating DNA synthesis. Two candidates for such negatively acting factors are the nuclear protein products of the retinoblastoma (Rb) and p53 genes. These genes have been designated tumor suppressor or antioncogenes (Green, 1989; Lane and Benchimol, 1990). The Rb or p53 genes are deleted or deregulated in many human cancers, and the Rb and p53 proteins are tightly bound by the transforming gene products of several oncogenic DNA viruses. It has been proposed that the interaction between Rb or p53 and these viral transforming proteins may circumvent the growth inhibitory effects of Rb and p53, resulting in the aberrant growth control associated with viral transformation (Sager, 1989). These data suggest that, in addition to the positive control mechanisms described earlier, negative controls also play a role in regulating the onset of DNA synthesis.

Thymidine Kinase (TK) as a Model for Late G_1 Regulation of Gene Expression

Concomitant with the onset of S phase, the expression of mRNAs and protein products for the replication-dependent histones and many of the enzymes associated with DNA synthesis increases (Baserga, 1985). Because expression of DNA biosynthetic enzymes, such as thymidine kinase (TK) and thymidylate synthase (TS), closely parallels the onset of S phase, the genes encoding these enzymes may be viewed as molecular models for regulatory events at the G_1/S boundary of the cell cycle (Pardee, 1989). In murine BALB/c 3T3 (clone A31) cells, both the onset of DNA synthesis and the induction of TK mRNA and enzyme activity have been shown to be regulated by a labile protein(s) (for review, see Knight et al., 1989). A pulse of cycloheximide given to serum-stimulated cells in mid- to late-G_1 delays the onset of DNA synthesis and the induction of TK and TS by an interval that is about 2 hr longer than the length of the pulse. Other studies have implicated IGF-I as the growth factor responsible for the coordinate regulation of these three events at the G_1/S boundary (for review, see Gudas et al., 1989). Considered together, these results suggest that the same or similar factors are involved in the control of all three processes. Thus, by undertaking a detailed investigation of the molecular mechanism(s) responsible for TK induction, we expect to gain insights into more global regulatory events that occur at the onset of DNA synthesis.

Following release from quiescence, there is a dramatic rise in TK mRNA near the G_1/S boundary that precedes the rise in TK enzyme activity by 2–3 hr (Fig. 1) (Gudas et al., 1989). The induction of TK mRNA at the G_1/S boundary is subject to multiple levels of control that differ qualitatively and quantitatively from those regulating the early response genes. For example, unlike the early G_1 inductions of c-*fos* and c-*myc*, accumulation of TK mRNA is absolutely dependent upon de novo protein synthesis. This induction of TK gene expression involves regulation at both the transcriptional and posttranscriptional levels (Knight, et al., 1989). As TK mRNA has a relatively long half-life (about 8 hr), its abundance does not fluctuate greatly in subsequent cell cycles. In addition to the multiple levels of control that regulate TK mRNA levels, recent evidence indicates that the amount of TK protein in cells is

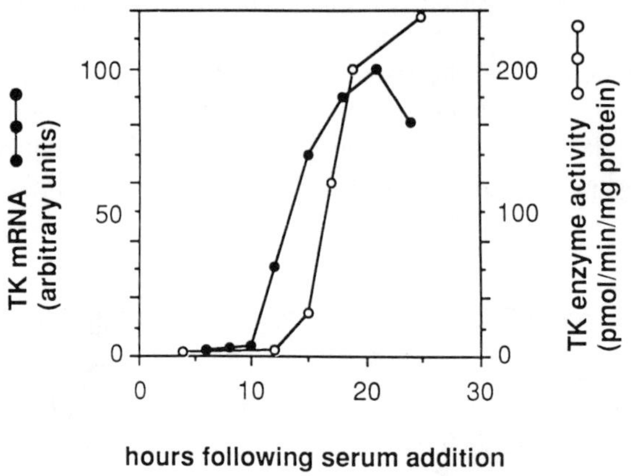

Fig. 1. Thymidine kinase (TK) mRNA and enzyme levels following serum stimulation of quiescent murine 3T3 fibroblasts. Cells made quiescent by serum deprivation for 48 hr were stimulated with fresh serum-containing medium and harvested for analysis at the times indicated.

also tightly regulated at the level of translation (Sherley and Kelley, 1988; Gross and Merrill, 1989).

Are there common regulatory signals that underlie the different ways in which TK expression is controlled? To address this question we have directed our efforts toward identifying the DNA and RNA sequences involved in cell cycle-dependent regulation of TK gene expression. As these sequence elements are defined, we are beginning to identify and purify the proteins that bind to them using a combination of molecular and biochemical techniques. In the process of characterizing the elements and factors responsible for each level of TK regulation, we hope to gain insight into the larger question of how the cell coordinates so precisely the timing of seemingly unrelated regulatory mechanisms. Furthermore, we hope to identify links between the regulation of TK and other genes expressed at the onset of S phase.

Transcriptional Regulation of TK Gene Expression

Several lines of evidence indicate that TK expression is regulated at the transcriptional level. Among these are the results of nuclear run-on transcription assays that have been performed using cell lines derived from two different mammalian species, mouse and monkey (Coppock and Pardee, 1987; Stewart et al., 1987). In both cell types, transcription increased as cells entered S phase from quiescence. Although the transcriptional increase accounted for only a portion of the total increase in mRNA, it is likely that TK promoter sequences play a role in the regulation of TK expression.

The 5′ sequences of TK genes from a number of species, including human, hamster, chicken, and mouse, have been compared. The analyses indicate that

TK promoter sequences have undergone extensive evolutionary divergence (Arcot et al., 1989). The human TK promoter contains two inverted CCAAT box elements, as well as a consensus TATA sequence. The chicken TK promoter, although significantly divergent from its human counterpart, also carries two inverted CCAAT boxes and one TATA box. The hamster TK promoter has one CCAAT box and one TATA element. In striking contrast, the mouse promoter contains neither of these elements (Lieberman et al., 1988; Seiser et al., 1989).

The presence or absence of CCAAT and TATA elements in the promoter regions of the various TK genes may influence the placement and/or efficiency of transcriptional initiation. Transcriptional start sites, mapped by ourselves and others, indicate that both the human and hamster TK genes rely upon one or two preferred sites of transcriptional initiation. In contrast, the mouse gene have over twenty different start sites that are distributed over a distance of about 200 bases upstream of the translational initiation site (Gudas et al., in preparation). In this regard the mouse TK gene resembles other traditional "housekeeping" genes, such as those encoding the enzymes dihydrofolate reductase (DHFR) and thymidylate synthase.

Perhaps one of the most intriguing questions to emerge from these studies is: how do these divergent TK promoter sequences respond to mitogenic stimulation with a similar induction of activity at the onset of S phases? We want to define the sequence elements within each of these promoters that are responsible for mediating the response, and ultimately, to identify and characterize the nuclear factors that interact with these sequences.

Functional analysis of critical TK promoter elements has been performed on both human and mouse genes using TK promoter/reporter gene fusion constructs transfected into cells. Results from a number of different laboratories show that human TK promoter sequences are sufficient to confer S phase specific regulation to a heterologous reporter gene (Kim et al., 1988; Travali et al., 1988). Both the CCAAT and TATA sequences are essential for the activity of this promoter (Arcot et al., 1989; Lipson et al., 1989). Our own studies of the mouse TK promoter indicate that, despite its lack of CCAAT and TATA consensus elements, it also contains sequences capable of conferring S phase specific regulation to a reporter gene (Fig. 2) (Fridovich-Keil et al., 1991). Using deletion analysis and site-specific mutagenesis we find that a consensus Sp1 site located within 100 bp of the TK translation initiation site is essential for efficient activity of the murine TK promoter. We are currently investigating the roles of other sequences in regulating TK gene expression in murine cells.

DNA regulatory sequences appear to modulate transcription via their interactions with nuclear proteins (Dynan and Tjian, 1985). This concept of transcriptional control mediated by the interactions of DNA sequences with specific binding proteins was first proposed over 30 years ago by Pardee, Ja-

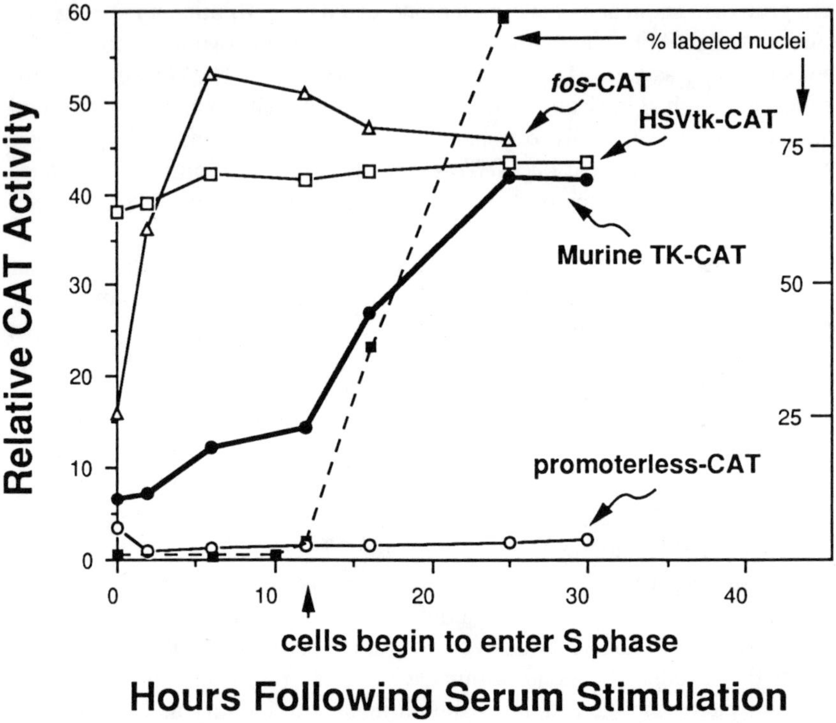

Fig. 2. S phase-specific induction of a heterologous gene driven by the murine TK promoter. 3T3 cells stably transfected with the indicated promoter/chloramphenicol acetyltransferase (CAT) fusion constructs were synchronized by serum starvation and serum restimulated. The dashed line indicates the percent labeled nuclei detected by autoradiography at several time points from one representative experiment. The solid lines represent relative CAT activity detected in extracts of cells harvested at the times indicated. HSVtk, herpes simplex virus thymidine kinase promoter.

cob, and Monod from their work with the *lac* operon of *E. coli* (Pardee et al., 1959). Applying this paradigm to the induction of TK gene expression, we might expect to find regulatory elements in the TK promoter that bind specific proteins in a cell cycle-dependent manner.

Indeed, in vitro analyses of specific DNA–protein interactions indicate that sequences in both the human and murine TK promoters do bind nuclear proteins in an S phase-specific manner. Knight, Gudas, and Pardee reported a striking, cell cycle-dependent shift in the pattern of protein binding to a fragment of human TK promoter sequence containing a CCAAT box element (Knight et al., 1987). As visualized by band-shift analysis, the DNA fragment was consistently bound by nuclear proteins throughout the cell cycle, but the apparent sizes of the DNA–protein complexes changed dramatically as cells traversed the G_1/S boundary.

Our more recent studies of the murine TK promoter indicate that, as reported in the human system, some of the proteins that bind to the promoter do so in an S-phase-specific manner (complex C in Fig. 3) (Dou et al., 1991). These changes are consistent with the idea that different sets of proteins interact with the TK promoter at different points in the cell cycle. The close temporal correlation between the increased rate of TK transcription in these cells and the observed band-shift changes suggests that the proteins in the DNA–protein complexes may play a role in the induction of transcription. Competition studies demonstrate that the proteins that bind to the mouse TK promoter are distinct from those that bind to the human TK promoter, consistent with the sequence divergence between the two promoters.

Using DNase 1 footprint analysis of a 170 base pair fragment of the murine TK promoter incubated either with crude nuclear extracts, or with purified Sp1 protein, we have identified an Sp1 binding site as well as two other binding sites, designated I and II (Fig. 3B). The S phase-specific binding activity in the mouse system appears to interact with all three sites (Dou et al., 1991). It is noteworthy that the site that binds proteins in a cell cycle dependent manner lies within the region that is critical for gene expression, as indicated by our genetic studies. We are now working to further characterize and isolate these DNA binding proteins so that we may ultimately clone the genes that encode them. These genes may constitute elements of the cellular regulatory machinery responsible for guiding cells across the G_1/S boundary of the cell cycle.

Posttranscriptional Regulation of TK Gene Expression

The steady state levels of TK mRNA increase 20- to 50-fold as cells traverse G_1 and enter S phase (Gudas et al., 1989). However, TK gene transcription increases only 3- to 4-fold in mouse cells and 6- to 7-fold in monkey CV-1 cells. Although these differences are significant, they cannot account for the large accumulation of TK mRNA at the onset of S phase. Furthermore, no differences in TK gene transcription were detected in chicken nuclei obtained from either dividing or nondividing cells (Groudine and Casmir, 1984). These results indicate that posttranscriptional mechanisms must contribute to the overall accumulation of TK mRNA at the G_1/S boundary, albeit to varying degrees dependent upon the species or cell type.

We have examined nuclear (Fig. 4A) and cytoplasmic (Fig. 4C) RNA from quiescent murine cells stimulated to proliferate by the addition of fresh serum-containing medium (Gudas et al., 1988). Very little TK mRNA was detected in the nuclei of cells harvested during G_0 or G_1. At the onset of S phase, however, a dramatic change in the processing of TK pre-mRNA was observed. In addition to the appearance of mature TK mRNA, a series of high-molecular-weight, nuclear precursors was detected. These high-molecular-weight TK

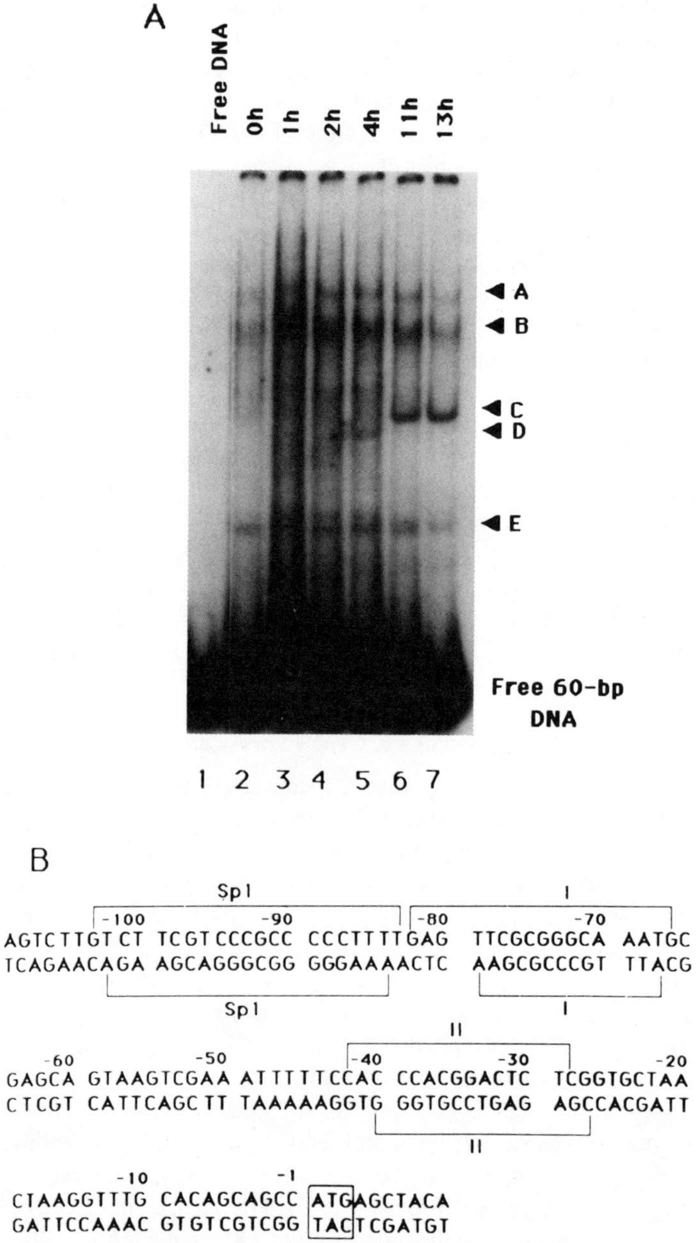
A
Free DNA
0h
1h
2h
4h
11h
13h
A
B
C
D
E
Free 60-bp DNA
1 2 3 4 5 6 7
B
Sp1
I
-100 -90 -80 -70
AGTCTTGTCT TCGTCCCGCC CCCTTTTGAG TTCGCGGGCA AATGC
TCAGAACAGA AGCAGGGCGG GGGAAAACTC AAGCGCCCGT TTACG
Sp1
I
II
-60 -50 -40 -30 -20
GAGCA GTAAGTCGAA ATTTTTCCAC CCACGGACTC TCGGTGCTAA
CTCGT CATTCAGCTT TAAAAAGGTG GGTGCCTGAG AGCCACGATT
II
-10 -1
CTAAGGTTTG CACAGCAGCC ATGAGCTACA
GATTCCAAAC GTGTCGTCGG TACTCGATGT

hnRNAs could be chased into mature mRNA in the presence of the RNA synthesis inhibitor actinomycin D, suggesting a precursor–product relationship. These results suggest that quiescent cells may be inefficient in processing certain mRNAs, such as TK, that are required during S phase, and that processing activity might be restored just prior to the onset of DNA synthesis. The failure to process TK hnRNA may somehow signal its degradation in the nucleus, accounting at least in part, for the low levels of TK mRNA detected in quiescent cells.

One possible site for the posttranscriptional regulation of TK mRNA could be at the level of cleavage and polyadenylation of TK pre-mRNAs. If there were a failure to properly cleave and polyadenylate TK pre-mRNAs during G_0 and G_1, we would expect to detect unprocessed run-on transcripts in the nucleus using a sensitive RNase protection assay. When such experiments were performed, no 3′ unprocessed TK pre-mRNAs were detected in the nuclei of cells prior to entry into S phase. As cells reached the G_1/S boundary, however, we began to detect 3′ unprocessed pre-mRNAs concomitant with the appearance of TK mRNA splicing intermediates. Our data, therefore, are consistent with the idea that cleavage/polyadenylation and splicing of pre-mRNAs are interdependent processes in the formation of mature TK mRNA (Gudas et al., 1990; Gudas et al., in preparation).

The ability to detect a series of TK nuclear pre-mRNAs has allowed us the opportunity to characterize the various splicing intermediates that are generated during the formation of a mature mRNA. Because the hybridization patterns obtained with total nuclear RNA and nuclear poly(A^+) RNA were virtually identical, 3′ end cleavage and polyadenylation most likely precede the subsequent splicing of TK pre-mRNAs. The murine TK gene contains seven coding exons and six introns. Random removal of introns from the primary transcripts of this gene should lead to a great number of intermediate TK pre-mRNAs. The fact that only a few discrete TK pre-mRNAs were visualized with a cDNA probe implies that rate-limiting intron excision occurs via a preferred and orderly manner. To test this hypothesis, we characterized the high-molecular-weight splicing intermediates using unique sequence DNA probes isolated from each of the six TK intervening sequences (Gudas et al., 1990). These analyses have allowed us to predict that intron removal from

Fig. 3. S phase-specific interactions between nuclear protein(s) and the murine TK promoter **(A)** The nuclear extracts used in these band-shift analyses were prepared from serum-starved and restimulated cells. Extracts prepared from cells harvested at the times indicated were incubated with a labeled 60 base pair (bp) fragment of DNA including the murine TK sequence −47 to −4 bp (see **B**). Bands representing complexed and free DNAs are indicated. **(B)** Partial sequence of the murine TK promoter. Brackets indicate sequences, labeled Sp1, I, and II, that are protected by proteins from DNase 1 digestion during footprint analysis. The TK translational initiation site, ATG, is boxed.

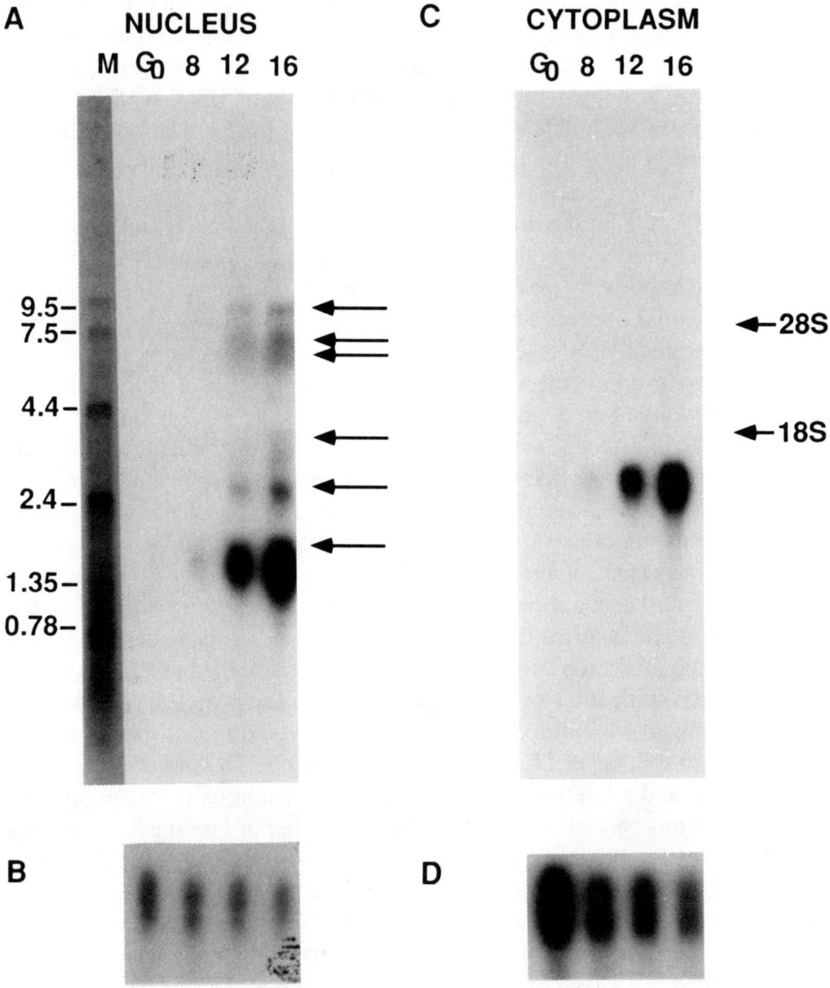

Fig. 4. Northern blot analysis of nuclear and cytoplasmic TK mRNAs after release from quiescence. Twenty-five micrograms of nuclear (1A) or 15 μg of cytoplasmic (1C) RNA obtained from cells that were either quiescent (G_0) or had been restimulated with fresh serum-rich medium for the times indicated, was analyzed by northern blot hybridization. **(A, C)** Filters hybridized with a murine TK cDNA probe. Arrows to the right of the nuclear samples indicate TK pre-mRNAs. The sizes of the marker RNAs (lane M) are indicated to the left of **A.** The positions of the 18 S and 28 S ribosomal RNAs are indicated to the right of **C.** To control for RNA loading and transfer, the membranes shown in **A** and **C** were washed and rehybridized with a probe for the β_2-microglobulin gene (**B** and **D**, respectively).

nascent TK transcripts occurs via two predominant routes. Moreover, our studies indicate that the same preferred order of intron removal from TK pre-mRNAs has been conserved in Chinese hamster cells.

Several investigators have transfected into recipient TK^- cell lines TK cDNA sequences that were linked to constitutive promoter elements (Knight et al., 1989). TK enzyme activity, and in several cases TK mRNA levels, were shown to be expressed from the transfected genes in a cell cycle dependent manner. These experiments suggest that at least some of the genetic elements capable of conferring S phase-dependent regulation on TK mRNA levels and enzyme activity are contained within the mature cytoplasmic mRNA. Regulated expression of TK enzyme activity may be explained by the enhanced translation of TK mRNAs during S phase and enhanced degradation of the protein upon cell division. The fluctuations in TK mRNA levels, under these conditions, must involve posttranscriptional controls working via mechanisms that have yet to be clearly defined.

Conclusions

The insights derived thus far from experiments performed in the Pardee laboratory and by other colleagues, all indicate that much can be learned from studying the regulation of S phase-specific genes in general, and TK in particular. The identification of critical DNA or RNA elements within the TK gene, combined with the characterization and study of proteins that bind them, is already beginning to shed light on the molecular mechanisms that underlie the progression of cells from G_1 into the S phase of the cell cycle.

Although much knowledge has already been gained concerning regulatory events at the G_1/S boundary, many questions remain unresolved. For example, How are the various levels of control of TK gene expression coordinated so precisely by the cell? Are other S phase specific genes regulated in a manner similar to TK? We have yet to determine how the controls placed on TK gene expression interplay with other regulatory cascades in the cell, such as those involving the retinoblastoma, cdc2, and p53 proteins. And finally, we have yet to resolve the question of how this regulation becomes deranged in cancer. HiNF-D, a transcription factor involved in the regulation of histone gene expression, recently was reported to be expressed in a cell cycle-dependent manner in normal cells but constitutively in transformed cells (Holthuis, et al., 1990). These results, together with our own, support our belief that an investigation of TK gene expression at the G_1/S boundary will provide valuable insights, not only into the regulation of normal cells, but also into the aberrant proliferation characteristic of cancer cells.

References

Arcot SS, Flemington EK, Deininger PL (1989): The human thymidine kinase gene promoter. J Biol Chem 264:2343–2349.

Baserga R (1985): The Biology of Cell Reproduction. Cambridge: Harvard University Press.

Coppock DL, Pardee AB (1987): Control of thymidine kinase mRNA during the cell cycle. Mol Cell Biol 7:2925–2932.

Dou QP, Fridovich-Keil JL, Pardee AB (1991): Inducible proteins binding to the murine thymidine kinase promoter in late G_1/S phase. Proc Natl Acad Sci USA, in press.

Dynan WS, Tjian R (1985): Control of eukaryotic messenger RNA synthesis by sequence-specific DNA-binding proteins. Nature (London) 316:774–778.

Fridovich-Keil JL, Gudas JM, Dou QP, Bouvard I, Pardee AB (1991): Growth-responsive expression from the murine thymidine kinase promoter: Genetic analysis of DNA sequences. Cell Growth Diff, in press.

Green M (1989): When the products of oncogenes and anti-oncogenes meet. Cell 56:1–3.

Gross MK, Merrill GF (1989): Thymidine kinase synthesis is repressed in nonreplicating muscle cells by a translational mechanism that does not affect the polysomal distribution of thymidine kinase mRNA. Proc Natl Acad Sci USA 86:4987–4991.

Groudine M, Casmir C (1984): Post-transcriptional regulation of the chicken thymidine kinase gene. Nucleic Acid Res 12:1427–1446.

Gudas JM, Knight GB, Pardee AB (1989): The cell cycle and restriction point control. In E.I. Frei (ed): The Regulation of Proliferation and Differentiation in Normal and Neoplastic Cells. San Diego, CA: Academic Press, pp. 3–20.

Gudas JM, Knight GB, Pardee AB (1988): Nuclear posttranscriptional processing of thymidine kinase mRNA at the onset of DNA synthesis. Proc Natl Acad Sci USA 85:4705–4709.

Gudas JM, Knight GB, Pardee AB (1990): Ordered splicing of thymidine kinase pre-mRNA during the S phase of the cell cycle. Mol Cell Biol 10:5591–5595.

Holthuis J, Owen TA, van Wijnen AJ, Wright KL, Ramsey-Ewing A, Kennedy MB, Carter R, Cosenza SC, Soprano KJ, Lian JB, Stein JL, Stein GS (1990): Tumor cells exhibit deregulation of the cell cycle histone gene promoter factor HiNF-D. Science 247:1454–1457.

Kim YK, Wells S, Lau Y-FC, Lee AS (1988): Sequences contained within the promoter of the human thymidine kinase gene can direct cell-cycle regulation of heterologous fusion genes. Proc Natl Acad Sci USA 85:5894–5898.

Knight GB, Gudas JM, Pardee AB (1987): Cell-cycle-specific interaction of nuclear DNA-binding proteins with a CCAAT sequence from the human thymidine kinase gene. Proc Natl Acad Sci USA 84:8350–8354.

Knight GB, Gudas JM, Pardee AB (1989): Coordinate control of S phase onset and thymidine kinase expression. Jpn J Cancer Res 80:493–498.

Lane DP, Benchimol S (1990): p53: Oncogene or anti-oncogene? Genes Dev 4:1–8.

Lieberman HB, Lin P-F, Yeh D-B, Ruddle FH (1988): Transcriptional and posttranscriptional mechanisms regulate murine thymidine kinase gene expression in serum-stimulated cells. Mol Cell Biol 8:5280–5291.

Lipson KE, Chen S-T, Koniecki J, Ku D-H, Baserga R (1989): S-phase-specific regulation by deletion mutants of the human thymidine kinase promoter. Proc Natl Acad Sci USA 86:6848–6852.

Murray AW (1989): The cell cycle as a cdc2 cycle. Nature (London) 342:14–15.

Murray AW, Kirschner MW (1989): Dominoes and clocks: The union of two views of the cell cycle. Science 246:614–621.

O'Farrell PH, Edgar BA, Lakich D, Lehner CF (1989): Directing cell division during development. Science 246:635–640.

Pardee AB (1974): A restriction point for control of normal animal cell proliferation. Proc Natl Acad Sci USA 71:1286–1290.

Pardee AB (1989): G1 events and regulation of cell proliferation. Science 246:603–608.
Pardee AB, Jacob F, Monod J, (1959): The gneetic control and cytoplasmic expression of inducibility in the synthesis of β-galactosidase of *E. coli*. J Mol Biol 1:165–178.
Rollins BJ, Stiles CD (1989): Serum-inducible genes. Adv Cancer Res 53:1–31.
Sager R (1989): Tumor suppressor genes: The puzzle and the promise. Science 246:1406–1412.
Seiser C, Knofler M, Rudelstorfer I, Haas R, Wintersberger E (1989): Mouse thymidine kinase: The promoter sequence and the gene and pseudogene structures in normal cells and in thymidine kinase deficient mutants. Nucleic Acid Res 17:185–195.
Sherley JL, Kelley TJ (1988): Regulation of human thymidine kinase during the cell cycle. J Biol Chem 263:8350–8358.
Stewart CJ, Ito M, Conrad SE (1987): Evidence for transcriptional and post-transcriptional control of the cellular thymidine kinase gene. Mol Cell Biol 7:1156–1163.
Travali S, Lipson KE, Jaskulski D, Lauret E, Baserga R (1988): Role of the promoter in the regulation of the thymidine kinase gene. Mol Cell Biol 8:1551–1557.
Yang HC, Pardee AB (1986): Insulin-like growth factor I regulation of transcription and replicating enzyme induction necessary for DNA synthesis. J Cell Physiol 127:410–416.

Perspectives on Cellular Regulation:
From Bacteria to Cancer, pages 279–296

Cell Cycle-Specific Control of Terminal Cell Differentiation

Andrew Yen

Department of Pathology, Cornell University Veterinary College, Ithaca, New York 14853

The pioneering work of Arthur Pardee in cell cycle regulation has provided impetus and insight to many of us. I first became aware of his work when I was a gradaute student. The experiments from his laboratory then had an elegance of logic that was to be a model for work I pursued as a postdoctoral fellow in his laboratory. It was a stimulating time to be working with him. Certainly, it was a formative time for a variety of ideas about cell cycle regulation. It was a unique education to be sharing in Art's thinking as he shaped the ideas that would have a profound influence on our own subsequent thinking and work, as well as on so many others in the field. Finally, any mention of memories of the time I was in his laboratory would be incomplete without note of the personal kindness he extended to me. I still remember his unsolicited offer of financial assistance when I first arrived in Boston with a nascent family. I imagine many have similar personal recollections.

Introduction

In mammalian cells the process of replication segregates into a series of obvious phases of distinguishable activities. The most obvious of these, mitosis or M, was known since the time of the pathologist Virchow. The discovery that DNA synthesis occurred during only a discrete portion, S phase, of the replicative cycle segregated the division cycle further and resolved the cell cycle into the four phases that we know now: G_1, S, G_2, and M. G_1 and G_2 are the periods anteceding and following DNA synthesis before mitosis, respectively. At the time they represented two gaps, hence G_1 and G_2, in knowledge of the cellular activity during these periods. The study of the orderly progression of cellular activities constituting the four phases of the cell division cycle has been a major focus in regulatory cell biology. Cell cycle studies have contributed greatly to our understanding of diverse pathological processes such as cancer and heart disease, as well as to fundamental biological processes such as development and immune function. Thus the study of the cell cycle has had far reaching consequences.

The G_1 phase has historically been of particular interest. This might come as no surprise. Most somatic cells are growth arrested with a G_1 DNA content, having arisen developmentally from proliferatively active cells. The primary exceptions are cells of the intestinal lining, hair follicles, and hematopoietic system. Of these, the hematopoietic system is perhaps most striking. It presents an ongoing developmental microcosm, complete with a range of pathologies, within a contained and well-defined system. With these exceptions, the disease of cancer has provided a striking "experiment of nature," indicating the consequences of dysfunction of the normally G_1 cell cycle regulatory functions. The proliferative activity that is the frequent hallmark of neoplasms points to a critical, G_1-specific, growth-regulatory function in G_1. Growth regulation apparently is largely exerted in G_1 or not at all. The study of the cellular pathology giving rise to neoplasms thus focused on the dysfunction of G_1-specific growth regulatory processes. These processes were historically studied using murine fibroblasts to great advantage.

G_1-Specific Growth Regulation by Serum

Murine fibroblast-like cells were a historically significant model in which neoplastic transformtion was studied because they showed G_1-specific growth arrest, whereas their transformed counterparts failed to do so. The untransformed fibroblasts grew in serum-supplemented medium and arrested growth with a G_1 DNA content, G_0, once the medium was depleted of factors essential for proliferation. In contrast, the transformed counterparts failed to growth arrest and continued to proliferate, exhibiting features that were classically associated with neoplastic transformation. These included abnormal morphology and an associated loss of contact inhibition and requirement for substrate adhesion. Thus the transformed cells failed to exercise growth regulation. The question I focused on when I joined Art Pardee's laboratory as a postdoctoral fellow was whether there was a discrete time in G_1 when cells became sensitive to growth regulation exerted by growth factors. The question was of obvious significance. If one knew when in G_1 the growth regulatory process occurred, then one could attempt to identify the cellular response in nontransformed cells that was aberrant in transformed cells.

Murine Swiss 3T3 fibroblasts were used for our studies because they were exquisitely sensitive to serum modulation of cell proliferation. The conventional means of determining events in the cell cycle relied on using synchronized cells. Exponentially growing cells could, for instance, be treated with a drug that inhibited progression beyond a certain point in the cell cycle, typically the G_1/S boundary, and a population of cells restricted to a discrete point in the cell cycle could thus be derived. These cells could then be released from the inhibition and allowed to continue the cell cycle. Alternatively, mitotic cells that had released their substrate adhesion could be harvested by mechanical shock and then recultured. These methods, however, were not well suited to the questions Art and I were interested in for three primaray reasons: (1) derivation of synchronized cells by pharmacological manipulation yielded cells that might have altered metabolism due to the drugs, (2) the timing of events could not be accurately resolved due to the heterogeneity of the cellular response to the synchronization procedure, and (3) there was significant cell-to-cell dispersion in cell cycle transit times, most of it in G_1, so that desynchronization was rapid and greatest in the phase of interest.

To circumvent these difficulties, we developed a technique using unsynchronized cells that were exponentially growing and not subject to pharmacological manipulation. The technique relied on mathematical modeling and the then relatively novel technique of laser flow cytometry. For a population of exponentially growing cells in steady state, the number density of cells can be approximated by the age-density formulation:

$$n(a,t) = n(0,0)\exp\{-k(a-t)\}, \qquad 0 < a < T$$

where a is age in the cell cycle, t is time of observation of the population, k is $(\ln 2)/T$, and T is the generation time or duration of the cell cycle. In this case, the integral of $n(a,t)$ over any domain of age a gives the number of cells in that age interval at time t. Obviously, if the age interval is from 0 to T, the integral of $n(a,t)$ gives the total number of cells in the population at time t. In exponentially growing cells, if the serum concentration was reduced from the typical 10 to 0.25%, then cells unable to progress through cycle without the higher serum concentration would be arrested, while those whose growth was not restricted by reduced serum would go on to mitosis. By using a drug to block mitosis (e.g., colchicine or colcimid, which inhibit microtubule polymerization), the number of cells sensitive or insensitive to serum for continued cell cycle progression could be determined from the number of cells blocked in G_1 or M. By staining the cells quantitatively for DNA content with a fluorescent dye, the number of serum-sensitive and -insensitive cells could be determined by flow cytometry, which measures fluorescence after excitation with a laser. Since the number of cells blocked by low serum equaled the integral of $n(a,t)$ over a from 0 to R, the restriction point or age in the cell cycle where they were inhibited, the value of R could be determined by solving the integral equations. [The calculation was done using a slightly more complicated form of $n(a,t)$ to allow for the variability in cell cycle transit times. However, the principle of the method is unchanged.] The calculated value for R in G_1 was 2 hr before the onset of S phase, given that G_1 was approximately 5.5 hr duration. Thus, as murine fibroblasts progressed through G_1, they encountered a point in their metabolism that responded to exogenous levels of serum. In the presence of adequate serum, they progressed to complete the cell cycle. If the serum concentration was too low, then they arrested in G_1 at the *R* point. This result meant that a regulatory function was localized to a discrete time in G_1.

G_0 versus G_1 Arrest

This finding suggested two other lines of inquiry. One was whether the G_1 regulatory point was a general restriction point for any essential growth factor. The other was what is the fate of a cell that has been arrested from further progression through the cell cycle. In the case of the first question, we examined the cellular response to other essential growth factors or nutrients. In particular, the removal of platelet-derived growth factor (PDGF) from the medium resulted in a G_1-specific growth arrest. The point of sensitivity was at the onset of G_1. Thus it was distinct from the restriction point for serum. Isoleucine was also found to be required for continued progression through G_1. The cells were sensitive to the level of exogenous isoleucine at a point in G_1 that was distinct from the other two. Thus there appeared to be a progression of metabolic steps sensitive to exogenous growth factors or nutrients that occurred in G_1. The finding that sensitivity to exogenous serum occurred only at a restricted point in G_1, 2 hr before DNA synthesis, was particularly interesting since this was the point at which the metabolism of the nontransformed cells and transformed cells diverged.

The other question was the nature of the arrest. The two simplest possibilities were that the cells either arrested in the course of G_1 transit and remained effectively in a prolonged G_1 state, or that they left the cell cycle at the time of arrest and entered a distinct quiescent state, G_1. At the time, these alternative were a subject of argument in the literature. We reasoned that if the arrested cells required a metabolic step to undergo DNA synthesis that cycling G_1 cells did not, then the arrested cells had to be in a state distinct from G_1. In collaboration with Robert Warrington, we found that recruitment of arrested cells into DNA synthesis by restoration of serum required a histidinol–sensitive step that actively cycling G_1 cells did not require. Thus the cells arrested in G_1 by low serum were not retarded in G_1 but had in fact left the cell cycle and were in G_0. In further studies, we found that different means of arresting cells in G_1 resulted in distinct G_0 states that were distinguishable by their requirement for recruitment to S phase. Thus cells could leave the G_1 phase at different points to assume distinguishable arrested states, all characterized by a G_1 DNA content.

Since G_1 serum sensitivity distinguished nontransformed from transformed cells, serum-mediated growth control was the focus of subsequent studies. We found that the arrested cells were biophysically distinguishable from proliferating cells by the light scattering properties of their nuclei, measured by laser flow cytometry. Arrested cells scattered less light into a forward angle cone along the axis of the incident laser beam, indicating that they were physically smaller. Furthermore their recruitment into S phase by restoration of serum depended on nuclear size. Subsequent work in the laboratory by others showed that a critical amount of protein synthesis was required prior to the serum-sensitive restriction point in G_1. Thus the synthesis of a critical regulatory protein to reach threshhold levels was implicated. Further experiments indicated that the critical protein was labile. The indication of a labile regulatory protein was a harbinger of the discovery of protooncogene products, such as the c-*myc* protein, which is labile and important in growth control.

G_0 Arrest and Differentiation

In a broader biological perspective, the question of G_0-specific growth arrest may be perceived as one of development. The growth-arrested cell is a proliferatively quiescent cell displaying a different phenotype than its precursor. This is especially evident in the case of cellular senesence, where proliferative capability is irrevocably lost. In this context, the control of differentiation and its relation to growth control become relevant. Growth regulation might then be perceived as a restricted case of growth/differentiation control.

Many in vitro systems exist in which to explore growth/differentiation control. Because cultured leukocytes can be grown in vitro easily, have had the benefit of decades of characterization to define the differentiation phenotypes, and can be related to an ongoing in vivo developmental process with pathological ramifications, they have clear advantages for studying the control of cell division and differentiation and their interrelationship. It was largely for these reasons that, when we expanded our interests to the more general question of growth/differentiation control, my laboratory focused on hematological cells.

Hematological neoplasias have certain differences distinguishing them from other tumors. The most striking is that the apparent tumor cell is not necessarily proliferatively active. For example, chronic lymphocytic leukemia is a disease with a large tumor cell burden, but almost all of the cells are in G_0. Tumors develop because the homeostatic mechanisms that regulate differentiation from immature to more mature cells are abnormal. Thus the normally coupled regulation of cell proliferation and differentiation is disrupted. This is in contrast to the simpler situation of release from growth inhibition discussed above. It is a distinction of hematological tumors from solid tumor neoplasms that must be recognized and can be exploited to study cell growth/differentiation regulation at a broader level.

Growth Regulation in Lymphocytes

Our early studies of leukocytes were restricted to growth regulation. The intention was to examine growth control without the influence of differentiation. We used a human lymphocytic cell line that was derived from a patient with leukemia and grew spontaneously in vitro. The cell line, SK-L7, was a lymphoblast that proliferated avidly in culture in serum-supplemented medium.

The initial studies examined the dependence of progression through the cell cycle on cell size. Since the density of the cell remains close to constant with progression through the cell cycle, this approximated the dependence of cell cycle transit on cell mass. Exponentially proliferating cells were analyzed by the pulse labeled mitosis technique. Cells in mitosis were segregated into cohorts of progressively increasing cell size, and the cell cycle kinetics of each cohort analyzed. Cells that were smaller at mitosis required longer to complete the cell cycle. The increase in time was due almost entirely to a longer G_1 period. Furthermore, the dispersion in size of G_1 cells was greater than that of cells going into S phase. Thus there appeared to be a critical cell mass that was required for cells to enter S phase. All of the size cohorts traversed the remainder of the cell cycle, S, G_2 and M, with similar kinetics. The lymphocytes thus behaved similarly to the fibroblasts described earlier with respect to a requirement for a critical amount of cell growth in G_1 phase, which determined subsequent entry into S phase.

We studied the dependence of cell cycle progression on growth and nutritional factors by culturing the cells in a device in which the cells were continuously fed serum and nutrients at a controlled rate. Under these conditions the cells grew exponentially until reaching a maximal cell density, whereupon growth reached a plateau and the cells accumulated in a G_1 DNA state. If the rate of medium exchange was increased, the maximal cell density increased accordingly. Thus the cells proliferated until they outgrew the nutritional supply. Like the fibroblasts, the cell cycle phase most sensitive to the availability of exogenous growth factors was G_1. The basic features of cell cycle regulation in response to exogenous factors thus appeared to be conserved between the earlier described fibroblasts and lymphoid cells.

There were, however, significant differences. The most interesting was that there was a maximal cell density that could not be exceeded regardless of how fast the medium was exchanged. The growth arrest did not appear to be mediated by nutritional exhaustion or, for that matter, the secretion of a negative regulator. Significantly, these plateau state cells were not arrested in G_1, as was evident from an analysis of their DNA content by flow cytometry, but were distributed throughout the cell cycle. This provided a hint that growth regulation in lymphocytes need not be restricted to a G_1-specific phenomenon.

Another interesting feature was revealed when we studied normal lymphocytes. Mononuclear cells (MNC), that is T and B lymphocytes and a small percentage of monocytes, isolated from peripheral blood can be stimulated with pokeweed mitogen (PWM). As a result, the lymphocytes proliferate and the B cells yield progeny that differentiate into plasmacytes. The plasmacytes were readily distinguished by their cytoplasmic immunoglobulin content, a hallmark of these differentiated cells. PWM stimulation of MNC has been used as an in vitro model of the humoral immune response. Using mathematical modeling, we found that the cells underwent two division cycles before yielding progeny that were refractory to further mitogenic stimulation. Thus the PWM signal presented to the initial cell population required two intervening division cycles before resulting in growth-arrested and differentiated cells. This was in contrast to the fibroblasts where the response occurred immediately, and did not require intervening division cycles.

Control of Terminal Differentiation

Recently, we sought an in vitro system that approximated an in vivo process, where growth and differentiation were coupled. The human promyelocytic leukemia cell line, HL-60, was used. HL-60 cells were originally derived from a patient with promyelocytic leukemia. The cells grow avidly in culture and retain the capability to undergo terminal differentiation along either the myeloid or monocytic lineages. Depending on the inducer, the cells will yield either mature granulocytes or monocytes. HL-60 cells are thus an uncommitted precursor cell with the same apparent terminal differentiation capabilities of promyelocytes in vivo. Agents such as retinoic acid and DMSO induce myeloid differentiation, whereas agents such as 1,25-dihydroxyvitamin D_3, sodium butyrate, or TPA cause monocytic differentiation.

Using retinoic acid as an inducer of myeloid differentiation, we asked whether the initial signal presented to the cell could originate at the plasma membrane, or required internalization. Retinoic acid is a ring with an alkyl chain and a COOH at carbon 15. The molecule was immobilized at the carboxy group to a substrate—either a plastic sphere or a tissue culture surface. The immobilized retinoic acid was effective at inducing terminal myeloid differentiation. Radiolabeled retinoic acid was used to confirm the localization of the immobilized retinoic acid. A membrane localized signal could, thus, initiate the metabolic cascade leading to terminal differentiation.

The cell cycle phase specificity of the seminal signal initiating the terminal differentiation was studied in several ways. The hope was to gain insight into the nature of the cellular process initiating this cascade. In the simplest case, one can begin with exponentially proliferating cells, expose them to retinoic acid, and observe the resulting kinetics of proliferation and differentiation. Onset of terminal differentiation, evidenced by G_0 arrest and phenotypic differntiation, occurred after a period corresponding to two division cycles. The kinetics were analyzed by a minimal mathematical model that assumed that there were two compartments of cells, proliferating and differentiating, such that, when proliferating cells passed some unspecified point in the cell cycle in the presence of inducer, the proliferation of the daughters is limited and progeny eventually leave the proliferating compartment and enter a differentiating cell compartment. By fitting the observed data, a differentiation control point was localized at approximately 2 hr past the start of S phase, that is

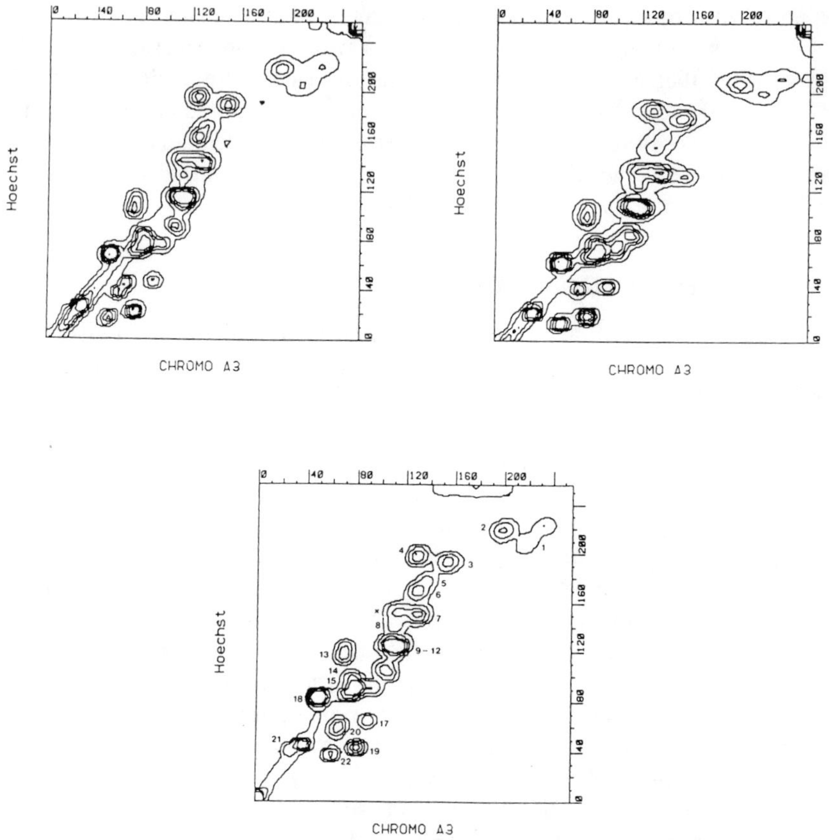

Fig. 1. Variants of HL-60 cells that are defective in their capability to differentiate in response to specific inducers of differentiation can also exhibit characteristic chromosomal aberrations. This figure shows the flow cytometric karyotype for (wild-type) HL-60 cells (upper, left) and for a differentiation-defective variant (upper, right). Chromosomes are seen in these isometric plots of Hoechst 33258 fluorescence, a measure of AT richness, and chromomycin A3 fluorescence, a measure of GC richness. A key giving the identity of the chromosomes observed is given (lower). It can be seen that the variant has lost a homologue of chromosome 20 and gained a chromosome near 7. The flow cytometric karyotypes represent 100,000 events. Thus even low levels of mosaicism can be detected with greater facility than conventional karyotyping. Furthermore the chromosomes can be preparatively isolated.

in early S phase. Similar analysis with DMSO-treated cells also indicated that the seminal signal initiating the terminal differentiation was in early S phase.

A simple experiment to determine whether the initial response to inducer is cell cycle phase dependent or independent is to use fluorescence-activated cell sorting to isolate cells enriched in different cell cycle phases. If there is no cell cycle phase dependence to the cellular response, initial cell populations enriched in different phases should respond to inducer with similar kinetics of growth arrest and differentiation. If the responses differ, then a cell cycle dependence is likely. When the experiment was done, the responses differed, indicating that the initial signal was cell cycle dependent. Furthermore, G_1-enriched cells responded slower than S phase-enriched cells by approximately the duration of G_1, consistent with a need to enter S phase before all can initiate a response to the inducer.

The Precommitment State

In investigating the metabolic cascade initiated by the inducer, we found that the response segregated into two basic steps, with an identifiable intermediate regulatory state. When exponentially proliferating cells were cultured in the presence of retinoic acid, the onset of G_1/G_0 growth arrest and phenotypic differentiation occurred after a period corresponding to two division cycles. In these cells, the division cycle was approximately 24 hr. If the cells were exposed to retinoic acid for one division cycle, washed and recultured in inducer-free medium, there was no differentiation or growth arrest at subsequent times. If retinoic acid was readded to the culture after an intervening division cycle in inducer-free medium, the onset of differentiation occurred in just one division cycle after readdition. The cells, thus, remembered the earlier exposure to inducer. If the intervening period was longer than several division cycles, then reexposure for the full, two-division cycles was again needed for onset of terminal differentiation. The memory state induced by the initial exposure was labile. Thus the initial dose primed the cells to terminally differentiation and ellicited a labile intermediate memory state, which we called "precommitment." Precommitment cells required only an abbreviated second exposure for onset of terminal differentiation. In other experiments, we found that the monocytic inducer 1,25-dihydroxyvitamin D_3, also induced precommitment. Thus the intermediate precommitment state appeared to be a general feature of either myeloid or monocytic differentiation.

The precommitment state can be shown to be differentiation lineage independent by a simple experiment. If cells exposed for one division cycle to either inducer, retinoic acid, or 1,25-dihydroxyvitamin D_3, were then exposed to the other inducer, the onset of terminal differentiation occurred in one division cycle of reexposure. But the differentiation phenotype that the cells exhibited was specified by the second inducer. Thus the character of the first inducer did not influence the differentiation lineage. The early events in the metabolic cascade, that is, those leading to precommitment, appeared to be differentiation lineage independent, whereas late events occurring subsequently specified the explicit differentiation lineage.

An obvious paradigm emerging from these data is that two signals must be presented to the cell to effect terminal differentiation. One is to ellicit the early events and another the late events. This was further explored in studies of the response of HL-60 cells to *cis–trans* isomers of retinoic acid. Certain isomers,

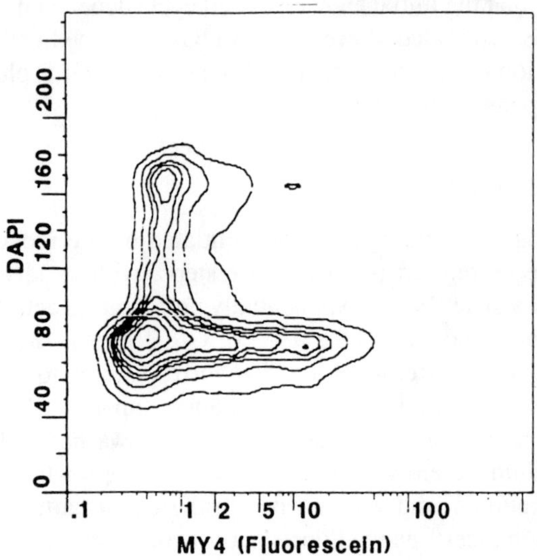
DAPI
0
40
80
120
160
200
.1
1
2
5
10
100
MY4 (Fluorescein)

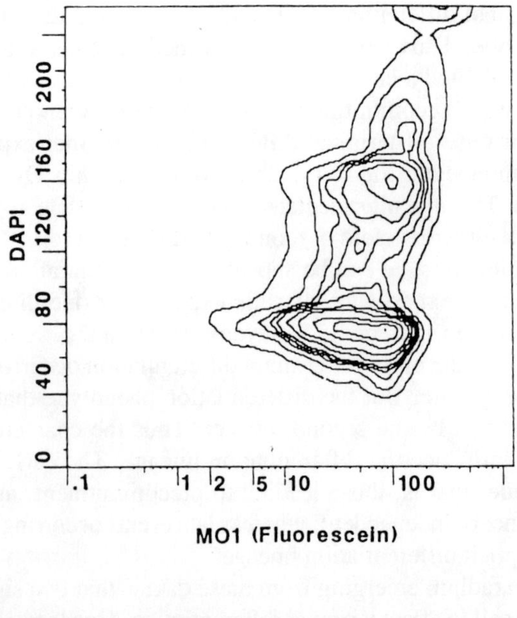
DAPI
0
40
80
120
160
200
.1
1
2
5
10
100
MO1 (Fluorescein)

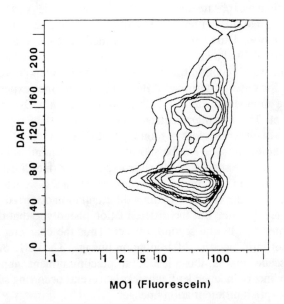

especially some di-*cis* isomers, were ineffective at eliciting terminal differentiation. By doing the type of sequential addition experiments described above, it was found that certain isomers were effective at eliciting only early events or late events. Characteristic early and late changes in the relative abundances of calcium-binding cytosolic proteins induced by retionoic acid or its isomers were consistent with the early or late capabilities of the isomers. Sequential application of isomers effective for early events, followed by an isomer effective for late events, reconstituted the normal kinetics of terminal myeloid differentiation. It thus appeared that different regions of the retinoic acid molecule were essential for eliciting either early or late events. In this way one inducer molecule might present two signals to the cell.

Given that terminal differentiation involves both G_0 arrest and phenotypic differentiation, it is natural to ask if these two features of the induced response are necessarily coupled in their regulation. It was found that control of proliferation and phenotypic differentiation could be uncoupled. Certain isomers or inducers resulted in G_0 growth arrest without any apparent phenotypic differentiation. Thus G_0 growth arrest was not irrevocably coupled to phenotypic differentiation. It is interesting to note that the converse situation has also been observed. There exist HL-60 variants that are defective in their capability to undergo terminal differentiation. In one variant, we found that the cells could be induced to express cell surface differentiation markers, but were not growth arrested. Normally, the expression of such markers is restricted to G_0 cells, but, in this case, bivariate flow cytometric analysis showed that cells throughout the cell cycle expressed the differentiation markers. Thus regulation of growth arrest and expression of differentiation specific phenotypic markers can be uncoupled.

Gene Amplification

If indeed the seminal cellular signal responding to the inducer were S phase-specific, then it might be possible to perturb S phase metabolism and trigger these early events. Years of pharmaceutical development in cancer chemotherapy provided a library of such S phase-specific agents. One of these is hydroxyurea, an inhibitor of ribonucleotide reductase. Exponentially proliferating HL-60 cells were exposed to a subcytotoxic dose of hydroxyurea for 20

Fig. 2. Expression of cell surface differentiation markers in differentiated HL-60 cells is usually restricted to G_0. However, in the case of the variant shown, expression of the differentiation specific marker is no longer restricted to G_1 DNA cells, but occurs for cells throughout the cell cycle. The differentiation-specific antigen was characterized by a fluoresceinated antibody, measured on the horizontal axis, and the cellular DNA content, measured by DAPI fluorescence, as shown in these bivariate isometric maps derived by flow cytometry. Both the normal HL-60 (upper) and variant (lower) cells were treated with 1,25-dihydroxyvitamin D_3 for 72 hr. The variant failed to growth arrest but did express the differentiation-specific antigen.

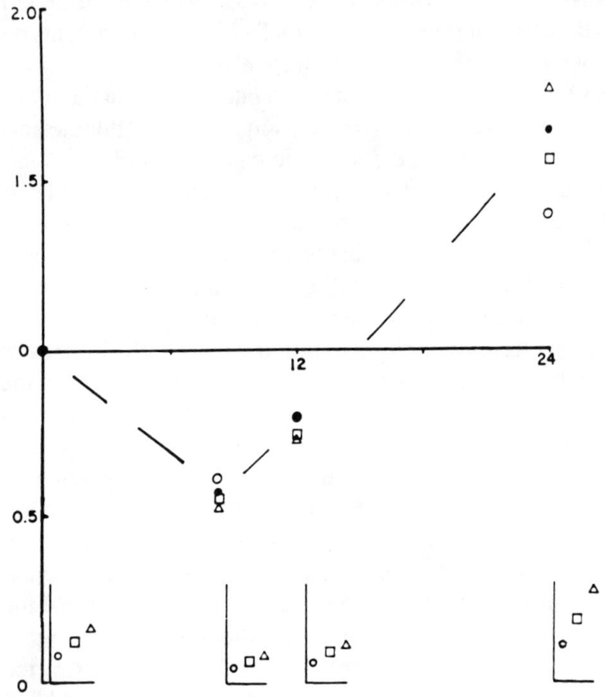
2.0
1.5
0
12
24
0.5
0

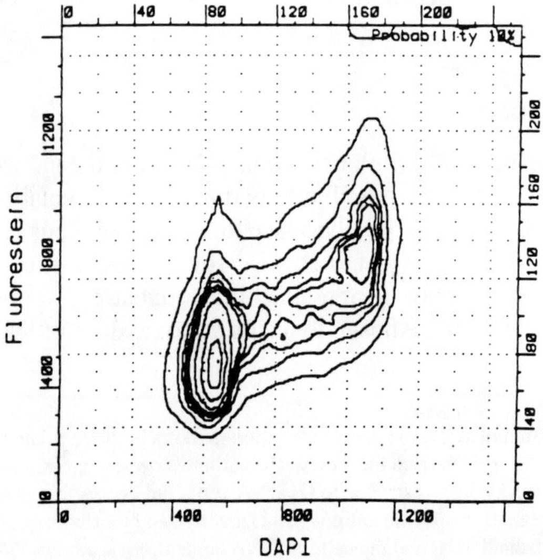
Probability 10%
Fluorescein
DAPI
0
40
80
120
160
200
400
800
1200

hr, washed free of the drug, and incubated in drug-free medium for an additional 6 hr. Retinoic acid was then added to the culture medium. The cells showed onset of terminal myeloid differentiation as evidenced by G_0 arrest and phenotypic differentiation within one division cycle. The hydroxyurea had thus primed the cells to differentiate and induced the precommitment state. If this model has validity, and the precommitment state is differentiation lineage independent, the same treatment should prime the cells to undergo monocytic differentiation as well. Using hydroxyurea priming and TPA as an inducer of monocytic differentiation, we found that this was the case. The priming effect is apparently relatively specific to hydroxyurea, since other S phase agents such as 5-fluorouracil and cytosine arabinoside failed to induce precommitment although they all affect DNA synthesis.

One of the striking features of hydroxyurea is that it induces gene amplification. It had been found originally by Robert Schimke and his co-workers, that the same pulse–release treatment of rodent cells, which coincidentally had similar cell cycle kinetics to HL-60 cells, caused an enhanced incidence of methotrexate-resistant cells. Methotrexate is a drug that inhibits the enzyme dihydrofolate reductase. It was found that the drug-resistant cells had an amplified dihydrofolate reductase gene. This gene is replicated in early S phase, and the suggestion was made that the hydroxyurea treatment could cause overreplication of early S phase genes. Since the seminal cellular event responding to inducer is an early S phase event and since hydroxyurea caused precommitment, DNA amplification may have a part in the early events leading to precommitment. Our data suggesting that some critical gene or genes is overreplicated in initiating the events leading to terminal differentiation provide a significant experimental challenge.

Fig. 3. The relative amounts of the c-*myc* gene product per cell can be measured for cells in each cell cycle phase in the course of induced terminal differentiation using flow cytometry. If HL-60 cells are stained with a primary antibody directed against the c-*myc* gene product, which is detected with a fluoresceinated secondary antibody, and then stained for cellular DNA, using DAPI, then a bivariate measurement of c-*myc* protein vs. DNA can be made (isometric plot shown in lower panel). For cells in each phase, G_1/G_0, S, G_2+M, the relative amount of c-*myc* protein can thus be determined. When this subline of HL-60 cells was cultured in the presence of retinoic acid, onset of terminal differentiation occurred at 48 hr. During the early events in this metabolic cascade, expression of the c-*myc* gene was down-regulated (upper panel, solid circle represents the mean of the total population). The graph shows the fractional change for the population as a whole (solid circle) and for G_1/G_0 cells (open circles), S phase cells (open square), and G_2+M cells (open triangles). The inserts show the relative amount of c-*myc* protein (vertical) for cells in G_1, S, and G_2+M (open symbols defined above). Expression of the c-*myc* protein is down-regulated in all cell cycle phases. Following the down-regulation, there is an increase in expression, also affecting all cell cycle phases. The increase is transient, returning to low levels of expression as the cells differentiate. In other sublines of HL-60, which also differentiate in response to retinoic acid, the early down-regulation of expression is not followed by a transient elevation in expression, but remains low.

c-*myc* and RB

An ongoing interest in the course of most of these studies has been the identity of genetic elements that mediate the early aspects of precommitment. We started our studies on genes having a high probability of involvement in the metabolic cascade leading to precommitment.

One highly implicated gene is the c-*myc* protooncogene. c-*myc* is the transforming agent of the avian myelocytomatosis virus. It can cause leukemia in animals. Its form or function has been found to be aberrant in a variety of hematological neoplasias. c-*myc* expression is typically positively correlated with recruitment to cell proliferation. Finally, it has been found translocated to an Ig locus in certain leukemias. It is likely that the normal function of c-*myc* is to participate in the control of leukocyte proliferation and differentiation.

Predictably, down-regulation of c-*myc* transcription had been found to be associated with terminal differentiation of HL-60 cells. We investigated whether the down-regulation was an essential component of the early metabolic cascade leading to precommitment. We found that early down-regulation of transcription was consistent for different means of inducing precommitment. Interestingly, the down-regulation was similar for cells in all cell cycle phases, although there was no modulation of cell cycle transit during these early events leading to precommitment. We used a bivariate flow cytometric assay that measured the c-*myc* protein and DNA content per cell simultaneously. These measurements showed that cellular c-*myc* protein content increased as cells progressed through the cell cycle, with cells at division having approximately twice the protein as cells initiating G_1. During early events, the amount of protein was down-regulated similarly for cells in all cell cycle phases. As would be expected of a short half-life regulatory protein, regulation of the protein levels mirrored that of the message in the cases we tested. Thus, an essential component of the early events leading to precommitment appears to be down-regulation of cellular c-*myc* protein levels mediated by transcriptional control.

Work from other laboratories indicated another relevant feature of c-*myc*. Data from studies of variants of HL-60 by George Klein and his co-workers indicated that it was not the level of c-*myc* expression that determined whether terminal differentiation could occur, but rather whether that level could be regulated by an inducer. The results suggested the possibility that if the absolute level of c-*myc* was not a determining factor, it might be the balance of *c-myc* and other gene products that determined subsequent differentiation. We had studied c-*myc* as an archtype of a class of growth regulatory nuclear protooncogenes, and now turned our attention to the possiblity of genes that might act in balance with c-*myc*. One class of genes that became of interest was the tumor suppressor or antioncogenes. The prototype is the RB or retinoblastoma susceptibility gene. RB is a recessive-acting gene, whose loss of function confers susceptibility to retinoblastoma and other tumors. Allowing that such a gene has a normal cellular function, it is likely to be in control of terminal differentiation. Using a primary antibody derived by Yuen Kai Teddy

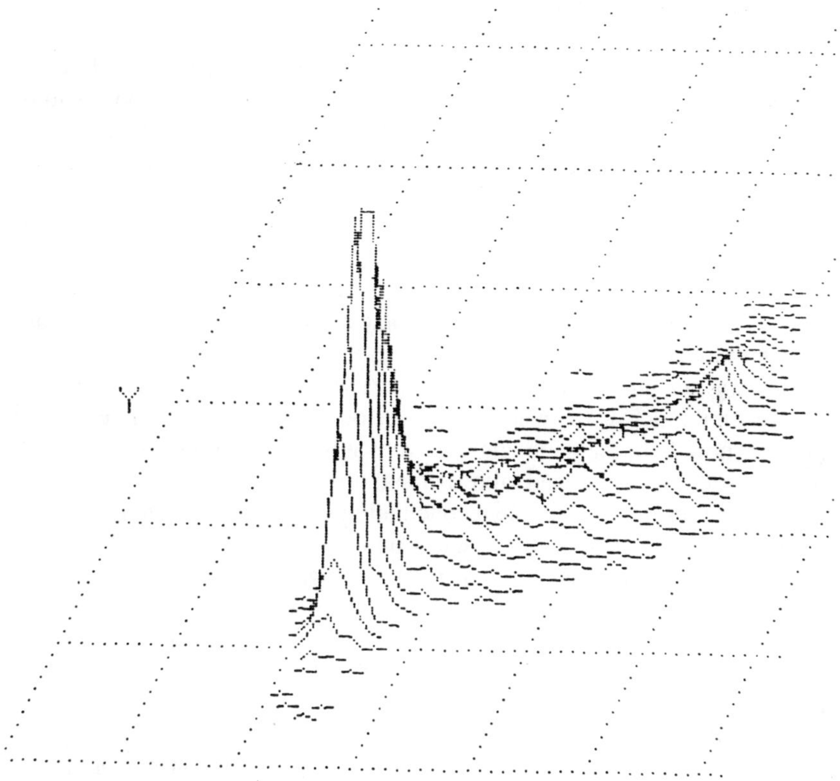

Fig. 4. The amount of RB protein per cell can be measured in bivariate analysis with cellular DNA. The amount of RB protein per cell continuously increases as cells progress through the cell cycle. The figure shows a contour map relating the number of cells (vertical axis) to their DNA (*x* axis) and RB protein (*y* axis) content. Cells in all cell cycle phases continue to accumulate the RB protein with no apparent threshold effect in protein levels as a function of cell cycle progression.

Fung's laboratory, the cellular levels of the 105-kDa RB gene product were measured during HL-60 differentiation. We found that retinoic acid caused an early down-regulation of RB protein levels within one division cycle of exposure to inducer, before any evidence of growth arrest or phenotypic differentiation. This early down-regulation occurred similarly for cells in all cell cycle phases. Furthermore, DMSO, another inducer of terminal myeloid differentiation, caused a similar down-regulation. If down-regulation of the RB gene product was an essential component of early events leading to precommitment, then it should be similarly regulated during induced monocytic differentiation. Pursuing this possibility, we examined if 1,25-dihydroxyvitamin

D_3, an inducer of monocytic differentiation, also regulated the RB gene product similarly. The early down-regulation of the RB gene product with respect to kinetics and cell cycle phases was similar to that seen for myeloid differentiation. Using a variant of HL-60 cells that could respond to retinoic acid or DMSO by undergoing myeloid differentiation and to 1,25-dihydroxyvitamin D_3 by undergoing monocytic differentiation, we found that down-regulation of the RB gene product occurred in all cases except in response to sodium butyrate, which induces monocytic differentiation and to which the variant cells were resistant. The results suggest of course that down-regulation of the RB protein is an essential feature of early events leading to precommitment. Thus tumor suppressor genes appear to play a role not unanticipated in control of terminal differentiation.

What was unanticipated in these findings was the direction of the regulation. Given that RB is the gene whose loss of function confers susceptibility to tumor formation, then one might naively anticipate that relatively high expression would be associated with terminal differentiation. This is counter to the observed result. One simple reconciliation of the data and our concept of the gene is to reconsider RB a ''status quo'' gene. In this case, the down-regulation or loss of function of RB might be regarded as permissive for cellular change from the proper differentiation state. Thus, in the case of HL-60 cells, down-regulation of RB is needed for conversion of these promyelocytes, normally a proliferatively active precursor cell, to a terminally differentiated phenotype. In this case it becomes of relevance to examine the relationship of RB regulation to that of c-*myc*, a presumed growth-promoting protooncogene. The HL-60 hematopoietic cells provide an experimentally convenient system to do this, an avenue we are now pursuing.

The paradigm for deriving terminal differentiation in these cells thus far indicated by the data can be stated as follows. The induced metabolic cascade leading to terminal differentiation can begin with a plasma membrane originated response to inducer. The seminal cellular response to the inducer is S phase-specific, and may involve a gene amplification event for an early S phase-replicated gene. The metabolic cascade occurs over a period corresponding to two division cycles and involves an intermediate regulatory memory state, precommitment. Early events leading to precommitment are differentiation lineage independent, while late events occurring after precommitment and before onset of terminal differentiation specify the explicit differentiated cell phenotype. An essential component of early events is the down-regulation of the c-*myc* gene product and the apparent coordinated down-regulation of the RB gene product.

If this paradigm has validity,then one might expect that it should be possible to find an agent that would elicit the cellular effect of precommitment by (1) priming the cells in a lineage-independent manner, (2) conferring a memory state with this priming, (3) be S phase-specific, (4) be ineffective at eliciting late events, (5) cause down-regulation of c-*myc*, and (6) be implicated with causing gene amplification. Indeed such an agent can be found. It is bromodeoxyuridine. Bromodeoxyuridine incorporation is S phase specific and it has been previously found to cause gene amplification. When cells were cultured for a period corresponding to one division cycle in bromodeoxyuridine, allowing unifilliary substitution, we found that it elicited precommitment with the characteristics above. A striking feature of this is that a priori one would not not expect a fraudulent nucleotide to cause regulation of the c-*myc* proto-oncogene. However, it is predicted by the model pursued here. Its experimental corroboration lends confidence to the validity of the general features of this evolving paradigm for control of terminal differentiation in these hematopoietic cells.

Acknowledgments

I am grateful to many colleagues in the laboratory, too many to name. Lynda King provided invaluable help in preparing this manuscript. This work supported in part by grants from the USPHS(NCI), American Institute for Cancer Research, Council for Tobacco Research, and the Cornell Biotechnology Program.

Bibliography

Guernsey DL, Yen A (1988): Retinoic acid induced modulation of c-myc not dependent on its continued presence: Possible role in precommitment for HL-60 cells. Int J Cancer 42:576–581.

Mihara K, Cao X-R, Yen A, Driscoll B, Murphee AL, T'Ang A, Fung YKT (1989): Cell-cycle dependent regulation of phosphorylation of the human retinoblastoma gene product. Science 246:1300–1303.

Yen A (1984): Control of HL-60 myeloid differentiation: Evidence of uncoupled growth and differentiation control, S-phase specificity and two-step regulation. Exp Cell Res 156:198–212.

Yen A (1989): HL-60 cells as a model of growth control and differentiation: The significance of variant cells. Hematol Rev 4:5–46.

Yen A, Albright K (1984): Evidence for cell cycle phase specific initiation of a program of HL-60 cell myeloid differentiation mediated by inducer uptake. Cancer Res 44:2511–2515.

Yen A, Chiao JW (1983): Control of cell differentiation during proliferation: I. Monocytic differentiation of HL-60 promyelocytes. Exp Cell Res 148:87–93.

Yen A, Fairchild DG (1982): T Cell control of B-cell proliferation uncoupled from differentiation. Cell Immunol 74:269–276.

Yen A, Guernsey DL (1986): Increased c-myc RNA levels associated with the precommitment state during HL-60 myeloid differentiation. Cancer Res 46:4156–4161.

Yen A, Lewin D (1981): Uncoupling lymphocyte proliferation from differentiation: Dissimilar dose-response relations for pokeweed mitogen-induced proliferation and differentiation of normal human lymphocytes. Cell Immunol 61:332–342.

Yen A, Pardee AB (1978a): Arrested states produced by isoleucine deprivation and their relationship to the low serum produced arrested state in Swiss 3T3 cells. Exp Cell Res 114: 389–395.

Yen A, Pardee AB (1978b): Exponential 3T3 cells escape in mid-G1 from their high serum requirement. Exp Cell Res 116:103–113.

Yen A, Pardee AB (1979): Role of nuclear size in cell growth initiation. Science 204:1315–1317.
Yen A, Riddle VGH (1979): Plasma and platelet associated factors act in G1 to maintain proliferation and to stabilize arrested cells in a viable quiescent state. Exp Cell Res 120:349–357.
Yen A, Stein LS (1981): Polyclonal mitogenesis of human lymphocytes by PWM: Two preprogrammed division cycles resulting in cells refractile to further mitogenesis. Cell Immunol 57:440–454.
Yen A, Fried J, Kitahara T, Strife A, Clarkson BD (1975a): The kinetic significance of cell size. I. Variation of cell cycle parameters with size measured at mitosis. Exp Cell Res 95:295–302.
Yen A, Fried J, Kitahara T, Strife A, Clarkson BD (1975b): The kinetic significance of cell size. II. Size distributions of resting and proliferating cells during interphase. Exp Cell Res 95:303–310.
Yen A, Fried J, Clarkson BD (1977): Alternative modes of population growth inhibition in a human lymphoid cell line growing in suspension. Exp Cell Res 107:325–341.
Yen A, Warrington RC, Pardee AB (1978): Serum stimulated 3T3 cells undertake a histidinol sensitive process which G1 cells do not. Exp Cell Res 114:458–462.
Yen A, Reece SL, Albright KL (1984a): Dependence of HL-60 myeloid cell differentiation on continuous and split retinoic acid exposures: Precommitment memory associated with altered nuclear structure. J Cell Physiol 118:277–286.
Yen A, Reece SL, Albright K (1984b): Membrane origin for a signal eliciting a program of cell differentiation. Exp Cell Res 152:493–499.
Yen A, Reece SL, Albright KL (1985): Control of cell differentiation during proliferation. II. Myeloid differentiation and cell cycle arrest of HL-60 promyelocytes preceded by nuclear structural changes. Leukocyte Res 9:51–71.
Yen A, Powers V, Fishbaugh J (1986a): Retinoic acid induced HL-60 myeloid differentiation: Dependence of early and late events on isomeric structure. Leukocyte Res 10:619–629.
Yen A, Freeman L, Powers V, Van Sant R, Fishbaugh L (1986b): Cell cycle dependence of calmodulin levels during HL-60 proliferation and myeloid differentiation: No changes during precommitment. Exp Cell Res 165:139–151.
Yen A, Freeman L, Fishbaugh J (1987a): Hydroxyurea induces precommitment during retinoic induced HL-60 terminal myeloid differentiation: Possible involvement of gene amplification. Leukocyte Res 11:63–71.
Yen A, Brown D, Fishbaugh J (1987b): Control of HL-60 monocytic differentiation. Different pathways and uncoupled expression of differentiation markers. Exp Cell Res 168:247–254.
Yen A, Forbes M, deGala G, Fishbaugh J (1987c): Control of HL-60 cell differentiation lineage specificity: A late event occurring after precommitment. Cancer Res 47:129–134.
Yen A, Brown D, Fishbaugh J (1987d): Precommitment states induced during HL-60 myeloid differentiation: Possible similarities of retinoic acid and DMSO induced early events. Exp Cell Res 173:80–84.
Yen A, Freeman L, Fishbaugh L (1988): Induction of HL-60 monocytic cell differentiation promoted by a perturbation of DNA synthesis: Hydroxyurea promotes action of TPA. Exp Cell Res 174:98–106.

Perspectives on Cellular Regulation:
From Bacteria to Cancer, pages 297–314

Regulation of DNA Replication in Mammalian Chromosomes

Joyce L. Hamlin

University of Virginia School of Medicine, Charlottesville, Virginia 22908

Introduction

I came to Art Pardee's lab in 1973 after having worked on the *Escherichia coli* lactose operon with Irving Zabin as a graduate student in the Molecular Biology Institute at UCLA. I had met Art a few years before at a Lac meeting at Cold Spring Harbor. As a student, I had always been impressed by the simple elegance of his work on bacterial regulatory and transport systems. I felt that to be in his laboratory as he was venturing into the world of animal cells would be a challenge and a great learning experience. In retrospect, I think that assessment was correct.

When I arrived at Princeton, Art was not yet back from sabbatical leave in England, and none of the lab group was actually working with animal cells. Within a few months, however, Art returned from London, where he had carried out the studies that culminated in the notion of the restriction or *R* point (Pardee, 1974), and his lab began in earnest its steady changeover to studies on mammalian cells. In the beginning, there was an interesting mix of bacterial and animal cell people that served to highlight for me the essence of Art's approach: he always seemed to get right to the heart of the problem, whether the question related to the simpler bacterial systems with which he was so familiar, or to the more complex and largely unfamiliar world of animal cells. He could quickly sense and identify the key questions and could develop myriad approaches for answering them. He is definitely a reductionist, and he is one of the most inventive people I have ever known from the point of view of the shear number of ideas he can conjure up and the graphic imagery that must fuel that invention. One of the important take-home lessons I learned from Art is that you can only ask (or rather answer) simple questions at each step in science, but that constant progression from one answer to the next question ultimately sketches the framework that constitutes the working knowledge of a subject. Second, I came to appreciate that the more ideas—whether good or even naive—that percolate through a laboratory, the greater the chances for learning new and exciting things (in Art's lab, of course, many, many of the ideas came directly from him). Since leaving Art's group to come to the University of Virginia, I have attempted to translate these two concepts into my own personal style of science, which I think is also earmarked by minimalism because of Art's influence (i.e., choosing the simplest system in which to address the questions at hand), and I have tried to foster an eclectic and broad-minded approach in my students so that we will always have many ideas afloat.

The DHFR Amplicon as a Model Replicon

The System

Nearly all of the ''animalists'' in Art's lab were in one way or another involved in studies on the G_1 period—the mysterious ''gap'' during which preparations were made for entry into S. For me, the whole cell cycle phenomenon was a mystery. You tickled the cell in G_0 or early G_1 with a hormone, for example, and after a 6- to 8-hr eclipse period, you finally got a phenotype (DNA replication) that could be measured by incorporation of radioactive thymidine. The cascade or train of events that occurs during that interval has, of course, occupied the attention of Art's lab and hundreds of others for many years, and is finally revealing its secrets in a dramatic and exciting way. But from the beginning, I was more fascinated by the S period itself. I wanted to understand the very last regulatory events that trigger traverse of the G_1/S boundary, and I wanted to know how the cell controls the temporal order of replication of defined sequences during the S period, presumably through the agency of the thousands of origins that had been detected in mammalian chromosomes in DNA fiber autoradiographic studies (Huberman and Riggs, 1968). The key question was whether origins of DNA replication in animal cells are fixed genetic elements, analogous to the origins of replication in simple microorganisms such as *E. coli* or SV40. If it turned out that they were, I wanted to devise a system in which to purify several origins known to fire at the same time in the S period. Presumably, this would allow us to determine what they had in common that caused them to fire synchronously with respect to one another.

My initial studies in Art's lab were dedicated to standardizing a reliable synchronizing regimen for CHO cells, i.e., getting all of the cells in a culture to behave as one so that the temporal events of the S period could be studied. The method we chose was adapted from an approach developed at Los Alamos National Laboratories (Tobey and Crissman, 1972). Cultures were arrested in G_0 by isoleucine starvation followed by release into complete medium containing hydroxyurea, an inhibitor of DNA replication. After 10 hr in the drug, which is long enough to allow even the slowest cell to traverse the G_1 period and to arrest at the G_1/S boundary, hydroxyurea is removed and the cells in the culture enter the S period in a synchronous wave (Hamlin and Pardee, 1976). Armed with this effective synchrony regimen, I performed some general studies on the effects of damaging early-replicating DNA on progress through the S period (such damage prevents cells from traversing S, for reasons that are not clear to me even now; Hamlin, 1978). But these studies did make it clear to me that nothing about mechanisms of entry into or orderly passage through the S period could be learned without analyzing defined origins of DNA replication.

About this time, Art told me about a seminar he had attended that was given by June Biedler from Sloan-Kettering Laboratories in New York. She and Barbara Spengler had shown that methotrexate resistance in Chinese hamster lung fibroblasts resulted in the appearance of expanded, homogeneously G-banding chromosome regions (HSRs; Biedler and Spengler, 1976). The implication was that increasing the level of methotrexate (MTX) in the culture medium over an extended time period selected cells that overproduced the target enzyme, dihydrofolate reductase (DHFR), by a DNA sequence amplification mechanism. The HSRs would then represent the multiple, tandem copies of the gene, which would be euchromatic (uniformly light-staining) in nature because they represented active chromatin. In the two or three years that followed, it was demonstrated convincingly that the HSRs in MTX-resistant CHO and Chinese hamster lung cells did, indeed, represent amplified DNA sequences containing DHFR genes (Melera et al., 1980; Milbrandt et al.; Nunberg et al., 1978).

Because the HSRs could represent as much as $2–4 \times 10^8$ bp of contiguous DNA sequence in a chromosome (Milbrandt et al., 1981), it was theoretically impossible for such a large region to be synthesized by replication forks emanating from an origin lying *outside* of the HSR. This followed from the fact that replication forks move at ~3 kb/min, which would require $\sim 10^5$ min (>1600 hr) to traverse a stretch of HSR 3×10^8 bp in length (the cells, in fact, were doubling every 18 hr). This meant that DNA replication must initiate from within the amplified units (amplicons) themselves. If all amplicons were similar to one another in sequence, and if origins were defined sequence elements, it followed that each amplicon must contain an origin of replication.

Initial studies on two of June Biedler's MTX-resistant cell lines (DC3F/A3 and DC3F/MQ19) were designed to determine the time of replication of the HSRs and whether or not initiation occurred at multiple loci within the HSR. Synchronized cells were pulse-labeled with [^{3}H]thymidine for 10 min at various times in the S period, the pulse was chased with cold thymidine, and metaphase cells were collected with colcemid and spread on microslides for autoradiographic analysis. The results were very dramatic. Replication in the HSRs initiated in the first 10 min of the S period, and silver grains were distributed uniformly all over the length of the HSRs (Hamlin and Biedler, 1981). This exciting finding suggested the possibility that initiation was occurring synchronously within the amplicons at the same time in the very early S period, and possibly at the same sequence in every amplicon (if origins were defined elements).

Drug-registant cell lines with these properties obviously represented an extremely good model system in which to study initiation of replication. When I moved to Virginia, the lab embarked on selecting a similar MTX-resistant CHO cell line (because CHOs are so easy to synchronize), and the line we subsequently developed (CHOC 400) has been the mainstay of our research program for the last 12 years.

Characterizing the Replication Pattern of a Defined Chromosomal Domain

The CHOC 400 cell line is resistant to 400 μg/ml MTX ($\sim 10^6 \times$ the LD_{50} of the starting CHO cell line), and contains ~1000 copies of the DHFR gene. The multiple copies of the amplicons are located in three different HSRs (one major, original site and two minor sites that resulted from chromosomal breakage and translocation of parts of the original HSR). Because of the high copy number, the restriction pattern of the DHFR amplicon can actually be visualized in ethidium bromide-stained agarose gels against the background smear of single copy restriction fragments (Milbrandt et al., 1981). By comparing each amplified fragment to size markers and by assessing whether each represented a single, double, or triple band, we were able to estimate that the minimal size of the repeated sequence had to be at least 135 kb. Since most of the known eukaryotic genes in those days were less than 15 kb in length, and the murine DHFR gene had been shown to be ~30 kb long (Crouse et al., 1982), it seemed likely that more than the DHFR gene had been amplified.

Even more importantly, the ability to visualize the restriction pattern of the amplicon suggested that we might be able to emulate an important experiment that had been performed by Marsh and Worcel to locate the bidirectional origin of replication of the *E. coli* chromosome (Marsh and Worcel, 1977). In their approach, a t_s initiation mutant was arrested at the beginning of the S period by incubating the culture at the nonpermissive temperature. The temperature was then lowered in the presence of [^{3}H]thymidine for a brief interval to allow initiation at the origin to occur. The DNA was harvested and was digested with a restriction enzyme, the digest was then separated on an agarose gel and was transferred to nitrocellulose, and the transfer was treated with a fluorophor. After exposure of the transfer to X-ray film, the most heavily labeled fragments (one of which eventually proved to contain the bonafide origin) were determined by densitometry.

In our application of this approach, CHOC 400 cells were arrested at the G_1/S boundary by isoleucine starvation followed by release into complete medium containing aphidicolin to inhibit the replicative DNA polymerases. When the aphidicolin was washed out and the cultures were pulsed for a brief interval with [^{14}C]thymidine, presumably only the origin-containing fragment and the immediately flanking fragments would be labeled (if the origin of replication in the DHFR domain is fixed). To assess the labeling pattern resulting from this experiment, the genomic DNA was purified, digested to completion with the restriction enzyme *Eco*RI, and the digest was separated according to size on an agarose gel (along with control DNA labeled during exponential growth). After transfer to DBM paper to immobilize the DNA, the transfer was exposed to X-ray film for several weeks.

From the resulting autoradiogram (Fig. 1A), it was quite apparent that, of the 40–45 *Eco*RI fragments arising from the amplified DHFR domain, two fragments 6.2 and 11.0 kb in length were preferentially labeled in the early S period (Heintz and Hamlin, 1982). This was a very exciting result because it showed for the first time that DNA replication initiates at preferred locations within this defined chromosomal domain, and suggested that replication initiation sites in mammalian chromosomes might be genetically fixed elements analogous to the origins of microorganisms. However, since there were *two* early-labeled fragments (ELFs), it was also possible that there were actually two origins of replication in the amplicon.

To determine whether there were one or two independent origins, and at the same time to isolate the fragments containing the presumptive origin(s), we ran a preparative agarose gel of an *Eco*RI digest of CHOC 400 DNA, excised the 6.2- and 11.0-kb ELFs from the gel, and labeled them in vitro with [^{32}P]dCTP. The labeled fragments were then used independently as specific hybridization probes on a genomic cosmid library that we had prepared from CHOC 400 DNA. If there were only one origin in the amplified DHFR domain, then the two ELFs would map together in the genome and would often hybridize to the same cosmids when used independently to screen duplicate filters containing recombinant clones. If the two fragments arose from two separate origins, however, they presumably would not be contained in the same cosmid. In fact, we found several cosmids that hybridized with both fragments, one of which (cS21) appeared to contain both the 6.2-kb and the 11-kb fragments in their entirety (Heintz et al., 1983). With the subsequent isolation of overlapping clones, the 6.2-kb ELF proved to contain two fragments 6.2 and 6.1 kb in length, which lie adjacent to one another in the genome (Fig. 2). In addition, the 6.2/6.1-kb and the 11-kb fragments are actually separated by 3.5- and 1.5-kb *Eco*RI fragments that, owing to their small size, were not detected as ELFs in the in vivo labeling studies (see map, Fig. 2). We concluded from these studies that there must be only one initiation locus in the DHFR domain, and that we had cloned it in the cosmid S21.

In the 11 years since we established the CHOC 400 cell line, we have cloned the DHFR gene itself, which is ~26 kb in length (Milbrandt et al., 1983), we have connected the gene to the initiation locus by chromosomal walking (Montoya-Zavala and Hamlin, 1985), and we have finally cloned the entire 240-kb major DHFR amplicon type from the CHOC 400 genome in a series of overlapping cosmids (Looney and Hamlin, 1987; see map in Fig. 2). This is the first complete amplicon to be cloned from any mammalian cell line. The multiple copies of this 240-kb amplicon are organized into alternating head-

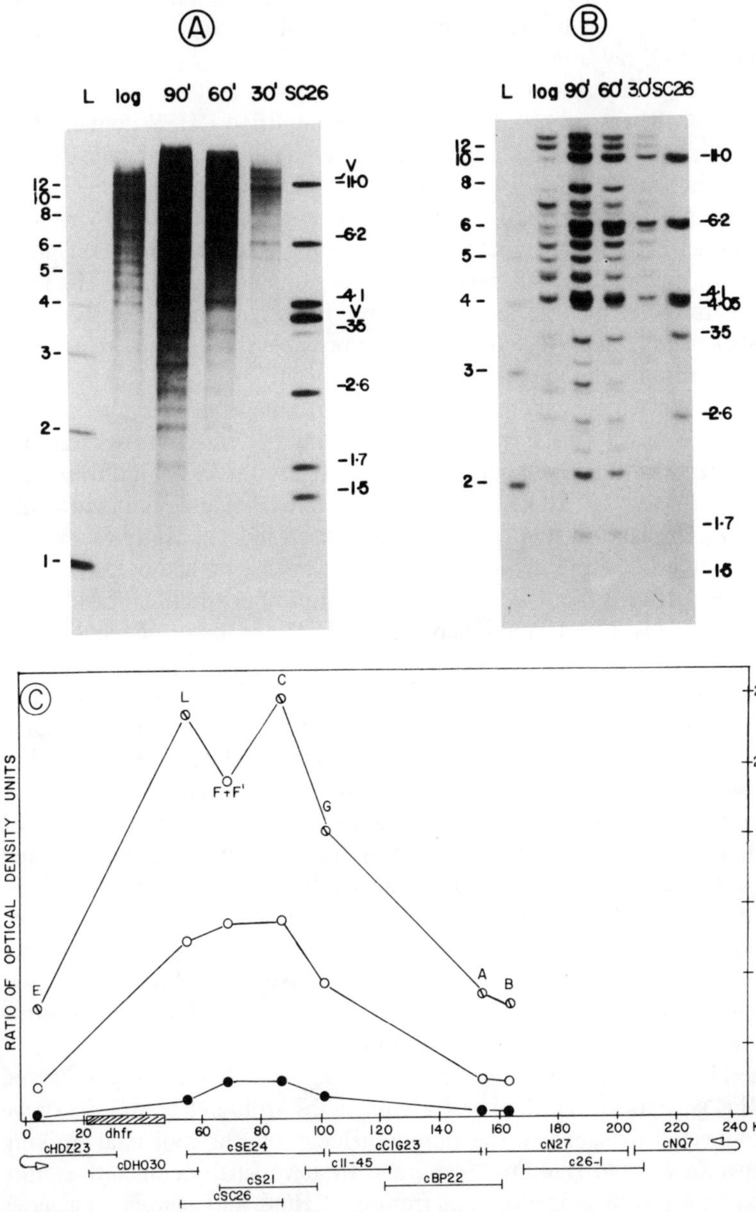
A
L
log
90'
60'
30'
SC26
12
10
8
6
5
4
3
2
1
V
11·0
6·2
4·1
V
3·5
2·6
1·7
1·5
B
L
log
90'
60'
30'
SC26
12
10
8
6
5
4
3
2
11·0
6·2
4·1
4·05
3·5
2·6
1·7
1·5
C
RATIO OF OPTICAL DENSITY UNITS
L
C
F+F'
G
E
A
B
0
20
dhfr
60
80
100
120
140
160
180
200
220
240 KB
4
8
12
16
20
24
cHDZ23
cSE24
cCIG23
cN27
cNQ7
cDHO30
cII-45
c26-I
cS21
cBP22
cSC26

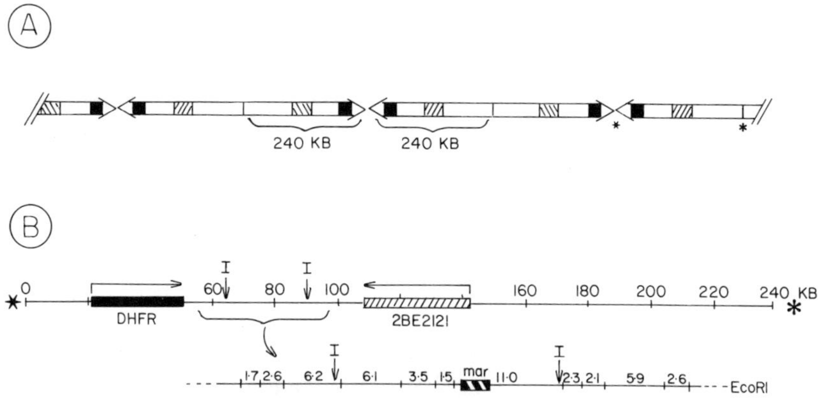

Fig. 2. Map of the 240-kb DHFR amplicon. **(A)** The palindromic arrangement of the major 240-kb type II amplicon in the CHOC 400 genome. The linear map in **B** shows the positions of genes and the two preferred zones of early labeling (the centers of which are labeled with an I). The direction of transcription of the two genes is indicated with arrows above the scale. The hatched box represents a prominent matrix-attachment region (*MAR*). A 40-kb region encompassing the initiation locus is expanded below to show *Eco*RI sites.

Fig. 1. Detection of early-labeled fragments in the DHFR amplicon. **(A,B)** CHOC 400 cells were synchronized at the G_1/S boundary by isoleucine starvation followed by release into complete medium containing aphidicolin. After 12 hr, the drug was removed and individual cultures were labeled immediately with [^{3}H]thymidine for the indicated times. The cultures were washed and returned to fresh medium containing cold thymidine for a total of 12 hr after initial release from aphidicolin. An exponential, control culture was also labeled for 12 hr with [^{3}H]thymidine. Genomic DNA was purified, digested with *Eco*RI, and separated on a 1% agarose gel along with an end-labeled l-kb ladder. In **A**, an end-labeled sample of the cosmid cSC26 (which contains most of the initiation locus) was also included. **(A)** The agarose gel was blotted onto GeneScreen Plus, and the transfer was sprayed with En^{3}Hance and exposed to X-ray film. Note that the 11-kb and 4.05-kb genomic *Eco*RI fragments in cSC26 are both attached to vector and migrate at 11.5 kb and 3.7 kb, respectively (V in the figure). The sizes of the marker fragments in the l kb ladder and in cSC26 are indicated. **(B)** *Eco*RI digests of the same samples analyzed in **A** were separated on 1.0% agarose, and the gel was subjected to two cycles of the in-gel renaturation procedure described in the text. The gel was then blotted to GeneScreen Plus, and the transfer was sprayed with En^{3}Hance and exposed to X-ray film. The transfer was then washed overnight with toluene to remove the En^{3}Hance, and the blot was hybridized with ^{32}P-labeled cosmid cSC26 and exposed to X-ray film. The signals obtained from all of the genomic digests after hybridization with the cSC26 probe were the same, but only the 90 min sample is shown (cSC26 lane). **(C)** The log-labeled and the 30, 60, and 90 min samples from the experiment in **B** were scanned on a densitometer and the relative peak heights of individual bands were recorded (note that the baseline went down to zero absorbance in between bands). The relative peak heights of individual bands in the early- and log-labeled samples were then expressed as the ratio ELF/log, and this ratio is plotted as a function of map position in the 240-kb amplicon. Only those bands that are known to represent single fragments were selected for analysis [with the exception of the two 6.2-kb *Eco*RI fragments (F + F′) that map together in the genome; see Fig. 2]. Closed circles, 30 min/log ratios; open circles, 60 min/log ratios; hatched circle, 90 min/log ratios.

to-head and tail-to-tail arrays in the genome (Looney and Hamlin, 1987; Ma et al., 1988). By cloning and mapping the entire amplicon, we were able to determine that the initiation locus represented by the 6.2/6.1-kb and 11.0-kb ELFs was located downstream from the DHFR gene (Fig. 2). Furthermore, since the amplicon appeared to have only one initiation locus, by definition we had cloned a replicon, whether or not its boundaries coincided with the two termini of the original replicon in the parental CHO cell. In recent years, we have identified a second, coamplified gene (2BE2121) that flanks the initiation locus on the downstream side (Foreman and Hamlin, 1989), resulting in the arrangement shown in Figure 2 (in which the bracket below the linear scale indicates the rough location of the 28-kb initiation locus). Note that both the DHFR and 2BE2121 genes are transcribed in a direction opposite to that of the replication forks emanating from the initiation locus. This interesting result is contrary to the suggestion that the leading strands of replication forks are polymerized in the same direction as transcription through genes (Smithies, 1982).

A major weakness of our in vivo labeling experiments was the extremely low resolution. Even though we had delimited the initiation locus to a region representing only ~10% of the 240-kb amplicon, this locus was still ~28 kb in length—six times the size of the entire SV40 genome! Part of the problem lies in the fact that we could not actually estimate the specific radioactivity of individual labeled bands relative to the same bands in exponential control cells because of the background smear of labeling from thousands of other single copy initiation sites firing at the same time. Therefore, we could not determine which of the ELFs was labeled first (i.e., had the highest specific radioactivity) in the early S period. Second, the resolution is necessarily limited by the size of the restriction fragments examined: if the fragment with the highest specific radioactivity after a brief pulse in early S were 4 kb in length, it would not be possible to determine the position of the origin with greater resolution than 4 kb.

Fortunately, we were able to eliminate background labeling by adapting a very clever in-gel renaturation method developed by Igor Roninson for the analysis of amplified DNA sequences in mammalian genomes (Roninson, 1983). In our application of this method, CHOC 400 cells were pulse-labeled with [^{3}H]thymidine in the early S period, and a restriction digest of the early-labeled DNA (along with a log-labeled control sample) was separated on an agarose gel by the usual method. However, before Southern blotting, the DNA digest in the gel was subjected to the in-gel renaturation procedure. The entire gel was soaked in a 0.3 N sodium hydroxide solution to denature the DNA fragments to single strands in situ. The restriction fragments in the gel were then renatured in an appropriate salt solution for a time long enough to allow the amplified bands (but not the single copy sequences) to reanneal, and the gel was then treated with a solution of S1 nuclease to rid of the single-stranded single copy fragments. The whole cycle was repeated once more to ensure that all background single copy fragments were removed by S1. The remaining amplified fragments that had reannealed during the renaturation step were then transferred to GeneScreen, and the transfer was sprayed with a fluorophor and placed next to X-ray film to determine the pattern of labeling.

The results from this analysis were quite beautiful: all of the background signal from early-firing single copy sequences had completely disappeared, resulting in easily quantifiable signals from the amplified bands that remained (Leu and Hamlin, 1989; Fig. 1B). Each lane on the film was scanned with a densitometer, and the resulting value for each band was compared to the corresponding band in the log-labeled control lane (expressed as the ratio ELF/log). When the ratio for each fragment was plotted as a function of map position in the amplicon, it was obvious that the 6.2/6.1-kb *Eco*RI doublet and the 11-kb *Eco*RI fragment were preferentially and equally labeled by 30 min into the S period (Fig. 1C). This result suggested that an origin might actually lie between these two, either in the 3.5- or 1.5-kb *Eco*RI fragments. Thus, the 6.2/6.1-, 3.5-, 1.5-, and 11-kb fragments again defined a 28-kb locus mapping downstream from the DHFR gene (Fig. 2). This result gave us tremendous confidence in our previous studies, and pointed out how important it was to have cloned and mapped the entire 240-kb amplicon.

However, the problem of low resolution was obviously not solved by this approach alone, since the details of labeling within this 28-kb locus were still not obvious owing to the large fragment sizes. In addition, much of the data was not useful because many of the bands on the gel are doublets and triplets, and the individual fragments in a given band could map at very different positions within the amplicon and therefore replicate at different times. An attempt to increase resolution by using enzymes that cut more frequently would only have resulted in a larger number of fragments and even more doublets and triplets for which the data would be unusable.

Clearly, what was needed was a method by which we could focus only on the 28-kb region that we knew contained the initiation locus, using restriction enzymes that would give us a unique spectrum of fragments containing no doublets or triplets. To do this, we developed a "hybridization enhancement" modification of the in-gel renaturation procedure. A cosmid that contains the entire initiation locus was first mapped with several restriction enzymes singly or in pairs to find the enzyme combination (*Bam*HI/*Hin*dIII in this case) that would give a unique spectrum of fragments 1–2 kb in length within this limited region. A 300-fold molar excess of a *Bam*HI/*Hin*dIII digest of the cosmid was then added to a *Bam*HI/*Hin*dIII digest of early-labeled genomic DNA from CHOC 400 cells, and the digest was separated on an agarose gel. The gel was then subjected to in-gel renaturation as described above. However, after the denaturation step, the DNA was allowed to reanneal only for a short time, i.e., long enough for those fragments driven into duplexes by excess cosmid to reanneal, but not long enough for other amplified bands or single copy fragments to rehybridize. After S1 treatment and another cycle of denaturation, renaturation, and S1 treatment, the remaining DNA was transferred to GeneScreen and was autoradiographed, and the relative labeling pattern of each band within the initiation locus was analyzed as described above. To our great surprise, this higher resolution analysis actually suggested that there were *two* rough zones of labeling within the previously identified 28-kb initiation locus, instead of the single, centered zone or site that we were expecting (Leu and Hamlin 1989).

Another quite different in vivo labeling protocol also suggested the presence of two initiation zones or sites in the DHFR amplicon (Anachkova and Hamlin, 1989). In this approach, CHOC 400 cells were synchronized at the G_1/S boundary by isoleucine starvation followed by release into complete medium containing aphidicolin for 12 hr. By this time, all cells should have arrived at the beginning of S and may actually have initiated replication at origins, but replication forks should be kept very close to origins by aphidicolin, which is a competitive inhibitor of dCTP. Prior to removal of the drug, the genomic DNA was cross-linked at 1–2 kb intervals with trioxsalen and UV light. At the same time that aphidicolin was removed, bromodeoxyuridine (BUdR—a dense analogue of thymidine) was added and the cells were allowed to enter S. One hour later, when replication forks should have advanced as far as possible before encountering a trioxsalen cross-link, the DNA was harvested, and the low-molecular-weight nascent DNA (termed X-DNA) was purified first by size on alkaline sucrose gradients and then by precipitation with anti-BUdR antibodies. This X-DNA, which should be centered around origins of replication, was then labeled with [^{32}P]dCTP and used to probe restriction digests of cosmids from the amplicon that had been separated on an agarose gel and transferred to GeneScreen. The fragments that hybridized the most intensely with the X-DNA were the 6.2/6.1-kb doublet and the 11.0-kb *Eco*RI fragment. This result supported the proposal that there were actually two initiation zones or sites within the 28-kb initiation locus mapping downstream from the DHFR gene.

To our great satisfaction, a report appeared from Howard Cedar's lab in which they had used our cosmid clones to develop an independent replicon mapping method, and had also found two origins of replication in the DHFR domain (Handeli et al., 1989). In their method, they isolated nascent DNA that was preferentially associated with that strand of the replication fork to which nucleosomes segregate in the presence of the protein synthesis inhibitor, emetine (purportedly the leading strand). This material was used as a hybridization probe on the separated single strands of M13 subclones to determine whether it hybridized specifically with the + or − strand of the genome. Since, according to viral models, leading strands should switch at origins of replication, the method has the potential to localize origins with high precision (in practice, the hybridization signal falls off near the origin for reasons that are not completely clear). The results of their analysis (which was actually done before they knew the results of our in-gel renaturation and cross-linking experiments) suggested the presence of two origins situated in the same approximate locations that we had previously detected.

The surprising finding of two initiation sites or zones separated by ~22 kb partially explained why many of our former strategies for increasing the resolution of replication fork mapping studies had failed (these are too numerous to recount here). Since there appeared to be no other initiation sites within the 240-kb DHFR amplicon, either (1) there are two separate origins of replication that happen to lie close to one another in this region of the genome, *or* (2) the "origin" in this region is much larger and more complex than the simple origins of microorganisms, which consist of a small *cis*-acting element that binds an initiation complex, resulting in helix destabilization and priming of nascent chains within a region of a few hundred base pairs.

At the same time that we were attempting to determine the precise location of initiation sites in the amplified DHFR domain, Pieter Dijkwel came to my laboratory to map the locations of matrix-attachment sites in the amplicon. A body of evidence had accumulated that genomic DNA is attached at approximately 100-kb intervals to a subnuclear proteinaceous scaffolding or matrix (Nelson et al., 1986). These permanent attachment sites have been suggested to represent specific DNA sequences and, in fact, the specific association of certain sequence elements located in the 5′ regulatory regions of genes had been demonstrated. There are two common ways of identifying specific matrix attachment regions (MARs). In the first, histones are extracted from isolated nuclei, which then allows the genomic DNA to be digested in situ with restriction enzymes. When the DNA that partitions with the matrix is compared to the DNA released by the restriction enzyme from the matrix, any preferential partitioning of a fragment with the matrix can be detected by Southern blotting and hybridization with probes from the region of interest. In a second method, the in vitro binding assay, histones are extracted from the nuclei, and the vast majority of DNA (>99%) is completely removed from the matrices by digestion to completion with micrococcal nuclease. These "bald" matrices are then incubated with end-labeled digests of the cosmids or plasmids of interest to determine whether any fragments in these mixtures selectively associate with the matrix.

When Pieter Dijkwel analyzed the amplified DHFR domain by these two methods, only two fragments from the entire 240-kb amplified DHFR domain demonstrated marked affinity for the nuclear matrix in both assays (Dijkwel and Hamlin, 1988). One of these maps near an interamplicon junction, but the significance of this, if any, presently escapes us. Interestingly, however, the second matrix-attachment site (MAR) is situated in the region between the two early-labeled peaks that we had identified (Fig. 2). This very provocative finding suggests a possible role for the MAR in initiation and raises the interesting hypothesis that the two zones of initiation may, in fact, be part of one complex origin, a critical component of which might be the centered MAR.

About the time that the studies on the MAR were being written up, two very novel and elegant two-dimensional replicon mapping methods were published that had the potential to give a more detailed picture of replication intermediates in this locus than we had been able to attain with our in vivo labeling protocols. While both of these methods had only been applied to genomes no more complex than those of yeast and of *Drosophila*, we felt that they had the potential for application to animal cell DNA.

In the first method (Brewer and Fangman, 1987), a restriction digest of genomic DNA is separated largely according to molecular mass on a low percentage agarose gel at low voltage in the absence of ethidium bromide. The lane is then excised, turned through 90°, and the DNA digest is separated in a high percentage gel at high voltage in the cold in the presence of ethidium bromide. In this second dimension, fragments separate according to both mass and shape. The power of this neutral/neutral gel system is illustrated in Figure 3. Regardlesss of whether a fragment contains a replication origin or is always replicated passively by replication forks emanating from a distant origin, the replication intermediates found in that fragment will range in size from slightly more than $1n$ (the unreplicated fragment) to almost $2n$ (the almost fully replicated fragment). However, depending on whether a fragment contains an origin or a single replication fork, the shape and migration of the replicating fragment in the second dimension of the gel will be different from a linear fragment of the same molecular mass. Therefore, a digest of genomic DNA isolated from either log or synchronized cells can be separated on such a gel, the DNA can be transferred to a filter by blotting, and individual fragments in a genomic region of interest can be analyzed by successive hybridizations with suitable probes that walk along the region.

When we applied this analysis to the amplified DHFR domain, the result was totally unexpected. Instead of finding two single origins of replication mapping in the center of each of the previously identified early-labeled peaks indicated in Figure 2, we found that initiations were occurring in every fragment that we examined within the 28-kb initiation locus: all of the patterns that we observed in gels suggested that each fragment from this locus was usually replicated passively by forks, but was also replicated sometimes by internal intiations. The most reasonable conclusion from these data was that initiation occurs randomly at many sites within the 28-kb initiation locus (Vaughn et al., 1990).

This was such a heterodoxic model and so different from what we expected that we wanted to use a completely independent method to analyze replication intermediates in this locus. We therefore turned to an independent neutral/alkaline two-dimensional replicon mapping method (Nawotka and

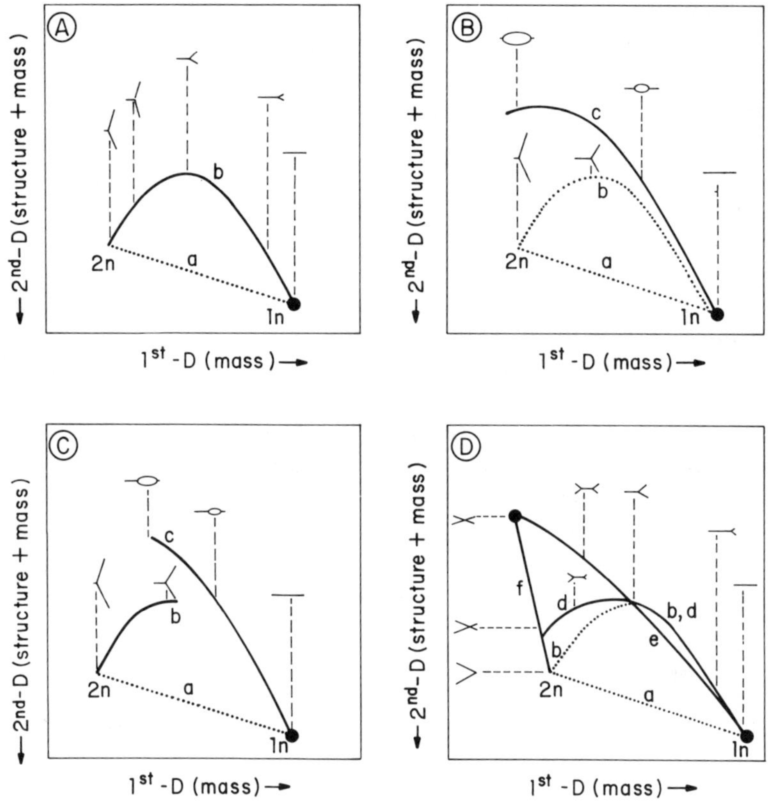

Fig. 3. Patterns of typical replication intermediates observed in the two-dimensional neutral/neutral replicon mapping method. Each panel shows an idealized autoradiographic image that would be obtained when a restriction digest of replicating DNA is hybridized with probes for fragments that contain different intermediates. **(A)** A complete simple Y or fork arc *(b)* resulting from a fragment that is replicated passively from an outside origin. Curve *a* represents the diagonal of nonreplicating fragments from the genome as a whole. **(B)** The pattern obtained when a fragment with a centered origin of replication is probed (curve *c*). Bubbles migrate more slowly at all extents of replication than do forks in a fragment of equal mass *(b)*. **(C)** The presence of an off-centered origin in a fragment gives rise to an incomplete bubble arc (*c*), which then reverts to the fork arc when the bubble expands beyond the nearest restriction site, resulting in a fork arc "break." **(D)** When two forks approach each other in a fragment either symmetrically or asymmetrically, curves *e* and *d* are obtained, respectively. If there is a fixed terminus in a fragment, the collected X-shaped structures would result in a concentrated spot somewhere on curve *f*. Recombination structures would also fall along curve *f* (Brewer and Fangman, 1987).

Huberman, 1988). In this method, a suitable restriction digest of genomic DNA from either log or synchronized cells is separated in the first gel dimension according to molecular mass under the same conditions as the neutral/neutral gel system described above. The lane of interest is then excised, turned through 90°, and electrophoresed in the second, alkaline dimension (0.3 N NaOH). The small nascent chains that are released from the $1n$ template by the denaturing agent migrate according to size, which will range from slightly less than $1n$ to almost zero, and trace a diagonal as shown in Figure 4. The direction of replication fork movement through a defined region of the genome (the DHFR locus, in this case) can be determined by hybridizing a transfer of such a gel successively to two small probes from either end of individual restriction fragments from the locus. If a particular fragment is replicated passively by forks, then the two hybridization patterns shown in Figure 4 (e.g., right-hand fragment) will be obtained, and the direction of fork movement can be determined unambiguously. Only if a fragment contains an origin, as shown in the central panel, will a probe from the middle of the fragment detect a diagonal that extends to almost zero fragment length.

When we scanned the 28-kb initiation locus by this approach, using either pairs of end fragments or centered probes, the results showed unambiguously that replication forks move in both directions in any given fragment from this region, and that initiations occur at virtually every region tested within the 28-kb locus, including fragments containing the MAR (Vaughn et al., 1990). Thus, the data from both the neutral/neutral and the neutral/alkaline two-dimensional replicon mapping methods supported the notion that initiation in the amplified DHFR domain occurs at many (probably random) positions within the previously defined 28-kb initiation locus.

Conclusions

How can our results from these two complementary two-dimensional replicon mapping methods be combined with our previous data into a unified model for the mechanism of replication initiation in this locus? The first question is how the two peaks of early labeling in this region relate to the random initiations that occur in the broad zone ecompassing both peaks. We are presently attempting to determine the relative number of initiations occurring in a given fragment relative to other fragments in the region, and it is possible that we will find a higher concentration of initiations per unit length of DNA in fragments centered over the two peaks of early labeling.

Another question is whether the two peaks of early labeling correspond to two separate origins separated by a matrix-attachment site, or whether the entire 28-kb locus (including the two peaks and the MAR) represents a single complex mammalian origin. To answer this question, we will have to learn much more about the role of the MAR (if any) in initiation, and we will also have to examine other initiation regions in the CHO genome. Using another

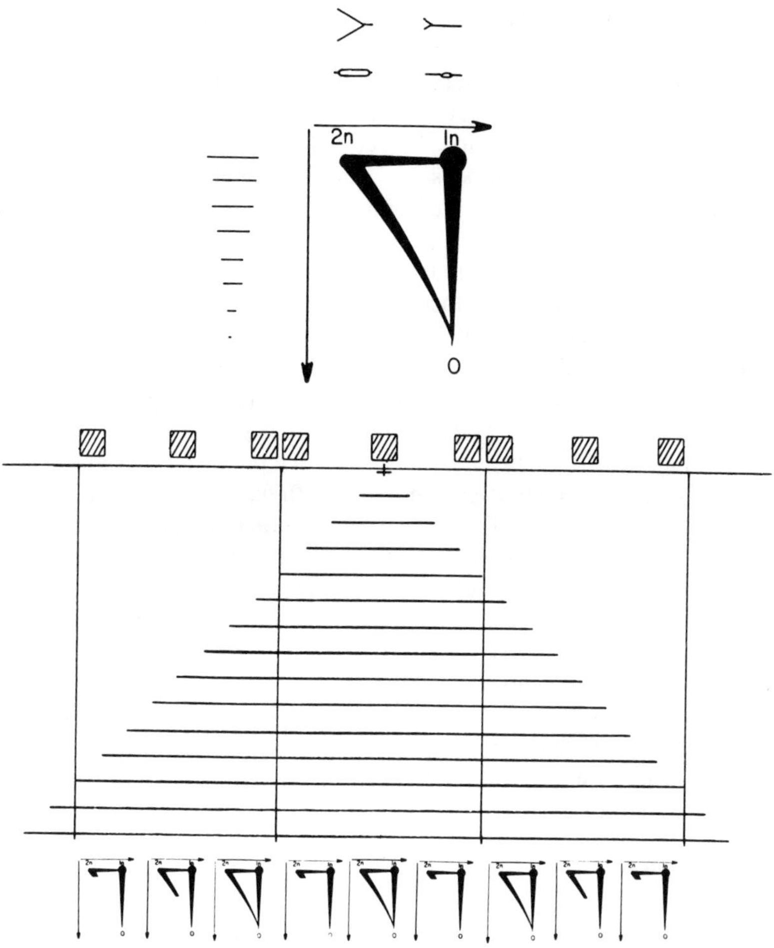

Fig. 4. Principle of the neutral/alkaline two-dimensional replicon mapping method. The upper diagram shows the idealized patterns of migration of double-stranded replication intermediates in the first (neutral) dimension and single-stranded nascent chains in the second (alkaline) dimension. The lower panel outlines the autoradiograms that would be obtained when each of the probes (hatched boxes) is used to illuminate its cognate restriction fragment. An origin of replication is positioned at the vertical hatchmark in the central fragment. Nascent strands of different sizes are indicated, and the expected autoradiograms resulting from each probe are shown below.

MTX-resistant Chinese hamster cell line with a much larger DHFR amplicon, we have recently identified and cloned a second (third?) initiation locus that lies ~250 kb upstream from the 28-kb initiation locus discussed above, and we have found a MAR in this new locus as well. We are presently attempting to determine whether there are one or two peaks of early labeling in this locus, and whether initiation also appears to be random and delocalized.

We would also like to determine whether this delocalized form of initiation is unique to amplified DNA. It is possible that the amplification process somehow may have deranged a normal process in which initiation occurs only in a small circumscribed region including and immediately surrounding a *cis*-regulatory "origin" of replication. In fact, we have recently had our first successes in applying the two-dimensional mapping methods to the single copy DHFR locus in CHO cells, and preliminary results suggest that initiation may also be delocalized in the single copy locus.

If it turns out that initiation, indeed, occurs at many random sites in a circumscribed zone of the DHFR locus in CHO cells (and in other single copy loci as well), does that mean that there are no *cis*-regulatory elements in mammalian chromosomes that are analogous to the origins of microorganisms, and that determine the location of initiation sites? To address this question, it would appear to be necessary to develop a phenotypic assay for origin function in which subfragments from the initiation locus can replicate autonomously (e.g., episomally) in animal cells after transfection. To date, we have not been able to demonstrate significant and reproducible autonomous replication of any fragments from the DHFR locus after transfection into CHO cells, and other laboratories have had similar difficulties with other presumptive mammalian origins of replication. However, it is possible that the two-dimensional replicon mapping methods described above could be applied to an analysis of fragments after they are transfected and *stably integrated* into mammalian chromosomes. Mutagenesis of appropriate subclones from the initiation locus, followed by assessment of the effect on replication initiation after transfection, could then be used to identify important *cis*-regulatory elements in mammalian origins.

Acknowledgments

I would like to thank Art Pardee for the support and guidance that he has given me over the years. I would also like to thank all of the members of my laboratory, past and present, each of whom has contributed in important ways to the success of our research program. These include (in alphabetical order): Boyka Anachkova, Jane Azizkhan, Sergei Bavykin, Kevin Cox, Pieter Dijkwel, Lorne Erdile, Pamela Foreman, Valerie Frazier, Mark Gray, Kay Greisen, Nicholas Heintz, Tzeng-Horng Leu, James Looney, Chi Ma, Jeffrey Milbrandt, Martin Montoya-Zavala, James Vaughn, and W. Carlton White. The work described here has been supported by grants from the NIH/NIGMS and the American Cancer Society.

References

Anachkova B, Hamlin JL (1989): Replication in the amplified dihydrofolate reductase domain in CHO cells may initiate at two distinct sites, one of which is a repetitive sequence element. Mol Cell Biol 9:532–540.

Biedler JL, Spengler BA (1976): A novel chromosome abnormality in human neuroblastoma and antifolate-resistant Chinese hamster cell lines in culture. J Natl Cancer Inst 57:683–695.

Brewer BJ, Fangman WL (1987): The localization of replication origins on ARS plasmids in S. cerevisiae. Cell 51:463–471.

Crouse GF, Simonsen CC, McEwan RN, Schimke RT (1982): Structure of amplified normal and variant dihydrofolate reductase genes in mouse sarcoma S180 cells. J Biol Chem 257:7887–7897.

Dijkwel PA, Hamlin JL (1988): Matrix attachment regions are positioned near replication initiation sites, genes, and an interamplicon junction in the amplified dihydrofolate reductase domain of Chinese hamster ovary cells. Mol Cell Biol 8:5398–5409.

Foreman PK, Hamlin JL (1989): Identification and characterization of a gene that is coamplified with dihydrofolate reductase in a methotrexate-resistant CHO cell line. Mol Cell Biol 9:1137–1147.

Hamlin JL (1978): Effect of damage to early, middle, and late replicating DNA on progress through the S period in CHO cells. Exp Cell Res 112:225–232.

Hamlin JL, Biedler JL (1981): Replication pattern of a large homogenously staining chromosome region in antifolate-resistant Chinese hamster cell lines. J Cell Physiol 107:101–114.

Hamlin JL, Pardee AB (1976): S phase synchrony in monolayer CHO cultures. Exp Cell Res 100:265–275.

Handeli S, Klar A, Meuth M, Cedar H (1989): Mapping replication units in animal cells. Cell 57:909–908.

Heintz NH, Hamlin JL (1982): An amplified chromosomal sequence that includes the gene for dihydrofolate reductase initiates replication within specific restriction fragments. Proc Natl Acad Sci USA 79:4083–4087.

Heintz NH, Milbrandt JD, Greisen KS, Hamlin JL (1983): Cloning of the initiation region of a mammalian chromosomal replicon. Nature 302:439–441.

Huberman JH, Riggs AD (1968): On the mechanism of DNA replication in mammalian chromosomes. J Mol Biol 32:327–337.

Leu T-H, Hamlin JL (1989): High resolution mapping of replication fork movement through the amplified dihydrofolate reductase domain in CHO cells by in-gel renaturation. Mol Cell Biol 9:523–531.

Looney JE, Hamlin JL (1987): Isolation of the amplified dihydrofolate reductase domain from methotrexate resistant Chinese hamster ovary cells. Mol Cell Biol 7:569–577.

Ma C, Looney JE, Leu T-H, Hamlin JL (1988): Organization and genesis of dihydrofolate reductase amplicons in the genome of a methotrexate-resistant Chinese hamster ovary cell line. Mol Cell Biol 8:2316–2327.

Marsh RC, Worcel A (1977): A DNA fragment containing the origin of replication of Escherichia coli. Proc Natl Acad Sci USA 74:2720–2724.

Melera PW, Lewis JA, Biedler JL, Hession C (1980): Antifolate-resistant Chinese hamster cells: evidence for dihydrofolate reductase gene amplification among independently-derived sublines overproducing different dihydrofolate reductases. J Biol Chem 255:7024–7028.

Milbrandt JD, Azizkhan JC, Greisen KS, Hamlin JL (1983): Organization of a Chinese hamster dihydrofolate reductase gene identified by phenotypic rescue. Mol Cell Biol 3:1266–1273.

Milbrandt JD, Heintz NH, White WC, Rothman SM, Hamlin JL (1981): Methotrexate-resistant Chinese hamster ovary cells have amplified a 135 kilobase pair region that includes the gene for dihydrofolate reductase. Proc Natl Acad Sci USA 78:6043–6047.

Mirkovitch J, Mireault M-E, Laemmli UK (1984): Organization of the higher order chromatin loop: Specific DNA attachment sites on the nuclear scaffold. Cell 39:223–232.

Montoya-Zavala M, Hamlin JL (1985): Similar 150-kilobase DNA sequences are amplified in independently derived methotrexate-resistant Chinese hamster cells. Mol Cell Biol 5:619–627.

Nawotka KA, Huberman JA (1988): Two-dimensional gel electrophoretic method for mapping DNA replicons. Mol Cell Biol 8:1408–1413.

Nelson WG, Pienta KJ, Barrack ER, Coffey DS (1986): The role of the nuclear matrix in the organization and function of DNA. Annu Rev Biophys Biophys Chem 15:457–475.
Nunberg JH, Kaufman RJ, Schimke RT, Urlaub G, Chasin LA (1978): Amplified dihydrofolate reductase genes are localized to a homogenously staining region of a single chromosome in a methotrexate-resistant Chinese hamster ovary cell line. Proc Natl Acad Sci USA 75:5553–5556.
Pardee AB (1974): A restriction point for control of normal animal cell proliferation. Proc Natl Acad Sci USA 71:1286–1290.
Roninson I (1983): Detection and mapping of homologous, repeated and amplified DNA sequences by DNA renaturation in agarose gels. Nucleic Acids Res 11:5413–5431.
Smithies O (1982): The control of globin and other eukaryotic genes. J Cell Physiol Suppl 1:137–143.
Tobey RA, Crissman HA (1972): Preparation of large quantities of synchronized mammalian cells in late G_1 in the pre-DNA replicative phase of the cell cycle. Exp Cell Res 75:460–464.
Vaughn JP, Dijkwel PA, Hamlin JL (1990): Replication initiates in a broad zone in the amplified dihydrofolate reductase domain in CHO cells. Cell 61:1075–1087.

Perspectives on Cellular Regulation:
From Bacteria to Cancer, pages 315–329

Allosteric Interactions Between the Enzymes of DNA Biosynthesis in Mammalian Cells

G. Prem Veer Reddy

Department of Obstetrics and Gynecology and Department of Biochemistry, Health Sciences Center, University of Virginia, Charlottesville, Virginia 22908

Introduction

I was peacefully in the pine trees of Corvallis, Oregon, doing my doctoral work with Chris Mathews. There was a sense of scientific excitement because my thesis work revealed a new, structural complex of enzymes (in T4 bacteriophage) that synthesized DNA and its precursors. Later, we would call the complex "replitase." I thought the scientific excitement would eventually wind down and I would finish my doctorate, return to Osmania University, in Hyderabad, India, and live out my days teaching nucleic acid biochemistry. Just then, the mists of the Pacific rain forests were suddenly shattered by a phone call from Art Pardee. Art called me to do postdoctoral work in his laboratory at Harvard and to extend my thesis work to mammalian cells. I could say he called me into the mayhem of Boston traffic, but that would be inaccurate; Boston traffic is child's play compared to India's. I could say he pulled me into the agony of writing grants and pushing manuscripts through stubborn reviewers. But that also would be inaccurate. Not only did the project develop beautifully under his guidance, but I received a postgraduate course from Art Pardee in seeing beyond the headaches and hard work to the intellectual excitement and beauty of the biochemistry of one of the most fundamental processes of life. Art, by providing an opportunity to work in his laboratory, has enabled me to combine my ambition to teach with scientific excitement. I have very good memories of that time and of Art bubbling with ideas and encouragement, a warm and caring person who has been my best friend in science. Because of this, I feel specially privileged to have an opportunity to contribute to this publication in Art's honor.

In this chapter I would like to present a view of the project I have pursued, with guiding support from Art. During the past decade, we have explored the idea of allosteric interaction between the enzymes of DNA replication. We would like to understand, first, how deoxynucleotides are being made available to the precisely timed and rapid process of DNA replication and, second, the events that immediately precede, and therefore could control, the onset of DNA synthesis during cell cycle. Inspired, in part, by Art's work on aspartate transcarbamylase, the term "allosteric" was originally coined in the laboratory of Jacob and Monod to describe the changes in the activity of an enzyme when it is bound by another substance (an effector) at a site distinct from the active site. In this chapter the phenomenon of "allosteric interaction" is invoked to explain how the interaction between enzymes of deoxynucleotide metabolism and DNA replication could influence the overall process of DNA synthesis in mammalian cells.

As suggested by Kornberg (1980), among others, proteins needed for DNA replication may bind to each other in vivo, forming a multiprotein replication complex with increased functional efficiency. Such a complex for DNA replication was successfully reconstituted in vitro using enzymes/proteins of T4 bacteriophage (Alberts, 1987). Complexes containing the enzymes of dNTP synthesis and DNA replication were also isolated from *Escherichia coli* infected with T4 phage (for a review, see Mathews et al., 1979) and, subsequently, a complex containing the enzymes of dNTP synthesis has been purified about 500-fold from T4 bacteriophage (Moen et al., 1988). The complexity of mammalian cells, and the lack of suitable mutants, made similar approaches to define associations between the enzymes of DNA synthesis in mammalian cells somewhat more difficult. Although my work began in the T4 phage system, in this chapter I will highlight some of the observations from mammalian systems that support the concept of functional association between the enzymes of deoxynucleotide synthesis and DNA replication and identify unresolved questions.

Physical Association Between the Enzymes of dNTP Synthesis and DNA Replication

Most of the enzymes involved in deoxynucleotide synthesis have been identified and characterized with respect to their in vitro properties (Fig. 1). Similarly, DNA polymerase(s), and several auxiliary proteins, are suggested to polymerize dNTPs in a manner tightly coupled to dNTP synthesis. However, it is not entirely clear how these two biosynthetic processes, i.e., dNTP synthesis and DNA replication, are so intricately coordinated to ensure an abrupt onset of DNA replication during the S phase of the cell cycle.

Through the years, many studies have correlated the increase in intracellular level of some of the key enzymes of deoxynucleotide metabolism with the onset of DNA replication by measuring enzyme levels and/or deoxynucleotides in extracts of cells progressing synchronously from G_1 to S phase. Although changes in the levels of enzymes and deoxynucleotides have been detected at different

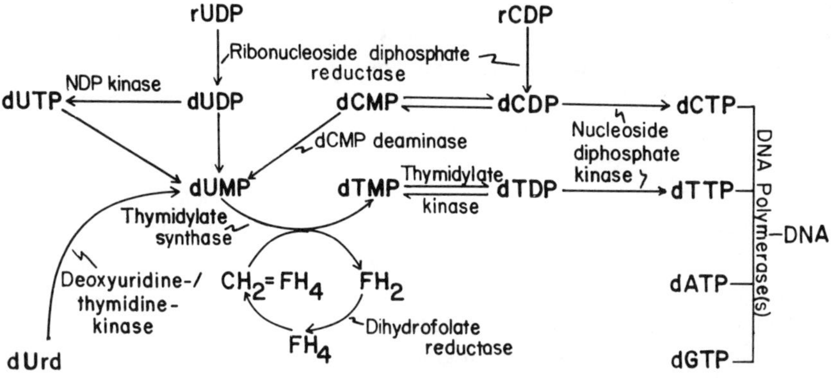

Fig. 1. Pathways of dNTP biosynthesis, showing key enzymes involved in TTP and dCTP synthesis from corresponding rNDPs and a distal precursor, dUrd. Synthesis of dATP and dGTP from corresponding rNDPs also involves ribonucleotide reductase and nucleoside diphosphate kinase.

stages of the cell cycle, a clear correlation with the onset of DNA replication has not been established. We, therefore, studied changes during the cell cycle of intracellular localization, rather than simply changes in the overall level, of these enzymes, particularly focusing on onset of DNA replication.

Reddy and Pardee (1980) first reported that at least seven enzymes associated with DNA biosynthesis, viz., DNA polymerase(s), ribonucleotide reductase, thymidylate synthase, dihydrofolate reductase, nucleoside diphosphate kinase, thymidylate kinase, and thymidine kinase, migrate into the nucleus and assemble into the multienzyme complex as Chinese hamster embryo fibroblast (CHEF/18) cells progress synchronously into S phase. Subsequently, $3' \rightarrow 5'$ exodeoxyribonuclease (associated with DNA polymerase δ), DNA topoisomerase II (Hammond et al., 1989), and nascent DNA that can serve as a template/primer for DNA polymerase activity (Noguchi et al., 1983) were also demonstrated to be integral components of this complex. This multienzyme complex, isolated from the nuclear lysate of S phase cells, was termed "replitase" to distinguish it from complexes such as the "replisome" (Kornberg, 1980), which contains exclusively the enzymes/proteins associated with DNA replication. Replitase, by contrast, contains the enzymes of both DNA replication and dNTP synthesis, organized in a supramolecular structure.

Ever since Baril et al. (1973) reported that DNA polymerase-α is aggregated with the other enzymes of dNTP synthesis, including ribonucleotide reductase, thymidylate synthase, and thymidine kinase in Novikoff tumor cells and in regenerating rat liver, there have been a number of confirmatory reports of such complexes in a variety of mammalian cells (Harvey and Pearson, 1988; Ayusawa et al., 1983; Wickremasinghe et al., 1983). Replitase complexes are best detected in cultures enriched in cells undergoing nuclear DNA replication. In studies with CHEF/18 cells, synchronized by release from isoleucine starvation, we observed that these enzymes are localized to the nucleus and assembled into the replitase complex only during progression into S phase (Reddy and Pardee, 1980). During G_1 phase, the enzymes of DNA replication and dNTP synthesis were absent from the nuclei, and could not be isolated in an aggregated form. Recovery of the complex fraction, specifically from the cultures containing S phase-enriched cells, was not an artifact of the isoleucine starvation method employed to synchronize the cells, because proliferating cells fractionated from a cell suspension from calf thymus also contained the replitase fraction in their nuclear lysate (Noguchi et al., 1983).

The finding that the enzymes migrate into the nucleus and assemble into a multienzyme complex when DNA replication starts abruptly suggests that the organization of the enzymes into a supramolecular structure may be a key event that initiates S phase. Temporal functioning of key enzymes of dNTP synthesis and DNA replication, could also be ascribed to the assembly of the enzymes into the replitase complex (Reddy, 1982).

Could the association of the enzymes into a supramolecular structure be an artifact of the experimental procedures employed during its isolation? The specificity of the association between the enzymes of deoxynucleotide synthesis and DNA replication is further established from the following observations:

1. Association of various enzymatic activities with the complex fraction seems to be in stoichiometry with the total DNA content of the cells. For example, in ribonucleotide reductase-overproducing Syrian hamster melanoma cells (Ashman et al., 1981) or in dihydrofolate reductase-overproducing CHEF/18 cells (Noguchi et al., 1983), only a specific fraction of these enzymes, in stoichiometry with the DNA content of the cells, was found associated with the replitase fraction.

2. Ayusawa et al. (1983) demonstrated that thymidylate synthase-negative mutants of murine FM3A cells could be transformed to thymidine prototrophs by transfecting human DNA into these cells. Thymidylate synthase expressed by the human genome was not integrated into the multienzyme complex of the transformed mouse cells, suggesting a species specificity of the enzymes involved in the macromolecular assembly of the complex.

3. There was no nonspecific association of cytoplasmic or mitochondrial proteins/enzymes with the complex fraction isolated from the nuclei of S phase cells (Noguchi et al., 1983; Reddy and Pardee, 1980).

4. As discussed below, the complex fractions prepared from S phase CHEF/18 cells (Noguchi et al., 1983) and from a human lymphoblastoid cell line (HPB-ALL) (Wickremasinghe et al., 1983) exhibited kinetic coupling between the enzymes, thereby limiting the accumulation of intermediary metabolites and providing a high concentration of precursors for DNA synthesis at the replication site.

Interactions between the enzymes of dNTP synthesis and DNA replication obviously support the idea that these enzymes are localized in a structured complex in the nuclei of S phase cells. Accordingly, nuclear, as well as cytoplasmic localization of ribonucleotide reductase, along with DNA polymerase, was observed in S phase-enriched CHEF/18 cells (Reddy and Pardee, 1980) and in actively replicating cells from regenerating rat liver (Youdale et al., 1984). By contrast, others reported to have found exclusively cytoplasmic localization of ribonucleotide reductase (Engstrom and Rozell, 1988; Kucera and Paulus, 1986). These claims, which utilized different cell systems, have been used as an argument against the universal nuclear localization of the key enzymes of dNTP synthesis and, therefore, against the idea of the replitase complex. With convincing evidence for the replitase complex from a variety of independent methods, we must therefore explain how these seemingly inconsistent data could have been obtained. As we improve the purification of the macromolecular structure containing the enzymes of dNTP synthesis and DNA replication, we should be able to better understand the biochemical and/or biophysical forces responsible for the protein–protein interactions in the complex, and whether such forces are affected by the fixation or fractionation techniques employed in these studies. This might account for the observed discrepancies in nuclear localization of the key enzymes of precursor synthesis in mammalian cells. Our early efforts to purify the complex already point to its labile nature (Noguchi et al., 1983; Wickremasinghe et al., 1983), possibly due to weak ionic forces involved in maintaining the integrity of the complex (Harvey and Pearson, 1988). Furthermore, it is not yet fully understood whether the enzymes are intranuclear or perinuclear during DNA replication, and if they are perinuclear, whether the nuclear matrix could serve as the venue for the assembly of the replitase complex and, therefore, for replication (Nelson et al., 1986). The existence of megacomplexes containing DNA polymerase-α and primase with the nuclear matrix of regenerating rat liver (Tubo and Berezney, 1987) favors such a possibility. Further progress in this direction will be as challenging and illuminating, if not more so, than it was during the period of purification and characterization of the ribosome.

Functional Association Between the Enzymes of dNTP Synthesis and DNA Replication

The primary physiological property of an aggregated multienzyme system, as opposed to soluble unaggregated enzymes, is its ability to regulate catalysis through the coordinated activation or inhibition of constituent enzymes. Sequestration of related enzymatic activities into functional units increases the efficiency of sequential reactions possibly by (1) decreasing the diffusion times for substrates, (2) minimizing competition by other pathways, (3) presenting substrates for the next reaction at effectively high concentrations, and (4) quickly utilizing potentially unstable intermediates in the final macromolecular biosynthesis (Srere, 1987). Some of these properties are manifested by complexes containing the enzymes of DNA precursor biosynthesis and DNA replication.

Cell Cycle-Dependent Regulation of the Enzyme Activities Associated With DNA Synthesis

Are the enzymes of dNTP synthesis catalytically active at all times during the cell cycle? If not, what controls their activities? In studies with synchronized CHEF/18 cells, it was observed that enzymes such as thymidylate synthase and DNA polymerase were present in the extracts of cells from all phases of the cell cycle. However, catalytic activity in intact cells was confined to the period of DNA replication (Reddy, 1982). A similar ''cryptic'' nature of thymidylate synthase activity was reported from studies with synchronized murine leukemia (L1210) cells (Rode et al., 1980; Matherly et al., 1989). It is intriguing to note that in cells that overproduce ribonucleotide reductase (Ashman et al., 1981) or thymidylate synthase (Danenberg and Danenberg, 1989), only a fraction of the total enzyme, comparable to that in normal cells, is catalytically active in intact cells. In ribonucleotide reductase-overproducing Syrian hamster melanoma cells (Ashman et al., 1981) and in thymidylate synthase-overproducing murine fibroblast cells (Reddy, unpublished results), deoxynucleotide levels remained almost unchanged relative to the levels in respective normal cells.

Is it possible that ''free'' deoxynucleotides could be responsible for the activation or inactivation of thymidylate synthase during the cell cycle? Lack of any significant difference in deoxynucleotide levels between thymidylate synthase-overproducing and normal cells tends to argue against such a possibility. In addition, in studies with permeabilized cells, where intermediary metabolites including dNTPs can freely diffuse in and out of the cells, thymidylate synthase exhibited functional properties similar to those observed with intact cells, ie., it was catalytically more active in S phase than in other phases of the cell cycle (Reddy, 1982). Moreover, the timing of the catalytic activation of thymidylate synthase and DNA polymerase in intact cells coincided with the timing (S phase) of their migration into the nucleus and their integration into the replitase complex (Reddy, 1982). These observations indicate that the catalytic activation of thymidylate synthase and other enzymes of deoxynucleotide metabolism could be a consequence of their allosteric interaction with the enzymes of DNA replication.

Inhibitor Evidence for Allosteric Interaction Between the Enzymes

If the catalytic activation of the enzymes in the replitase complex is a consequence of allosteric interaction, as proposed above, then it would be expected that allosteric interactions could also lead to inhibition. In particular, an inhibitor of one enzyme could lead to inhibition of a second unrelated enzyme of the complex by allosteric interaction. These inhibitors would have no effect on the second enzyme when it is separated from the complex. One example of this is the inhibition of thymidylate synthase by inhibitors of ribonucleotide reductase, DNA polymerase, or DNA topoisomerase (Reddy and Pardee, 1983). The inhibitory effect of hydroxyurea (HU), an inhibitor of ribonucleotide reductase, and aphidicolin, an inhibitor of DNA polymerase-α, on thymidylate synthase activity was also observed by other investigtors (Chiba et al., 1984; Rode et al., 1985; Nicander and Reichard, 1985). However, these investigators suggested alternative explanations for the inhibition by HU and aphidicolin of [^{3}H]dUrd flux through thymidylate synthase in intact cells. These alternative explanations were largely based on studies with nonsynchronized, exponentially growing cells, in which DNA polymerase was completely inhibited, while thymidylate synthase flux was partially inhibited by these antimetabolites. One plausible alternative suggestion is that the decrease in thymidylate synthase flux was due to the feedback inhibition by deoxynucleotides, accumulated when DNA polymerase reaction is inhibited by HU or aphidicolin (Chiba et al., 1984). However, our recent measurements of the specific activities of deoxynucleotides, particularly of dUMP and dTTP, in synchronized S phase CHEF/18 cells clearly demonstrates that neither the accumulation of thymidine nucleotide(s) nor the decrease in the specific activity of dUMP in HU or aphidicolin treated cells is large enough to account for the decrease in thymidylate synthase activity. Furthermore, it was observed that the inhibitory effect of HU on DNA replication or on thymidylate synthase activity in vivo could not be circumvented by providing all four deoxynucleosides in S phase-enriched CHEF/18 cells (Plucinski et al., 1990) or in thymus cells (Scott and Forsdyke, 1980). These observations cannot be explained by pool alterations, and point to inhibition by allosteric interaction between the enzymes within the replitase complex.

One reason why these results point so directly to the allosteric control within the replitase complex is because the cultures employed in these studies are highly enriched with S phase cells. Many of the studies in the literature are carried out on exponentially growing cells with the internal controls in the replitase complex obscured by the majority of the cells that are out of S phase and consequently do not have fully assembled replitase complexes. Confusion also stems from the failure of investigators to carefully distinguish between "replication-active" and "replication-inactive" deoxynucleotide pools as defined by Leeds and Mathews (1987).

Kinetic Coupling Between the Enzymes of DNA Synthesis in Isolated Replitase

The functional association among the enzymes of dNTP synthesis and DNA replication in a multienzyme complex in mammalian cells is evident from their ability to rapidly incorporate radioactivity in vitro into DNA from distal precursors such as [^{14}C]CDP (Noguchi et al., 1983). During the incorporation of labeled rCDP into DNA (rCDP→dCDP→dCTP→DNA), the sequestered enzymes in the replitase complex limit the local accumulation of intermediary nucleotides (dCDP and dCTP) and prevent a large excess of unlabeled dCDP or dCTP from interfering with the incorporation of labeled CDP into DNA. In this reaction,the fact that the incorporation of ribonucleoside diphosphates into DNA occurred subsequent to their conversion to deoxynucleotides by ribonucleotide reductase is evident from the sensitivity of the reaction to HU, and also from the requirement for dithiothreitol for optimal incorporation of radioactivity from rCDP into DNA (Nogushi et al., 1983). Wickremasinghe et al. (1983) also reported that the complex of DNA polymerase and DNA precursor synthesizing enzymes from a lymphoblastoid cell line (HPB-ALL cells) could incorporate the distal precursors [^{3}H]thymidine or [^{3}H]dTMP into DNA at rates comparable to those observed using an immediate precursor [^{3}H]dTTP. During this incorporation, the concentration of [^{3}H]dTTP formed from the distal precursors was so low that if it had to diffuse through the reaction mixture prior to its incorporation into DNA, it would have supported only 9% of the observed rate. It is, therefore, suggested that the [^{3}H]dTTP formed from the distal precursors was maintained at a high concentration in the vicinity of DNA polymerase by kinetically coupled enzymes in the complex.

Phenomenon of Deoxynucleotide "Channeling"

The process by which a distal precursor is transformed into a dNTP, the ultimate substrate of DNA replication, through sequential reactions, without allowing intermediary metabolites to build up and mix with free pools, is referred to as "channeling" of deoxynucleotides. As described above, this was observed with the multienzyme complex of deoxynucleotide metabolism and DNA replication isolated from mammalian cells. Channeling was further studied using permeabilized cells, where enzymes maintained in a physiological milieu are accessible to exogenously added distal precursors. In cells permeabilized with lysolecithin, it was reported that ribonucleoside diphosphates incorporated just as effectively as deoxynucleoside triphosphates into acid-precipitable and alkali-resistant material (Castellot, 1980). In a reaction mixture optimized for ribonucleotide reductase by inclusion of dithiothreitol, rNDPs incorporated 50% more efficiently than dNTPs in permeabilized S phase-enriched CHEF/18 cells (Reddy and Pardee, 1980, 1982). This incorporation was sensitive to allosteric effectors, viz. dATP and dTTP, and to an inhibitor (HU) of ribonucleotide reductase. More importantly, and consistent with the phenomenon of channeling, [^{14}C]dCDP or [^{14}C]dCTP formed from [^{14}C]rCDP was not diluted by prior addition of excess cold dCDP or dCTP (Reddy and Pardee, 1982). Furthermore, with intact replitase and with the reaction mixture optimized for the incorporation of rNDPs, the proximal precursors of DNA replication (dNTPs) incorporated poorly into DNA, although dNTPs could be made to incorporate at normal rates merely by inactivating ribonucleotide reductase with HU (Reddy and Pardee, 1980). These observations are summarized in the model presented in Figure 2. According to this model distal precursors, e.g., rNDPs, have much greater access to the site of DNA replication than do proximal precursors, the dNTPs. In fact, in vitro, dNTPs are readily incorporated into DNA only when the replitase complex is chemically impaired.

Why are distal precursors so efficiently incorporated? Is it because of their proximity to DNA polymerase, as mentioned above? It is critical that questions such as these be answered in order to better understand deoxynucleotide

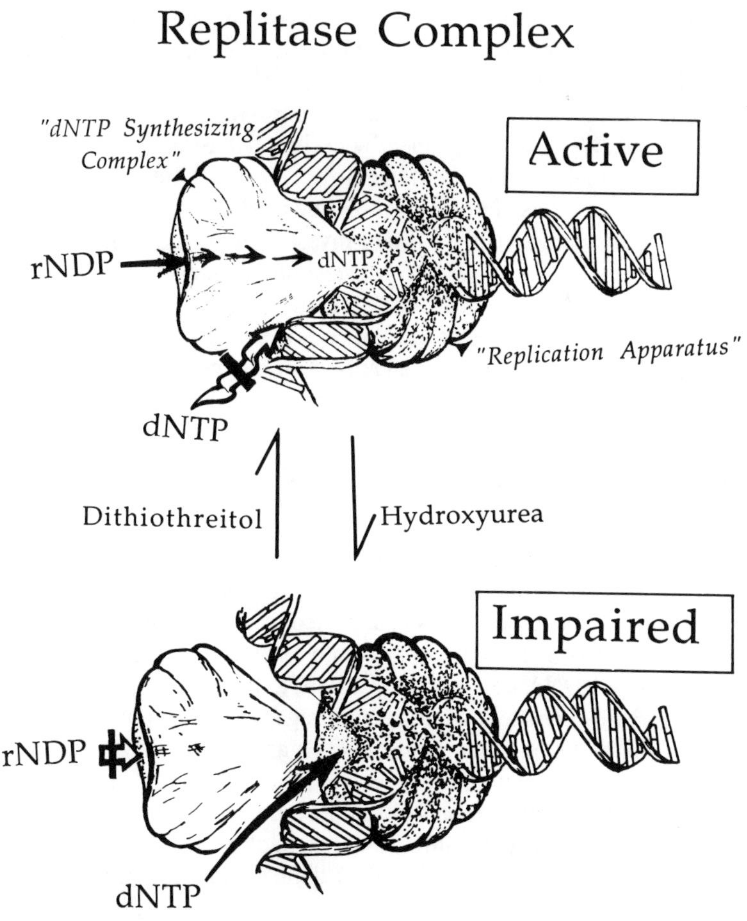

Fig. 2. A hypothetical model of replitase complex based on the data derived from the kinetics of incorporation of rNDPs and dNTPs into DNA of permeabilized S phase CHEF/18 cells.

channeling and the relationship between in vitro incorporation and the incorporation that occurs in intact mammalian cells.

Ayusawa et al. (1983) presented genetic evidence for enzyme specificity in the channeling of rNDPs into DNA in permeabilized murine FM3A cells. As an alternative to permeabilized cells, one can microinject the distal precursors into the cell to overcome the permeability barrier. However, rNDPs microinjected into mouse cells did not incorporate into DNA (Wawra, 1988). These studies lacked proper controls to determine what fraction of the microinjected cells were actually in S phase and how the intracellular milieu might limit the accessibility of microinjected rNDPs (as compared to dNTPs) to the site of semiconservative DNA replication (as opposed to repair synthesis).

While the model depicted in Figure 2 sets the stage for future investigations, it should be noted that the in situ system employed in these studies required high concentrations of dithiothreitol to stabilize ribonucleotide reductase and rNDP channeling into acid-precipitable and alkali-resistant material. Although the nucleic acids synthesized from rNDPs in this system exhibit the properties of DNA, as determined by cesium trifluoroacetate equilibrium density gradient analysis (Reddy et al., 1986), it is not clear whether artificially high concentrations of dithiothreitol used might possibly influence DNA synthesis in permeabilized cells. Further investigations of this important phenomenon of channeling must address the questions about whether experimental conditions might artifactually influence the experimental results. There is, however, no doubt that channeling exists, since it is shown clearly in in vivo experiments. The important questions that remain are the mechanism of channeling and its significance to the overall process of DNA synthesis.

Deoxynucleotide Compartmentation for DNA Replication in Mammalian Cells

The channeling process clearly leads to a functional compartmentation of deoxynucleotides in the microvicinity of their utilization. The idea of deoxynucleotide compartmentation is also supported by the fact that the overall intracellular concentrations of dNTPs are quite low (Jackson, 1984), relative to the apparent K_m of DNA polymerase (Castellot, 1980). The low steady-state levels of dNTPs also create a paradox with regard to the supply and demand of dNTPs in mammalian cells. Since dNTPs are polymerized at an overall rate of about 4×10^6 nucleotides per second [considering that there are on the average about 5×10^4 replication forks in operation during the entire S phase, and that each fork moves at the rate of about 80 nucleotides per second (Reddy, 1989)], the entire pool of dNTPs could be depleted in a fraction of a minute once replication starts (Jackson, 1984). How then does the cell sustain a constant and balanced supply of all four dNTPs to allow rapid replication of DNA? Furthermore, dNTPs, once formed from ribonucleotides, play a quite limited

TABLE 1. Channeling of Exogenous [^{3}H]Deoxyuridine Directly to DNA Synthesis Without Mixing With the Endogenous Deoxynucleotide Pools in CHEF/18 Cells

	(cpm/pmol)
Specific activity of exogenous [^{3}H]deoxyuridine	2460
Specific activity of endogenous nucleotides	
dUMP	263
dTTP	258
Specific activity of thymidine residues in newly synthesized DNA as predicted by total mixing of endogenous pools (average of dUMP + dTTP)	260
Actual specific activity of thymidine residues in newly synthesized DNA as calculated from DNA synthesis rate[a] and measured radioactivity[b]	2363[c]

[a]The DNA synthesis rate is derived from measuring the total DNA (6.5 μg/10^6 cells) and the total time (about 480 min) it takes to duplicate this DNA (S phase) in CHEF/18 cells. Considering that the average molecular weight of deoxynucleotide residue in DNA is 330, the rate of DNA replication in CHEF/18 cells during the entire S phase is 41 pmol of dNMP incorporated 10^6 cells^{-1} min^{-1} or 10.2 pmol of any one deoxynucleotide (dTMP, dAMP, dCMP, or dGMP) incorporated 10^6 cells^{-1} min^{-1}.

[b]When S phase CHEF/18 cells are incubated with [^{3}H]deoxyuridine (2460 cpm/pmol), 24,108 cpm are incorporated into newly synthesized DNA per minute.

[c]Therefore, the specific activity of dTMP in newly synthesized DNA is 24,108 cpm ÷ 10.2 pmol = 2363 cpm/pmol. The data presented in this table are adopted from Reddy (1989).

role outside their function as DNA precursors. Therefore, dNTPs are expected to be in the vicinity where DNA polymerase reaction occurs.

The most direct evidence for the compartmentation of deoxynucleotides for DNA replication in mammalian cells came from studies with S phase-enriched CHEF/18 cells (Reddy, 1989). In these studies it was observed that the radioactivity from exogenous deoxyuridine was readily incorporated into DNA without any detectable time lag, even though the cells contained a large excess of endogenous dUMP and dTTP pools. As seen from Table 1, exogenous radioactive precursor labels only 10% of the endogenous precursor pool (based on specific activity) under conditions where the DNA is made almost exclusively from the exogenous source. This demonstrates two important properties of DNA synthesis. The first is the phenomenon of channeling, described above. The second is the existence of separate functional compartments.

In addition to functional compartmentation, evidence for physical compartmentation of deoxynucleotides was seen when radioactive deoxyuridine-labeled S phase-enriched CHEF/18 cells were fractionated into cytoplasts and karyoplasts by cytochalasin B enucleation (Reddy, 1989). In these studies it was observed that more than 80% of each of the radioactive intermediates derived from [^{3}H]dUrd partitioned with the karyoplasts of S phase cells. Particularly, the proximal precursor of DNA replication, [^{3}H]dTTP, was almost completely confined to the karyoplasts. Leeds and Mathews (1987) suggested that there could exist two types of dNTP pools in mammalian cells: replication-active pools and replication-inactive pools. Replication-active pools are small, rapidly labeled from exogenous radioactive precursors, and support DNA replication. Replication-inactive pools, on the other hand, are relatively large, slowly labeled, and are not associated with scheduled semiconservative DNA synthesis; rather, they could be involved in DNA repair or recombinational processes. In this regard [^{3}H]dUMP and [^{3}H]dTTP labeled from [^{3}H]dUrd in studies with S phase CHEF/18 cells could represent replication-active pools and are present primarily in the karyoplast (Reddy, 1989). By contrast, large endogenous pools of these nucleotides could represent replication-inactive pools and therefore could be largely localized in the cytoplasmic compartment of the cell, as observed by Leeds et al. (1985).

The existence of deoxynucleotide compartments for DNA synthesis in mammalian cells was first shown in the kinetic studies of Fridland (1973) in human lymphoblast cells. These studies suggested that DNA replication is supplied by a small dTTP pool that becomes labeled rapidly with exogenous radioactive thymidine or deoxyuridine and turns over rapidly, but whose presence is obscured by a much larger, slowly replenished pool of dTTP. Furthermore, the radioactivity from deoxyuridine incorporated into DNA 2-fold more efficiently than that from thymidine, even when the dTTP pool was labeled to the same specific activity with both the radioactive precursors. Nicander and Reichard (1983) found kinetic evidence for the presence of separated dCTP pools, with preferential DNA synthesis from the pool labeled from cytidine in 3T6 cells. However, they could not find a similarly compartmentalized pool of dTTP in these cells, possibly because they used amethopterin to block the de novo pathway for thymidylate synthesis. In disagreement with the observations of Nicander and Reichard (1983), Chiba et al. (1984) did not find a separate dCTP pool with preferential DNA synthesis from the pool labeled with cytidine in mouse L1210 leukemia cells. This discrepancy could be partly due to the failure of Chiba et al. (1984) to take into account the specific activity of intermediary nucleotide pools.

Compartmentation of purine nucleotides for DNA synthesis was clearly demonstrated in studies with thymus cells (Scott and Forsdyke, 1980) and with mouse T lymphoma (S-49) cells (Nguyen and Sadee, 1986).

How does a mammalian cell maintain a distinct compartment of deoxynucleotides for DNA synthesis? This question is particularly interesting in view of the lack of any physical barriers that limit metabolites like dNTP from freely diffusing in the intracellular environment, and the lack of a physical diffusion barrier between the nucleus and the cytoplasm. One possible explanation for functional compartmentation might be the aggregation of the metabolic enzymes of a specific pathway into units or the presence of several related enzymatic activities in a single protein (multienzmic protein). This type or organization would increase the efficiency of sequential reactions, thereby leading to the compartmentation of the corresponding metabolites (Srere, 1987). The replitase complex described in this chapter could represent such a functional unit.

Conclusions

In recent years, most of the enzymes involved in DNA precursor biosynthesis have been identified and characterized. However, the intracellular organization of these enzymes, nature of the controls of nuclear DNA replication, and details of the supply of deoxynucleoside triphosphates (dNTPs) to the site of DNA replication remain obscure. Our investigations have revealed that the enzymes of DNA replication and DNA precursor biosynthesis interact with each other within the structure of a multienzyme complex which we have called "replitase." The replitase complex in mammalian cells could be responsible for the compartmentation of dNTPs in the vicinity of DNA replication. Our findings of enzyme rearrangement and complex formation at the time of DNA replication indicate that the organization of the enzymes into a supra molecular structure may be the key event that initiates S phase.

Successful purification of the replitase complex should aid in understanding both the process of the initiation of DNA synthesis and the controls of synthesis rate. Knowledge of the interactions within the replitase should clarify much of a literature that is presently confusing. Purified replitase should provide an in vitro system in which much more detailed studies can be carried on about DNA synthesis, the most important single biochemical process of mammalian cells. Beyond this, in a more practical realm, a greater understanding of the initiation process and controls of normal DNA synthesis may give valuable insight into how these controls break down in cancer cells. This knowledge may suggest novel cancer treatments and may also suggest systematic preventive medicine to make the initiation of cancer less likely.

Acknowledgments

This work is supported by National Institute of Health Grant CA-39445.

References

Alberts BM (1987): Prokaryotic DNA replication mechanisms. Phil Trans R Soc London Ser B 317:395–420.

Ashman CR, Reddy GPV, Davidson RL (1981): Bromodeoxyuridine mutagenesis, ribonucleotide reductase activity, and deoxyribonucleotide pools in hydroxyurea-resistant mutants. Som Cell Genet 7:751–768.

Ayusawa D, Shimizu K, Koyama H, Takeishi K, Seno T (1983): Unusual aspects of human thymidylate synthase in mouse cells introduced by DNA-mediated gene transfer. J Biol Chem 258:48–53.

Baril E, Baril B, Elford H, Luftig RB (1973): In Mechanism and Regulation of DNA Replication. A.R. Kolber and M. Kohiyama, Eds. New York: Plenum, pp 275–291.

Castellot JJ (1980): Lysolecithin-permeabilized animal cells as a tool for studying cell growth and metabolism. In R. Baserga, C. Croce, and G. Rovera (eds): Introduction of Macromolecules into Viable Mammalian Cells. New York: Alan R. Liss, pp 297–324.

Chiba P, Bacon PE, Cory JG (1984): Studies directed toward testing the "channeling" hypothesis—ribonucleotides → DNA in leukemia L1210 cells. Biochem Biophys Res Commun 123:656–662.

Danenberg KD, Danenberg PV (1989): Activity of thymidylate synthase and its inhibition by 5-fluorouracil in highly enzyme-overproducing cells resistant to 10-propargyl-5,8-dideazafolate. Mol Pharmacol 36:219–223.

Engstrom Y, Rozell B (1988): Immunocytochemical evidence for the cytoplasmic localization and differential expression during the cell cycle of the M1 and M2 subunits of mammalian ribonucleotide reductase. EMBO J 7:1615–1620.

Fridland A (1973): DNA precursor in eukaryotic cells. Nature New Biol 243:105–107.

Hammond RA, Miller MR, Gray MS, Reddy GPV (1989): Association of 3′→ 5′ exodeoxyribonuclease activity with DNA replitase complex from S-phase chinese hamster embryo fibroblast cells. Exp Cell Res 183:284–293.

Harvey G, Pearson CK (1988): Search for multienzyme complexes of DNA precursor pathways in unifected mammalian cells and in cells infected with Herpes simplex virus type I. J Cell Physiol 134:25–36.

Jackson RC (1984): A kinetic model of regulation of the deoxyribonucleoside triphosphate pool composition. Pharmacol Ther 24:279–301.

Kornberg A (1980): DNA Replication. San Francisco: Freeman.

Kucera R, Paulus H (1986): Localization of the deoxyribonucleotide biosynthetic enzymes ribonucleotide reductase and thymidylate synthase in mouse L cells. Exp Cell Res 167:417–428.

Leeds JM, Mathews CK (1987): Cell cycle-dependent effects on deoxynucleotide and DNA labeling by nucleoside precursors in mammalian cells. Mol Cell Biol 7:532–534.

Leeds JM, Slabaugh MB, Mathews CK (1985): DNA precursor pools and ribonucleotide reductase activity: Distribution between the nucleus and cytoplasm of mammalian cells. Mol Cell Biol 5:3443–3450.

Mathews CK, North TW, Reddy GPV (1979): Multienzyme complexes in DNA precursor biosynthesis. Adv Enzyme Reg 17:133–156.

Matherly LH, Schuetz JD, Westin E, Goldman ID (1989): A method for the synchronization of cultured cells with aphidicolin: Application to the large scale synchronization of L1210 cells and the study of the cell cycle regulation of thymidylate synthase and dihydrofolate reductase. Anal Biochem 182:338–345.

Moen LK, Howell ML, Lasser GW, Mathews CK (1988): T4 phage deoxyribonucleoside triphosphate synthetase: Purification of an enzyme complex and identification of gene products required for integrity. J Mol rec 1:48–57.

Nelson WG, Pienta KJ, Barrack ER, Coffey DS (1986): The role of the nuclear matrix in the organization function of DNA. Annu Rev Biophys Biophys Chem 15:457–475.

Nguyen BT, Sadee W (1986): Compartmentation of guanine nucleotide precursors for DNA synthesis. Biochem J 234:263–269.

Nicander B, Reichard P (1983): Dynamics of pyrimidine deoxynucleoside triphosphate pools in relationship to DNA synthesis in 3T6 mouse fibroblasts. Proc Natl Acad Sci USA 80: 1347–1354.

Nicander B, Reichard P (1985): Relations between synthesis of deoxy nucleotides and DNA replication in 3T6 fibroblasts. J Biol Chem 260:5376–5381.

Noguchi H, Reddy GPV, Pardee AB (1983): Rapid incorporation of label from ribonucleoside diphosphates into DNA by a cell-free high molecular weight fraction from animal cell nuclei. Cell 32:443–451.

Plucinski TM, Fager RS, Reddy GPV (1990): Allosteric interaction of components of the replitase complex is responsible for enzyme cross-inhibition. Mol Pharmacol 38:114–120.

Reddy GPV (1982): Catalytic function of thymidylate synthase is confined to S phase due to its association with replitase. Biochem Biophys Res Commun 109:908–915.

Reddy GPV (1989): Compartmentation of deoxypyrimidine nucleotides for nuclear DNA replication in S phase mammalian cells. J Mol Rec 2:75–83.

Reddy GPV, Klinge EM, Pardee AB (1986): Ribonucleotides are channeled into a mixed DNA-RNA polymer by permeabilized hamster cells. Bichem Biophys Res Commun 135:340–346.

Reddy GPV, Pardee AB (1980): Multienzyme complex for metabolic channeling in mammalian DNA replication. Proc Natl Acad Sci USA 77:3312–3316.

Reddy GPV, Pardee AB (1982): Coupled ribonucleoside diphosphate reduction, channeling and incorporation into DNA of mammalian cells. J Biol Chem 257:12526–12531.

Reddy GPV, Pardee AB (1983): Inhibitor evidence for allosteric interaction in the replitase multienzyme complex. Nature (London) 303:86–88.

Rode W, Scanlon KJ, Moroson BA, Bertino J (1980): Regulation of thymidylate synthetase in mouse leukemia cells (L1210). J Biol Chem 255:1305–1311.

Rode W, Jastreboff MM, Bertino J (1985): Thymidylate synthase inhibition in cells with arrested DNA synthesis is not due to an allosteric interaction in the replitase complex. Biochem Biophys Res Commun 128:345–351.

Scott FW, Forsdyke DR (1980): Isotope-dilution analysis of the effects of deoxyguanosine and deoxyadenosine on the incorporation of thymidine and deoxycytidine by hydroxyurea-treated thymus cells. Biochem J 190:721–730.

Srere PA (1987): Complexes of sequential metabolic enzymes. Annu Rev Biochem 56:89–124.

Tubo RA, Berezney R (1987): Identification of 100 and 150S DNA polymerase-α-primase megacomplexes solubilized from the nuclear matrix of regenerating rat liver. J Biol Chem 262:5857–5865.

Wawra E (1988): Microinjection of deoxynucleotides into mouse cells. No evidence that precursors for DNA synthesis are channeled. J Biol Chem 263:9908–9912.

Wickremasinghe RG, Yaxley JC, Hoffbrand AV (1983): Gel filtration of a complex of DNA polymerase and DNA precursor-synthesizing enzymes from a human lymphoblastoid cell line. Biochim Biophys Acta 740:243–248.

Youdale T, Frappier L, Whitfield JF, Rixon RH (1984): Changes in the cytoplasmic and nuclear activities of the ribonucleotide reductase holoenzyme and its subunits in regenerating liver cells in normal and thyroparathyroidectomized rats. Can J Biochem Cell Biol 62:914–919.

**Perspectives on Cellular Regulation:
From Bacteria to Cancer, pages 331–340**

From Molecular Evolution to Body and Brain Evolution

Allan C. Wilson

Division of Biochemistry and Molecular Biology, University of California, Berkeley, California 94720

This article is a salute to Arthur Pardee on the occasion of his 70th birthday. Both my scientific career and my personal life owe much to this eminent scientist. If it were not for Art's laboratory, I would not have met Leona Greenbaum, my wife of 32 years!

An intellectually exciting atmosphere greeted me upon my arrival in Art's lab at Berkeley in June 1957. The laboratory was on the top floor of a rather new building that housed the Departments of Biochemistry and Plant Biochemistry as well as the Virus Laboratory. During my 4-year stay, our floor was also occupied by the laboratories of W. Stanley, H. Fraenkel-Conrat, G. Stent, A. Knight, and H. Rubin. To me it was a veritable cradle of molecular biology. In the two years before my arrival, this floor had witnessed such notable discoveries as (1) the dissociation and reconstitution of a virus, (2) the infectivity of a pure viral nucleic acid, (3) the production of specific amino acid changes by specific chemical mutagenesis of isolated nucleic acid encoding a protein, (4) feedback inhibition of enzyme activity by the end product of a biosynthetic pathway, (5) repression of the synthesis of enzymes in a biosynthetic pathway by an end product, (6) incorporation of base analogues (e.g., bromouracil) into nucleic acids and the consequent generation of mutations, and (7) ultracentrifugal demonstration of the existence of ribosomes in bacteria. Four of these landmark discoveries were made by Art and his associates. Following these remarkable achievements came others such as the demonstration that the binding site of the feedback inhibitor was separate from the active site of the enzyme and the discovery that enzyme induction was due to the inhibition of a repressor by the inducer. However, Art's colleagues were slow to comprehend the significance of these findings. Sensing their incomprehension, Art accepted an offer of a full professorship at Princeton and left Berkeley in the same year that I completed the Ph.D. degree (1961).

Perhaps one reason for their incomprehension was Art's modesty. For students like me, who were unsure of our ability to succeed in research, this trait was attractive. Moreover, he encouraged my aspiration to help (after I left his laboratory) bring molecular biology and natural history together in the molecular study of evolution.

Molecular Biology and Evolutionary Time

During the late 1950s and the 1960s, comparative studies of mammalian proteins led to three major discoveries, which are reviewed by Kimura (1983), Nei (1987), and Li and Graur (1991), as well as by Wilson et al. (1977,1987) and Wilson (1985). The first one was that the evolution of genes is dominated by changes that are functionally inconsequential or nearly so. Insulin provided the earliest example of this phenomenon when Sanger's group found that functionally equivalent insulins from different species differed in amino acid sequence.

The second discovery was that these changes accumulate relentlessly at nearly constant rates in proteins such as hemoglobin, cytochrome *c*, and albumin. We now know, as the result of extensive comparative studies of genes during the 1970s and 1980s, that the average rate of evolutionary change is remarkably uniform at functionally equivalent sites in vastly different creatures. The silent sites in codons may be taken as an example. The average rate of divergence due to base substitution at these sites is nearly 1% per million years for genes in enteric bacteria and in the nuclei of higher plants and metazoan animals.

The rates of evolutionary substitutions are uniform enough to allow proteins and nucleic acids to be used as dating devices. By measuring the extent of divergence in primary structure between specific proteins or nucleic acids of two species, one can estimate the time elapsed since they had a common ancestor, provided that the average rate of change in the macromolecule under study is known from studies on other species.

The third discovery was the neutral theory, which explained how mutations with no advantage over the wild type could be fixed, i.e., spread through a population at the expense of the wild type, and why the number of neutral mutations fixed per unit time might be rather uniform from species to species.

The vision of being able to put knowledge of biological diversity on a time scale transforms the way in which one can study evolution. Evolutionary genetics used to function for the most part without a time scale. This situation was epitomized by the *Drosophila* workers who sought to understand the genetic basis of evolutionary change without having any idea of when their favorite species (*D. melanogaster* and *D. simulans*) last shared a common ancestor. So, they were unable to measure rates of change. Yet my upbringing in the Pardee lab had indelibly imprinted on my mind the necessity of measuring rates of biological processes in order to test ideas about the mechanisms underlying those processes (Wilson and Pardee, 1962). For me, molecular evolutionary clocks were a godsend that provided the time scales needed for quantitative analysis of evolutionary processes.

Brain Evolution on a Time Scale

To illustrate how evolution can be studied quantitatively with respect to time, I shall deal first with the brain. The study of brain evolution has a long history but almost no attention was given to a temporal analysis of brain evolution (Jerison, 1973). Figure 1 introduces us to a way of analyzing how the size of the brain has changed in relation to body size during the evolution of land vertebrates. It shows a genealogical tree relating modern people to apes, other mammals, reptiles, and amphibians. The order of branching of the lineages is well known from classical studies of bodily traits in both living and fossil forms as well as from comparative studies of macromolecules. The **a** symbol refers to the common ancestor of amphibians and other land vertebrates, while

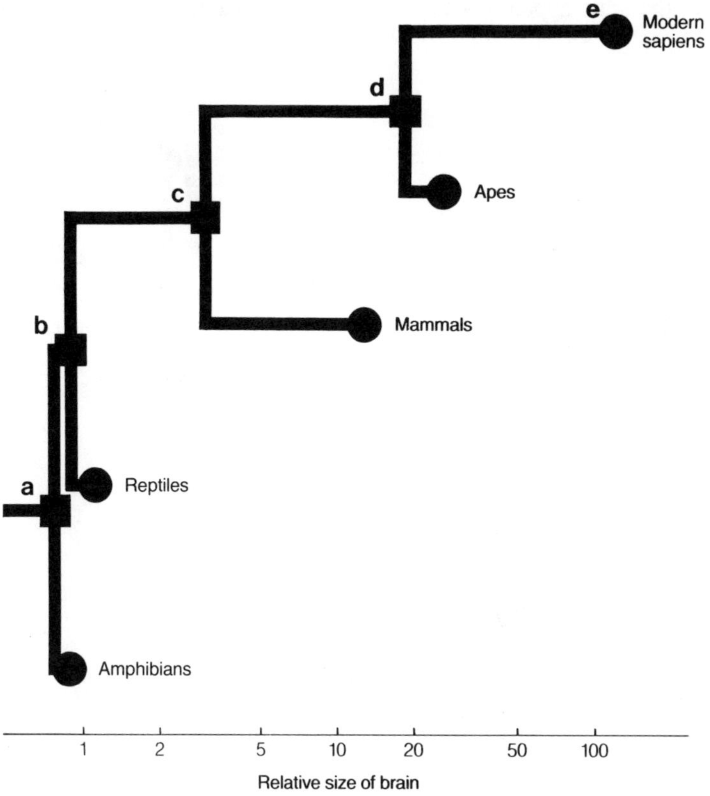

Fig. 1. Tree showing how the size of the brain has changed in relation to that of the body along the lineages leading from an early amphibian (**a**) to modern four-legged vertebrates. Relative brain size (R) is given by the equation $R = yx^{-0.67}$, where y is brain volume in cubic centimeters and x is body weight in kilograms. The brain values for modern animals are from Wyles et al. (1983).

b stands for the common ancestor of reptiles and mammals, and so on. The scale below the tree gives the relative sizes of the brain in the five sorts of living animals. This scale also provides interpolated estimates of relative brain sizes for the four ancestors (**a–d**). The principle used to make these interpolations is that of minimizing the variation in rates of change in relative brain size. Although no recourse was made to the fossil record in deriving these estimates, they are in satisfactory agreement with the fragmentary evidence available from fossils.

The times when the four ancestors lived are also rather well known from a combination of both fossil and molecular evidence. These times and the relative brain size estimates are plotted in Figure 2, which shows that, on the lineage leading from the early amphibian ancestor **a** to modern people, the brain has grown in relative size by 100-fold. Furthermore the evolutionary growth of the brain appears to have increased exponentially at the least.

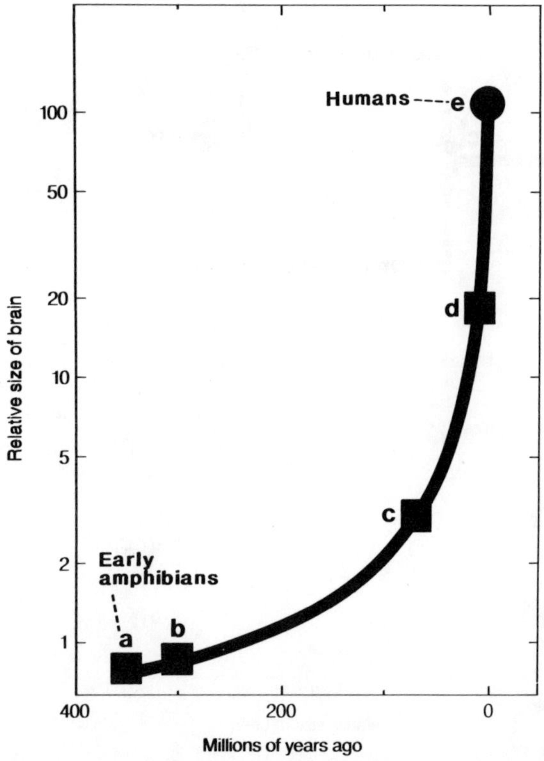

Fig. 2. Relative brain size at various times on the lineage leading from early amphibians (**a**) to modern people (**e**). The *R* values are from Figure 1 and the times are based on Romer (1966) and Carroll (1988) as well as on molecular comparisons.

TABLE 1. Distribution of Learning Ability

	Invertebrates		Vertebrates		
Type of learning	Most	Social	Cold	Warm	Humans
Social					
1. Teaching	−	−	−	−	+
2. Imitation	−	−	−	±	+
3. Primitive[a]	−	±	±	+	+
Individual	+	+	+	+	+

[a]Social facilitation and local enhancement (Thorpe, 1963).

A curve of this shape implies that the process is autocatalytic, i.e., that the brain has driven its own evolution on the lineage leading to humans. The following model explaining how this could happen takes account of the fact that evolution occurs in populations and that the social propagation of new behaviors can occur not only in humans but in many other species of big-brained animals as well (Table 1):

1. An advantageous new habit arises by chance within an individual.
2. Among the other individuals in the population there is genetically based variation in the capacity to detect, evaluate, and copy the new habit. Hence, the new habit spreads nongenetically and unevenly through the population.
3. Selection favors those individuals bearing ''brain'' mutations, which improve the capacity to catch on to the new habit.

A consequence according to the model is that each time a new habit arises and spreads through a population, it selects for improvement in the capacity to catch on to discoveries made by others. Since both innovation and catching on require the ability to imagine one's body in a new relation to the environment, the selection process to which I refer is likely to enrich the population as regards the ability both to innovate and catch on. One then has the possibility of a positive feedback loop operating on brain mutations.

Among the many kinds of mutations that could improve the brain's ability to innovate and catch on, some will produce more neurons or dendrites and thus raise the relative size of the brain. The model enables one to understand in principle how the brain's relative size could go up exponentially with time. To explain why the curve is apparently hyperexponential in Figure 2, we may ask whether there are multiple synergistic positive feedback loops or whether more points might show that the process is not as nonlinear as it seems to be in Figure 2.

Rates of Bodily Change

The model proposed for explaining the brain's role in its own evolution leads one to expect a correlation between relative brain size and the rate of bodily evolution. That such a correlation exists was pointed out by Wyles et al. (1983) and the results are summarized in Figure 3.

The reasons for expecting this correlation are as follows. New habits are expected to arise and spread by social learning through populations more often in big-brained than in small-brained animals (Table 1). Such social-propagation events are known to take place frequently in many songbirds and social mammals (especially primates) but rarely, if at all, in less brainy species. Whenever a species adopts a new habit, the body engages the environment in a new way. In this circumstance, selection favors mutations that improve the body's performance of the new act. Brainy species will thus subject themselves frequently to new selection pressures that fix ''body'' mutations as well as brain mutations.

Figure 3 shows that the rules of body evolution are not the same as those for molecular evolution. Whereas body evolution is driven by natural selection (emanating to a large extent from the brain in higher animals), molecular evolution caused by base substitutions seems to be due largely to random fixation and thus depends chiefly on time rather than relative brain size.

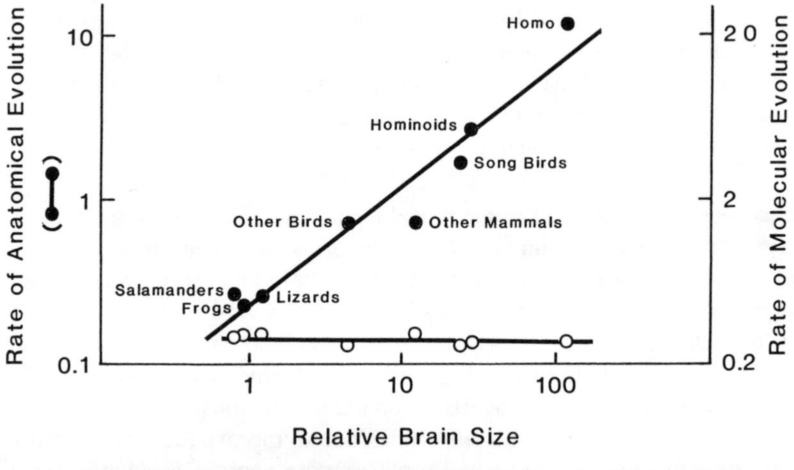

Fig. 3. Rates of change in body plan and at noncoding sites in nuclear DNA for eight taxonomic groups whose relative brain sizes are known. The body and brain values are from Wyles et al. (1983). Extent of change in body plan per unit time is based on measurements of eight linear traits (representing parts of the head, the fore and hind limbs, and the trunk) for pairs of species whose time of divergence is known. A value of 1 corresponds to an average divergence of 1% in relative trait length per million years. Relative trait length is the length of a trait divided by the sum of the lengths of the eight traits (Wyles et al., 1983). The molecular values refer to percent divergence (owing to base substitutions) per million years and are from diverse sources and are approximate only (Helm-Bychowski and Wilson, 1988, and unpublished compilation).

Genetic Basis of Body and Brain Evolution

The results in Figure 3 prompt the question of what kinds of genetic change underlie evolution at the organismal level. The figure also draws attention to the likely value of studying these problems in the creatures whose bodies and brains have changed fastest, namely the big-brained birds and mammals, and especially humans. These are the creatures in which the ratio of adaptive genetic change (driven by positive selection) to neutral change will presumably be highest. Thus the signal associated with adaptive evolution is less likely to be drowned by the noise of neutral change in brainy species.

A popular idea has been to ascribe the adaptive evolution of organisms chiefly to regulatory mutations, which can affect the amount of a gene product drastically without affecting its structure (Wilson et al., 1977; Dickerson, 1988; MacDonald, 1990). An example, which also illustrates the brain's role in adaptive evolution, is provided by those human populations whose ancestors tamed cattle and became dependent on dairy farming, thereby subjecting themselves as adults to lactose in the diet. Within the short period of 5,000 years, the people of northern Europe, who seem to have been especially dependent on dairy products, underwent virtual fixation for a regulatory mutation that causes β-galactosidase to be expressed permanently in the small intestine. By contrast, in human populations that did not depend heavily on dairy products as adults, the galactosidase gene is usually expressed in the gut only during infancy, as is the case in most mammals. Unfortunately, the basis for this regulatory polymorphism affecting human β-galactosidase concentrations is unknown at the DNA level (Boll et al., 1991).

Lysozyme Evolution in Foregut Fermenters

Lysozyme offers a similar opportunity to investigate the genetic basis of evolutionary adaptations. In conventional mammals, this enzyme is expressed in macrophages, where it helps to fight invading bacteria. By contrast, in those mammals with a fermentative foregut, lysozyme expression occurs mainly in the lining of the true stomach. On secretion into the stomach fluid, the enzyme may help to digest the many bacteria flowing from the foregut into the stomach. The recruitment of lysozyme as a major digestive enzyme in advanced ruminants (i.e., cows, sheep, deer, antelopes, and their relatives) has involved four types of genetic event (Irwin and Wilson, 1990):

1. A major regulatory change affecting the tissue specificity of gene expression.
2. Multiplication of the number of lysozyme genes from one in conventional mammals to an array of 10 genes in ruminants.
3. Recombination between the exons of nonallelic genes in the array.

4. Amino acid replacements in the protein, suiting it for surviving and functioning in the presence of acid, pepsin, and a troublesome fermentation product, diacetyl (which modifies arginine residues and is responsible for the smell of butter).

We wonder whether intergenic recombination had a role in the adaptive evolution of ruminant stomach lysozymes. Tree analysis suggests that these lysozymes have disobeyed the rule of the molecular clock during ruminant evolution. The primary structure changed quickly at the outset of ruminant evolution and slowly later (Jollès et al., 1989). The way in which intergenic recombination could have contributed to such departures from the normal rate of amino acid change in this protein is illustrated in Figure 4. Such recombination can, in principle, accelerate adaptive evolution by quickly assembling double mutants that are advantageous.

Since lysozyme is one of the most thoroughly known proteins in terms of three-dimensional structure, there is hope that we may be able to retrace the pathway of its evolution in ruminants. By site-directed mutagenesis coupled with structural and functional analysis, we may be able to find out how a macrophage lysozyme turned into a stomach lysozyme. This approach has already been used in a preliminary way to reconstruct pathways of lysozyme evolution in birds (Malcolm et al., 1990). Further development of the approach may enable us to find out whether double mutants were required to get from one adaptive state to another during the evolution of ruminant stomach lysozymes. Questions like these, as well as questions about the molecular basis of the evolutionary shifts in gene regulation that took place during the evolution of foregut fermenters from conventional mammals, occupy attention in my laboratory at present.

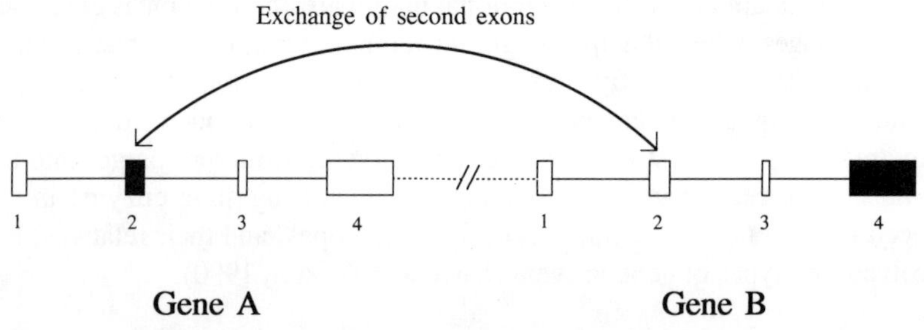

Fig. 4. Scheme showing the exchange of information between the second exons of nonallelic lysozyme genes. Such an exchange can generate a doubly mutant lysozyme gene or a gene lacking either mutation. The boxes refer to coding blocks, empty boxes being wild type and solid boxes being mutant. The actual lysozyme genes of ruminants have longer introns than is evident here and the lysozyme genes are far apart (Irwin and Wilson, 1990, and unpublished).

Summary

1. Molecular biological ways of studying evolution have revealed evolutionary clocks.

2. These clocks provide an approximate time scale for organizing knowledge of biological diversity.

3. The time scale stimulates the desire to measure extent of biological change at many levels of biological organization and thus to estimate rates of evolution.

4. Rates of evolutionary change in body plan are found to correlate with relative brain size, which has risen at an increasing rate on the lineage leading from early land vertebrates to humans.

5. A model explaining these results says that in big-brained species the chief driving force for body and brain evolution is not the changing environment but the brain. The brain invents new ways of exploiting the environment and, as the new life-styles spread nongenetically through populations, they select for body and brain mutations.

6. The genetic basis of adaptive changes in the body and brain may reside mainly in regulatory mutations assisted by gene duplications and intergenic recombination. Lysozyme gene evolution in those mammals having a fermentative foregut provides a model system for linking gene evolution to evolution at the organismal level.

References

Boll W, Wagner P, Mantei N (1991): Structure of the chromosomal gene and cDNAs coding for lactase-phlorizin hydrolase in humans with adult-type hypolactasia and persistence of lactase. Am J Hum Genet, in press.

Carroll RL (1988): Vertebrate Paleontology and Evolution. New York: Freeman.

Dickerson WJ (1988): On the architecture of regulatory systems: evolutionary insights and implications. BioEssays 8:204–208.

Helm-Bychowski KM, Wilson AC (1988): Temporal calibration of nuclear DNA evolution in phasianoid birds: Evidence from restriction maps. Proc Int Ornithol Congr 19:1896–1904.

Irwin DM, Wilson AC (1990): Concerted evolution of ruminant stomach lysozymes. Characterization of lysozyme cDNA clones from sheep and deer. J Biol Chem 265:4944–4952.

Jerison HJ (1973): Evolution of the Brain and Intelligence. New York: Academic Press.

Jollès J, Jollès P, Bowman BH, Prager EM, Stewart C-B, Wilson AC (1989): Episodic evolution in the stomach lysozymes of ruminants. J Mol Evol 28:528–535.

Kimura M (1983): The Neutral Theory of Molecular Evolution. Cambridge: Cambridge Univ. Press.

Li W-H, Graur D (1991): Fundamentals of Molecular Evolution. Sunderland, Massachusetts: Sinauer Assoc.

MacDonald JF (1990): Macroevolution and retroviral elements. BioScience 40:183–191.

Malcolm BA, Wilson KP, Matthews BW, Kirsch JF, Wilson AC (1990): Ancestral lysozymes reconstructed, neutrality tested, and thermostability linked to hydrocarbon packing. Nature (London) 345:86–89.

Nei M (1987): Molecular Evolutionary Genetics. New York: Columbia Univ. Press.

Romer AS (1966): Vertebrate Paleontology, 3rd ed. Chicago: Univ. of Chicago Press.

Thorpe WH (1963): Learning and Instinct in Animals. London: Methuen.

Wilson AC (1985): The molecular basis of evolution. Sci Am 253 (4):164–173.
Wilson AC (1988): Time scale for bird evolution. Proc Int Ornithol Congr 19:1912–1917.
Wilson AC, Pardee AB (1962): Regulation of flavin synthesis by *Escherichia coli*. J Gen Microbiol 28:283–303.
Wilson AC, Carlson SS, White TJ (1977): Biochemical evolution. Annu Rev Biochem 46:573–639.
Wilson AC, Ochman H, Prager EM (1987): Molecular time scale for evolution. Trends Genet 3:241–247.
Wyles JS, Kunkel JG, Wilson AC (1983): Birds, behavior, and anatomical evolution. Proc Natl Acad Sci USA 80:4394–4397.

Epilogue: Understanding Growth Control—Quantal Leaps or Continuous?

The above question may be a rhetorical one for a molecular biologist or cell biologist working on growth control. If a pediatrician or a fermentation microbiologist were to answer this question, he would consider growth as a continuous phenomenon. For a DNA replication specialist working with the cell cycle, however, growth will be a phenomenon that proceeds in quantal leaps. In this field, as much as in astronomy, where the debates were raging between the Ptolemaic and the Galileic concept of the world, the answer depends very much on the relative position of the scientist considering the question.

If a science historian were asked the same question about the development of the entire field over the last three decades, namely whether knowledge about mechanisms of growth control has advanced in quantal leaps or continually, his answer would again depend on the exact field he would be considering. If he is a bacterial physiologist, he would probably acknowledge continuous but slow growth of our knowledge in the field of growth and chromosomal replication. If he is a cancer researcher, however, he will have to admit that we are making little progress and that there may have been certain small increments or leaps, but that convincing leaps will have to follow if we want to understand the defects of growth regulation of tumor cells in vivo.

The outcome of the answer will of course also depend on the expectations of the analyst and to whom he talks. If he has Art Pardee's tenacity to stick to a great goal, he will see in each increment of knowledge in this field a solid contribution continually adding up to help us reach the final goal. While talking to the new students and postdoctoral fellows, he will, like Art, pass on his infectious enthusiasm and thereby convince them to enter the field of growth control for the remainder of their careers.

Many of us have a biphasic psyche, a fact that Goethe describes as man having two souls in his body ("Zwei Seelen wohnen, ach, in meiner Brust!"). Art Pardee's enthusiasm in the company of graduate students would easily be matched by his total modesty among peers, where Art would be among the first to admit the molecular intricacies of the field. The big leaps are yet to come, and since the field belongs to the great challenges in biology of today, it was and is the right field for Art Pardee to work in and to convey his enthusiasm to his students, collaborators, and postdoctoral fellows.

It is the small steps, if underpinned with sufficient evidence, that are considered solid science publishable in the *Journal of Biological Chemistry* and similar journals and that continually contribute to the growth of science. They are our daily bread and guarantee the flux of grant money. They are not satisfactory, however, to that part of Art Pardee's soul that is aristocratic in spirit. That part of his personality loves the challenges, looks for terra incognita, and tries to tackle the key questions of tomorrow.

With this type of approach and motivation he successfully nailed down the repressor concept experimentally (the well known PaJaMo, i.e., Pardee-Jacob-Monod experiment). He was similarly successful with the concept as well as with the evidence that feedback-controlled enzymes have a metabolic regulatory site separate from the catalytic site. Both breakthroughs were not announced in Broadway style but in the usually modest Pardee style. Several other solid firsts followed. No wonder that such a scientist will pick up a similarly challenging project: the molecular basis of growth control.

After the year 2000, it will be interesting to look back at the present decades to find out whether it was the continuous growth of color, shape, and structure of the mosaic that eventually led to a holistic view of the picture or whether it was one or a few leaps in understanding growth control that resolved the problem at the molecular level. For the time being, we all have to content ourselves with the small steps and with Churchill's motto that "Success is failure after failure after failure but not losing the motivation."

If by stepping back you cannot decipher the basic framework and the meaning of a mosaic to which you have added your 477th color stone out of a thousand, you have to return to the picture, fill in another 23 color stones, and step back again. What Louis Pasteur once called the prepared mind may only then see the contours of the mosaic, i.e., see the motif, achieve the breakthrough, and thereby make the quantal leap forward hoped for over decades.

Whether the next one or two decades bring this single breakthrough in growth control at the molecular level, or whether it will be a stepwise unveiling of the mosaic, remains to be seen. All of us—students, collaborators, and long-time friends alike—hope that Art will still be actively contributing to the solution, and we wish him all the best and dedicate this book to him in gratitude.

Max Burger
January 1991

Index